123 Robotics Experiments for the Evil Genius

123 Robotics Experiments for the Evil Genius

MYKE PREDKO

McGraw-Hill
New York   Chicago   San Francisco   Lisbon
London   Madrid   Mexico City   Milan   New Delhi
San Juan   Seoul   Singapore   Sydney   Toronto

The McGraw·Hill Companies

Library of Congress Cataloging-in-Publication Data

Predko, Michael.
    123 robotics experiments for the evil genius / Myke Predko./
    p. cm.
    ISBN 0-07-141358-8
    1. Robotics   I. Title: One hundred twenty-three robotics exeriments for the evil genius.
    II. Title: One hundred twenty-three robotics experiments for the evil genius.   III. Title.

TJ211.P73   2003
629.8'92—dc22   2003066532

1 2 3 4 5 6 7 8 9 0   QPD/QPD   0 9 8 7 6 5 4 3

P/N 141359-6
PART OF
ISBN 0-07-141358-8

*The sponsoring editor for this book was Judy Bass and the production supervisor
was Pamela A. Pelton. It was set in Times Ten by MacAllister Publishing Services, LLC.*

Printed and bound by Quebecor/Dubuque.

 This book is printed on recycled, acid-free paper containing a minimum of 50 percent recycled
de-inked fiber.

McGraw-Hill books are available at special quantity discounts to use as premiums and sales promo-
tions, or for use in corporate training programs. For more information, please write to the Director
of Special Sales, McGraw-Hill Professional, Two Penn Plaza, New York, NY 10121-2298. Or contact
your local bookstore.

Contents

Contents

Myke's Rules of Robotics

Throughout this book, I will be keeping to my 10 rules of robotics:

1. Start small.

2. Design everything together.

3. Jerkiness in a robot is not a selling point.

4. Protect your drivetrains from the environment.

5. Keep the robot's center of mass in the center of the robot.

6. The faster a robot runs, the more impressive it is.

7. Object detectors should detect objects far enough away from the robot so that it can stop before damaging the object or itself.

8. Complexity adds weight.

9. Weight adds weight.

10. If the robot isn't doing anything, it shouldn't be expending any energy.

Section One
Introduction to Robots

When you think of the term "robot," what comes to mind? The following are some definitions that attempt to explain what a robot is:

A true robot is a machine that can be "taught," programmed like a computer, to make different kinds of motions and perform a variety of jobs. . . . Machines that do one job only and cannot be "retrained" are not true robots, either.

The New Book of Knowledge, 1998

Robotics A field of engineering concerned with the development and application of robots, as well as computer systems for their control, sensory feedback, and information processing. There are many types of robotic devices, including robotic manipulators, robot hands, mobile robots, walking robots, aids for disabled persons, telerobots, and microelectronomechanical systems.

The McGraw-Hill Encyclopedia of Science & Technology, 8th Edition

A **robot** is a mechanical device that operates automatically. Robots can perform a wide variety of tasks. They are especially suitable for doing jobs too boring, difficult, or dangerous for people. The term robot comes from the Czech word *robota*, meaning *drudgery*. Robots efficiently carry out such routine tasks as welding, drilling, and painting automobile body parts.

The World Book Encyclopedia, 1995

A **robot** is a machine that performs a task automatically. The robot's actions are controlled by a microprocessor that has been programmed for the task. The robot follows a set of instructions that tell it exactly what to do to complete the task.

World Book's Young Scientist, 2000

robot /'ro:bot/n. **1** a machine with a human appearance or functioning like a human. **2** a machine capable of carrying out a complex series of actions automatically. **3** a person who works mechanically and efficiently but insensitively.

The Canadian Oxford Dictionary, 1998

Humans are the ultimate generalists, with a form designed by millions of years of evolution to respond to a very wide variety of circumstances. The science and technology of robotics is usually concerned with building machines to perform a much smaller number of tasks within a specific set of problems, such as inspection or assembly parts on production lines. Such robots generally have a much simpler form. They often consist of a jointed arm with a gripper or other devices that work like a hand and a microprocessor that functions like a brain.

Encyclopedia of Technology and Applied Sciences, 1994

Robot "A reprogrammable, multifunctional manipulator designed to move material, parts, tools or specialized devices through various programmed motions for the performance of a variety of tasks."

Robot Institute of America, 1979

Now, here we go on a more-detailed examination and explanation of robots, which, to coin a definition, are fully automated machines which may respond to external stimuli as well as to internal commands which have been prerecorded. It is important to note that we have here the term "robot," which is different from android, or droid for short, or from humanoid, another term associated with these machines.

The Complete Handbook of Robotics, 1984

Robot Any mechanical device that can be programmed to perform a number of tasks involving manipulation and movement under automatic control. Because of its use in science fiction, the term *robot* suggests a machine that has a humanlike appearance or that operates with humanlike capacities; in actuality modern industrial robots have very little physical resemblance to humans.

AP Dictionary of Science and Technology

Robot

(1) A device that responds to sensory input.

(2) A program that runs automatically without human intervention. Typically, a robot is endowed with some artificial intelligence so that it can react to different situations it may encounter. Two common types of robots are *agents* and *spiders*.

Webopedia

A robot is a machine designed to execute one or more tasks repeatedly, with speed and precision. There are as many different types of robots as there are tasks for them to perform.

what is ? com

A robot has three essential characteristics:

- It possesses some form of mobility.
- It can be programmed to accomplish a large variety of tasks.
- After being programmed, it operates automatically.

Australian Robotics and Automation Association

1. A robot may not injure a human being, or, through inaction, allow a human being to come to harm.

2. A robot must obey the orders given it by human beings except where such orders would conflict with the First Law.

3. A robot must protect its own existence as long as such protection does not conflict with the First or Second Law.

Isaac Asimov

Clearly, no one single definition encompasses what a robot is and how it is supposed to work. Different people have widely different and often conflicting ideas of what a robot is and what it isn't. Many different types of robots exist, each one meeting some of the definitions above.

In the following pages, I will investigate some of the different types of robots and introduce you to many of the skills and much of the knowledge to create your own robots.

Just remember that if you create a robot to take over the world and it fails, when the authorities come, you've never heard of me or this book.

Experiment 1
Toilet Paper Roll Mandroid

In the 1950s, scientists determined that the most likely body shape an alien from space would have would be a biped, the same as humans. A biped consists of two arms and two legs arranged symmetrically around a vertical line. The reasoning behind this conclusion was largely based on the scientists' familiarity with their own bodies: Recognizing that humans, as the result of hundreds of millions of years of evolution, are capable of carrying out an astonishingly varied range of tasks. Carrying this line of logic further, it was felt that aliens, evolved to the point where they can create machines similar to ours, must have bodies that are similar to ours.

This line of thinking is essentially what people go through when they are thinking about how robots should look. If you were asked what a robot should look like, you would probably think first of a biped robot like the Terminator or Robby the Robot. Using the logic of the scientists of the 1950s, building robots using the same body types as we have makes a lot of sense because we are comfortable with using our bodies to move around and manipulate objects.

As this book is about robots, I'm sure you want to start designing and operating robots. Because we have a successful form to follow (our own), let's start designing and building a simple biped robot out of toilet paper rolls, pipe cleaners, and some glue. Once this robot has been built, you can perform the book's first experiment for yourself—seeing how a biped robot would transition from standing straight up to walking forward. Once we have done this, we can start experimenting with other actions humans perform.

The mandroid for this experiment consists of a skeleton of used toilet paper rolls that are connected using pieces of pipe cleaner that have been glued to the inside of the toilet paper rolls. If the toilet paper rolls are analogous to the bones in your body, then the pipe cleaners are the connective tissues and your joints. In the plan view (Figure 1-1), note that I have specified locations for the pipe cleaners so that the skeleton can move (or articulate) the same way that your body can. Because we are following a human model very closely, we can expect to be successful and be able to go on to other experiments with this robot, such as having it walk over to an object and pick it up.

Note that for the different pipe cleaner joints, you'll see that I took into consideration their placement so that the robot would be capable of moving in the same manner as a human being.

To build the robot, I cut 10 2.5-inch-long (6.4 centimeters) pieces of pipe cleaner and collected 10 old toilet paper rolls. To cut pipe cleaners, I used a set of wire clippers—don't use scissors (especially ones that are important to other people). I shouldn't have to say this, but you should wait for toilet paper rolls to become available; don't expedite the process of getting bare toilet paper rolls. I don't want to get any angry emails from parents saying that one day they walked into their bathroom and found enough toilet paper lying on the floor to fill 10 rolls.

Once you have the cut pipe cleaners to size using a sharp knife, cut one of the rolls into two smaller cylinders, each piece 1 inch (2.5 centimeters) long. The longer piece will become the "back" of the robot with the small cylinder being the robot's "pelvis." On another toilet paper roll, cut a ring about 0.75 inches (2 centimeters) long; this will become the robot's "head."

With the other eight rolls of toilet paper, you are ready to start assembling the robot using some kind of paper or wood glue. Model airplane cement, epoxies, and contact cements are not appropriate for this task. You may want to try using a cyanoacrylate such as Krazy Glue to hold down the pipe cleaner pieces before using the paper or wood glue. Personally, I would discourage doing this as you will probably end up gluing yourself to short pieces of pipe cleaner and empty toilet rolls. After other people see this, it will be hard for them to think of you as an "Evil Genius" with any kind of seriousness.

On each toilet paper roll, put down a 1-inch (2.5-centimeter) bead of glue along the inside to affix one

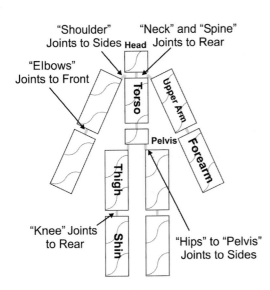

"Shoulder" Joints to Sides Head   "Neck" and "Spine" Joints to Rear

"Elbows" Joints to Front

Torso   Upper Arm

Pelvis   Forearm

Thigh

"Knee" Joints to Rear   Shin   "Hips" to "Pelvis" Joints to Sides

"Torso" and "Pelvis" cut from a single roll.
"Pelvis" is 1" (2.5 cm) Long. "Head" is cut from
a single roll and is 1.2 "(3 cm) long.

Figure 1-1   Toilet paper robot plan

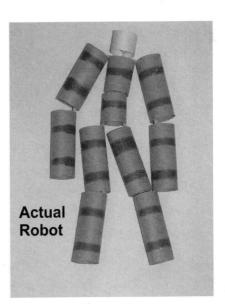

Actual Robot

end of a piece of pipe cleaner. Push 1 inch (2.5 centimeters) of pipe cleaner into the glue, leaving the other 1.5 inches (3.8 centimeters) outside the roll. When you have pushed the pipe cleaner down into the glue, put some glue on top of the pipe cleaner to make sure that it is secure. Once this is done, leave the rolls of toilet paper and pipe cleaners to dry for a day.

Next, repeat the process by gluing 1 inch (2.5 centimeters) of the exposed pipe cleaner into another toilet paper roll using exactly the same process and leave it to dry for another day. There will be 0.5 inches (1.25 centimeters) of pipe cleaner joint between the two rolls. Once the pieces have dried, glue them to the torso, one side at a time, to avoid having the glue run. When you are finished (a couple of days or so from when you started), you will have a model robot that looks like the one in Figure 1-1.

As I indicated, I would like to experiment with trying to get a robot to change from standing straight up to walking forward. With the glue on your robot dry, try to get it to stand up.

Chances are you will end up with a heap of seemingly loosely connected empty toilet rolls, similar to the pile I ended up with (Figure 1-2). You will also have the vexing problem of determining how to support the robot so that you can start experimenting with how you will make it walk.

Looking at this heap of paper, glue, and pipe cleaner, you can draw a few conclusions. The first one is that pipe cleaners are not rigid enough to support the collection of toilet paper rolls in a set position. You might be thinking of materials you could replace the pipe cleaners with, but I want to tell you to avoid going through the effort. Even if you had a robot that could support itself standing up, it is very difficult to come up with the motion required to walk without

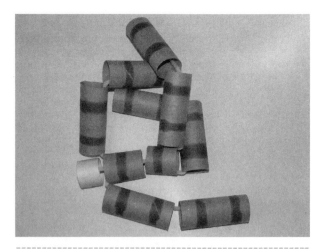

Figure 1-2   *Not an auspicious beginning to building your own robots*

the robot falling over. Remember that it took you a year or so to program yourself to stand up and walk forward, and in your case, you had all the necessary equipment to begin with. Walking forward is just one aspect of the problem—along with it, you will have to figure out how to turn and walk over uneven terrain. Stairs are a particularly vexing problem for walking robots.

Designing a mobile biped robot that can stand up and walk like a human being is considered by many roboticists as the "Holy Grail" of robotics—it is an incredibly challenging task in which large, well-funded companies and laboratories are just beginning to have success. With this in mind, I would like to change the way you look at robots so you start looking at them from the bottom up, gaining the necessary skills to build the different components used in a robot. One day you may build a robot that looks and acts like a human, but for right now, let's set our sights at the bottom.

Experiment 2
Pipe Cleaner Insect

In the first experiment in this book, I gave you some idea how difficult it is to create a robot based on the human form. I mentioned that there would be problems getting the robot to walk, but didn't go into detail because I didn't know of a way to get the robot to even stand reliably. Before starting to work on actual robots, I think that it is important to come up with a platform that is stable (it can stand up reliably) and then investigate how the robot can move or manipulate objects.

When looking for ideas or a better understanding of how to approach a problem, you will often find the answer by looking at nature and seeing how different animals (and even plants) respond to the same challenge. If we want to look at a stable platform capable of movement and moving objects, then we will probably look through different multilegged animals. This should be an obvious simplification of the robot platform. As a child, you were able to learn to crawl on all fours much sooner than you were able to walk.

Looking at animals that can walk around on all fours and manipulate objects similarly to humans, the obvious animal to me is the elephant. It can move around on its legs and manipulate objects using its trunk. The problem with the elephant (and any four-legged animal) is its dynamic and unstable motion. As the elephant walks, it transfers mass between its different legs so that it never quite falls over. Implementing this motion in robots is not terribly difficult, but it has the problem of the robot falling over if it were to stop abruptly or with one leg left in the air.

As a simple test of this statement, get down on your hands and knees and crawl across the floor, stopping abruptly with one arm or leg in the air. Depending on what you are holding up when you stop, you will either fall onto your side or your face. Initially, it might be difficult for you to fall over; you will automatically compensate for the lifted limb and move your body's center of gravity so that you are stable on three limbs. You might want to conduct this experiment over a gym mat to make sure you don't hurt yourself.

By going to a lower life-form with four legs, we have solved the problem of being unstable when standing up, but we still have a problem of movement. Let's look for a lower life-form that can move on multiple legs, but is always stable. The obvious animal that meets this requirement is the bug. If you observe the motion of an ant (cockroaches are too fast), you will see that at all times, the ant has at least three legs on the ground. As I show in Figure 1-3, when an ant moves forward, two legs on one side and one leg on the other are pushing it forward while the other three legs are getting into position to take over and move the insect forward.

The legs must be hinged and driven in such a way that they can move up and down and back and forth. Moving the lower leg up and down pushes the insect off the ground, and moving back and forth is used to either propel the robot or move the leg into position to propel the insect. Figure 1-4 shows a mechanical analog of an insect leg with the side-to-side motion provided by a hinge joint on the side of the insect, and the up-and-down motion accomplished by moving the lower leg up and down.

When referring to robot arms and legs, each dimension the limb can move in is called a *degree of freedom*. Although this simple insect leg only has two degrees of freedom (up/down and backwards/forwards), you will find that other robots have limbs with as many as eight degrees of freedom to allow them to perform complex tasks.

The insect is always stable (its center of gravity is always in the center of at least three legs), and if it were to stop for any reason, it wouldn't fall over because it is always stable, unlike the four-legged animal. Along with forward movement being easily implemented, changing direction for an insect is also quite simple. This is why insect-based robots (sometimes called *insectoids*) are more popular than ones based on cats, dogs, or elephants.

You can very easily investigate the properties of the insect-based robot by building a simple model for

Move 1

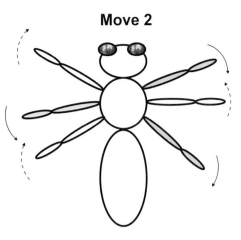

Shaded Legs on Surface, Pushing Insect Forward. White Legs off Surface Moving Forward to Push Next.

Move 2

Legs Previously off Surface. Lowered to Surface and Pushing Insect Forward. Legs Previously on Surface Lifted up and Moved Back to Position to Position to Drive Insect.

Figure 1-3 *Insect movement*

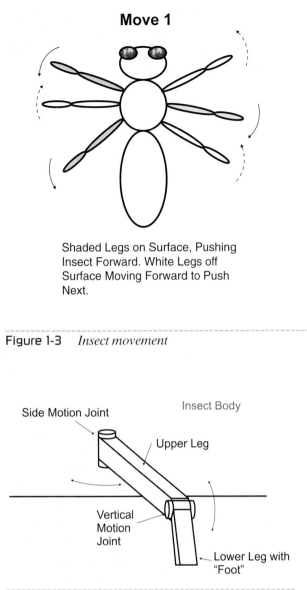

Figure 1-4 *Insect leg*

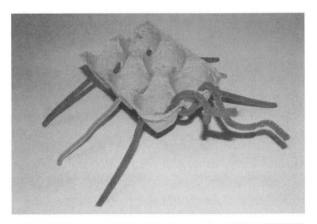

Figure 1-5 *The completed robot insect*

yourself. For my prototype, I used half of the bottom of an egg carton and, using white glue, mounted some pipe cleaner "legs." To create the legs, I poked holes in the side of the egg carton (in each well where eggs are stored), and pushed a pipe cleaner through the hole. Once I had the six legs (made out of the three pipe cleaners), I evened out the amount of pipe cleaner that was present on each side and then glued the legs into the egg carton wells. The antennae that you can see in Figure 1-5 are strictly for decoration.

You will find that it will take a day for the glue to dry. Once it has dried, you can start experimenting with the motion of the legs. Using Figures 1-3 and

1-6 showing the motion for an insect, you can demonstrate how an insect moves forward quite easily.

Turning the insect is accomplished by moving the legs in a similar manner as moving forward, but instead of moving all three legs in the same direction, the single leg on one side moves in the opposite direction, causing a differential force on the insect and turning it. This can be very easily demonstrated with your pipe cleaner and egg crate ant.

Reviewing the model robot you've just built, you should see two of areas of concern. The first is the need for the robot to support itself. Depending on

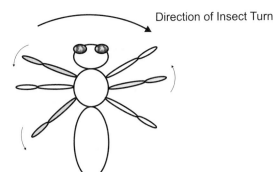

Direction of Insect Turn

Shaded legs on left side push to the rear while the shaded leg on the right side pushes forward. The white legs are above the surface, preparing to continue the turn.

Figure 1-6 *Insect turning*

how the robot is implemented, this could be a major concern as the weight of the robot may be more than what the legs (and the mechanisms that drive them) can handle. The second is the apparent complexity of the robot—you are probably feeling like there could be an easier way of developing a mobile robot.

Experiment 3
LEGO Mobile Robots

In the previous pages, I have looked at some different types of legged mobile robots. In presenting these different types, I have also noted that they have some fairly significant concerns that go with them in terms of complexity. Thinking about it, you may come to the conclusion that basing robots on some kind of life-form is not the best way to design them, and maybe we can look somewhere else for inspiration.

In our modern society, many different moving devices do not follow the human, animal, or insect form that was discussed previously in this section. For example, virtually 100 percent of the cars on the road are built using the same platform consisting of four wheels, two of which are driving and two are steering. For most modern (front-wheel drive) cars, the steering wheels are also the driving wheels.

Making the use of the car platform very attractive is the ability of many different model *remote-control* (R/C) cars to be converted into a robot format. Later in this book, I will discuss my experiences trying to convert a prebuilt car product into a base for a mobile robot.

If you want to build a car platform from scratch, you will discover two different problems that have to be overcome when you want to turn the vehicle. The first one has to do with the steering wheels. As can be seen in Figure 1-7, the "right" wheels (closest to the axis of the turn) will have to be at a sharper angle than the "left" wheels. (Actually, when you look at Figure 1-7, you will see that all four wheels are actually turning through a slightly different radius curve.) Most cars have an offset built into the steering gear pieces (known as *linkages*) that automatically turn the wheels at the required angle. This can be built into a robot, but you will have to work through the proper angles of the linkages to be successful.

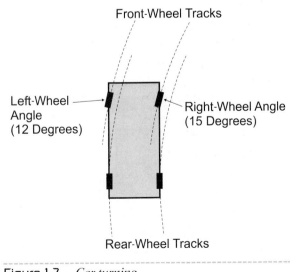

Front-Wheel Tracks

Left-Wheel Angle (12 Degrees)

Right-Wheel Angle (15 Degrees)

Rear-Wheel Tracks

Figure 1-7 *Car turning*

Robot designers use two common solutions to this problem. The first is to just use a single steering wheel. By doing this, the robot base is known as a *tricycle* for obvious reasons. The second solution to the different angle on the steering wheels is to mount both of them on a single bar, like a toy wagon. Using a few pieces of LEGO, you can build a model of a car robot with wagon steering, like the one shown in Figure 1-8.

For the examples in this experiment, I bought a LEGO Creators kit, which costs less than $10. If you have any LEGO toys in your home, chances are you will have enough parts to build the example robot models in this section without having to buy anything else. What you need is four wheels with axles attached to LEGO blocks, a vertical hinge (or small, round pieces of LEGO that will allow a block to pivot), and a few LEGO blocks to hold the model robot together.

This method works reasonably well, but can result in the robot tipping over when turning sharply, and it can have difficulty running over rough surfaces. The second issue with the car-based mobile robot is the difference in speeds between the driving wheels when the robot is turning. In the diagram of a car in a turn (Figure 1-7), you can see that the inside wheels have a smaller radius than the outside wheels. This smaller radius means that the inside wheels have to go a shorter distance in the same amount of time as the outside wheels. In a car, the solution to this problem is to build in a *differential*, which is a special type of gearbox that changes the speed of the different driving wheels depending on the angle of the turn. A differential could be built into a robot, but a much simpler solution is to just drive one wheel.

For many robot designers who want to create a steered robot, the tricycle platform, with the steered wheel being driven (the other two wheels are allowed to turn freely) is the optimum choice.

For other robot designers, after looking at the extra complexity of turning the wheels, they question the need for steering wheels at all. In Figure 1-9, I have built a simple LEGO robot with only two wheels.

This is the simplest robot that you can build and it is known as a *differentially driven* robot because the two wheels should be capable of moving independently so the robot can turn. With this robot model,

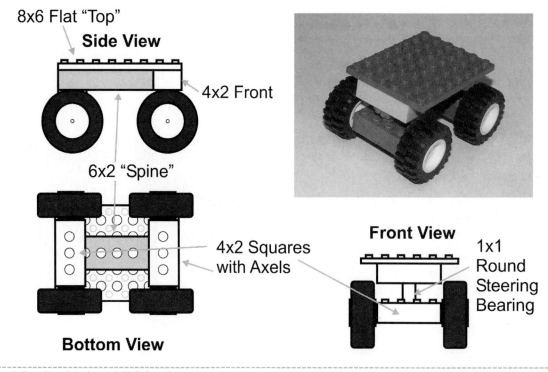

Figure 1-8   Design for a LEGO robot wagon

Top View

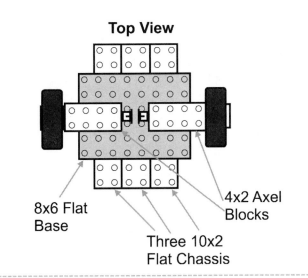

8x6 Flat
Base

4x2 Axel
Blocks

Three 10x2
Flat Chassis

Figure 1-9   *Two-wheel, differentially driven Lego robot*

you can place your fingers on the wheels and see how the robot moves with the wheels both going in the same direction, as well as turning the robot by turning the wheels at different rates or even turning the wheels in different directions.

If you look at different robots on the Internet, you will discover that 90 percent or more of the different robots that have been built at home follow this format. Of the different robot body types investigated so far, this is the simplest and least expensive to implement—its only real problem is that it does not handle changing or uneven surfaces well.

It is very easy to develop circuits and software to control the motion of the two wheels of this robot, and I will be using the differentially driven robot as

the base for the actual mobile robots I work through later in the book.

An evolutionary step for the differentially driven robot is the *tracked differentially driven* robot, which uses tracks like a military tank or bulldozer. This type of robot handles uneven surfaces very well (which is why it is used for tanks and bulldozers), but it can have a lot of resistance to movement. This is especially true when you are trying to turn the robot.

To minimize the robot's turning resistance, I feel the optimal solution is to have two driving wheels on each side of the robot (both linked to each other). You can build this robot very simply using the LEGO pieces you've used for the previous example bases as shown in Figure 1-10.

Top View

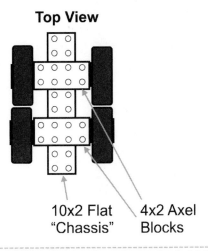

10x2 Flat
"Chassis"

4x2 Axel
Blocks

Figure 1-10   *Design for a four-wheeled LEGO cart*

Experiment 4
Cardboard Arm

So far I have discussed mobile robots, robots that can move about under their own control. Another area of robotics research is the operation of robot arms. These robots are usually designed to operate from some kind of base (it can move) and pick something up and move it to another location. In this experiment, you will build a simple model robot arm that is similar to the Canadarm on the space shuttle.

Saying that a robot arm can just pick up and move objects probably seems like an overly simplistic definition of a robot arm. You know that if you were to go into an automobile factory, you would see robots painting, welding, driving bolts, and assembling parts. None of these operations involve simply picking up and moving an object. If you were to examine the robots used to assemble cars, you would discover that each one is customized for the job that it has to do. The welding robot is designed with its "hand" (also known as an *end effector* or *gripper*) as a set of welding tongs. The painting robot's gripper is actually a paint sprayer. The assembly robot has an air wrench for its end effector. In each of these cases, specialized hardware is added to the robot for it to perform the desired function.

In some manufacturing settings where one robot is used to perform multiple tasks, the tools are designed to be picked up by the robot arm's gripper and then the robot moves the tools to the work area. By implementing the manufacturing process using this methodology, the robot arm is working in a manner similar to a human.

In Figure 1-11, I have outlined how to build a simple robot arm model in much the same way you did with the toilet paper roll biped robot in the first experiment. The difference in this robot is that I used the core from a roll of fax paper, cut it into two pieces, and using four pieces of 2.5-inch-long (6.4-centimeter) pipe cleaner, I glued them together as shown with a base joint, elbow joint, and two U-shaped pieces for the gripper.

Like the toilet paper roll biped robot, I put 1 inch (2.5 centimeters) of white glue into the rolls, placed the pieces of pipe cleaner into them, and let them dry for a day or so.

When the glue had dried, I then held down the pipe cleaner base joint and moved the arm around, similar to what is shown in Figure 1-12, and tried to

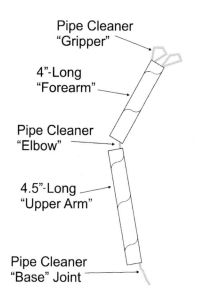

Pipe Cleaner "Gripper"

4"-Long "Forearm"

Pipe Cleaner "Elbow"

4.5"-Long "Upper Arm"

Pipe Cleaner "Base" Joint

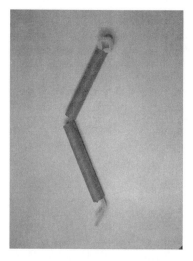

Figure 1-11   *A model robot arm that you can build*

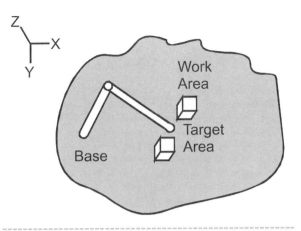

Figure 1-12   *Three-dimensional robot arm*

move to objects and see what was involved in the process.

In many real robot arms, this is similar to the process that is used to record or program their movements. For these robot arms, a human guides the arm (either directly or using some kind of remote control) to the various locations where the arm is needed. This method is quite fast and reasonably precise.

The problem with this method is that it is not practical for many robot applications. The classic example of something that cannot be programmed this way is the Canadarm used with the special shuttle. This robot arm cannot support itself under Earth's gravity, let alone carry any kind of payload. In this case, the motion of the arm must be worked out mathematically, usually by computer.

To illustrate what I mean, let's look at a two-dimensional robot arm that consists of a long upper arm and a short forearm (Figure 1-13). As I have drawn the arm, the upper arm can turn about 45 degrees from the shoulder while the upper arm can turn 180 degrees around the elbow. This range of motion has resulted in the strangely shaped *work envelope* that is drawn under the arm on the diagram.

When a robot arm is described, each direction a part can move in is described as a *degree of freedom*. For the robot arm shown in Figure 1-13, it has two degrees of freedom: one at the shoulder and one at the elbow. Your arm has seven degrees of freedom (three for your shoulder, one for your elbow, two for your wrist, and one for your hand opening and closing). The more degrees of freedom a robot arm can

move means increased complexity of the joints and the requirement for stronger actuators (the mechanisms that move the parts of the arm).

The *work envelope* defines all the places that the end of the robot arm can reach. To specify the "X" position within the work envelope for the two axis of freedom, you could use the formula below:

```
ArmX = (UpperArmLength x Cosine
        (UpperArmAngle)) +
        (ForeArmLength x Cosine
        (ForeArmAngle + UpperArmAngle))
```

Calculating the "armY" position is exactly the same, except sines are used rather than cosines. Note that to get the correct position at the end of the arm, the angle the forearm is at must also take into account the upper arm's angle. Hopefully, this formula does not scare you off—I gave it because I wanted to show that the position at the end of the arm can be fairly easily calculated using basic trigonometry.

I'm sure that my definition of a robot arm was disappointing to you—especially when you compared it to what your own arm and hand can do. This is a good opportunity for you to consider your arm and hand and think about just how amazing they are. Your brain is able to command your arm and hand to move to a specific location in space, in a specific manner, without having any visual feedback.

For Consideration

When I am asked by people how they can create their own experiments, I am surprised at how few people understand that in a well-designed experiment, the results are seldom a surprise. The stereotypical cloaked scientist mixing chemicals at random before he comes up with something interesting is a myth that many people believe. A properly designed experiment is used to test a theory or hypothesis, not to see what happens when something is done arbitrarily.

To illustrate what I mean, consider the experiments performed 150 years ago by many reputable scientists trying to discover what air is made up of. As you are probably aware, a bit more than

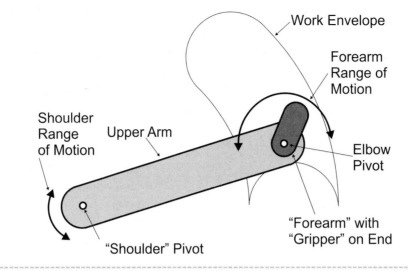

Figure 1-13   *Horizontal robot arm motion*

three-quarters of air is made of nitrogen, 20 percent or so is made of oxygen, and the remaining 1 or 2 percent consists of different trace gases. The experiments performed were made to understand what the trace gases were in the hopes of finding something that was valuable or new.

How the different gases were extracted was quite ingenious. Air is actually a solution (like salt water) and the different gases in it are distributed throughout and cannot be removed with techniques like centerfuging or heating. It actually took a hundred years of work before a method of extracting the different gases in air was demonstrated.

You may know that if you compress a gas enough, it will become a very cold liquid. Using this knowledge, air was compressed to the point where it liquefied, and then it was held at different temperatures to let the different gases boil off, allowing the scientists to collect what boiled off or what was left after everything else was allowed to escape. For example, nitrogen boils at −195.8 Celsius (C), oxygen at −182.8 C, carbon dioxide at −78.5 C, chlorine at −34.5 C, hydrogen at −252.7 C, and helium at −268.6 C. Just as a reference point, absolute zero is at −273 C.

The problem with the experiments was that the scientists performing them did not think about what was going to be produced by the experiment, so they tested them by opening the container the gasses were collected in and breathed them in. The assumption made was that because air was safe for people to breath, the different gasses that made it up must also be safe.

Unfortunately, this assumption is wrong. Mixed together in air, the different gases are quite benign, but by themselves, they can be very dangerous. Today we know that chlorine is a deadly corrosive poison, hydrogen explodes on contact with flame (remember light bulbs hadn't been invented when these experiments were taking place), pure oxygen will cause small fires to burn intensely, and large concentrations of carbon dioxide will stop a person's breathing.

It shouldn't be surprising that many of the scientists performing these experiments died (often horribly) without comprehending what was happening to them. The composition of the Earth's atmosphere and the properties of the different gasses that made it up weren't understood until the experiments were reviewed, and it was guessed that the constituent parts of air could be dangerous. With the guess (which is called a *hypothesis* or *theory*) that potentially dangerous gases were in the air, scientists were able to develop experiments that protected themselves (and their laboratories) and allowed them to better study the different gases.

To avoid getting into the trap of developing experiments that could fail (like the first one in this book), I am going to make sure the rest of the experiments presented in this book have the following six parts:

1. Statement of experiment's purpose. This is a simple statement, more than a title but not a

complete description of what's going to be done.

2. Theory behind the experiment. This is a statement of what the expected results are and why.

3. Bill of material (equipment or apparatus found in the "Parts Bin" and "Tool Box"). This is the list of equipment needed for the experiment.

4. Procedures. These are the different tasks needed to carry out the experiment. I will include assembly drawings and circuit diagrams for the experiment in this section.

5. Results (or observations). This is quite simply what I saw (and I would expect you to see). I will include photographs of the completed experiments as well as quantitative measurements of what was seen.

6. Conclusions. This is a discussion of any lessons learned and suggestions of what can be done next.

These six parts will be similar to how a high school science teacher would like experiments to be organized. Formatting the experiments will help you to organize your thoughts and understand exactly what you want to get out of the experiment. It also allows other people to repeat the experiment and test the results. Finally, the results should support the conclusion.

You may find that your teacher wants you to write up experiments in a specific format. Remember that if you want to get good marks in the course, write up your experiments in the format your teacher wants. Telling your teacher that you're just following the format I've used in this book isn't going to cut much ice with them.

An important tool for any experimenter is a neat, well-laid-out notebook. Going back to what your teachers have been telling you, they have probably been harping on the importance of a notebook for years. Notebooks are invaluable for storing ideas, circuit diagrams, formulas, or plans for the future. As you work through this book, I recommend that you keep a notebook alongside it and use it to save important points, the observations of your own experiments, or ideas that you can use later.

The experiments in this section are somewhat willy-nilly and I'm sure they don't measure up to what you are expecting for the book. Rather than calling them experiments, I would be more comfortable calling them model tests because they do not fit into my idea of what an experiment is and how it is conducted. Model is an excellent term for what was done in this section because the models demonstrate the concerns regarding different robot types. The purpose of this section is to show you that you can test out your own ideas for robots simply and inexpensively.

As I go forward in this book, I will be endeavoring to make the experiments much more rigorous and robust, and I'll reflect how experiments are actually conducted when robots are being developed. Whereas the experiments in this chapter took minimal thought to create, the experiments throughout the rest of the book will have much more up-front work done on them to make sure that they work reliably and can be used as a basis for later experiments or robot projects.

Section Two
Robot Structures

At the start of the previous section, I demonstrated that creating a robot based on the human form was difficult ("nontrivial" in engineering terms). The first robot experiment started with a somewhat skeleton-like structure that was connected with flexible joints. When the robot was stood up, it promptly collapsed in a heap. I explained that this form of robot is unstable and requires a series of muscles to keep it upright along with the skeleton and joints. After doing this, I then worked through a number of different forms that would be better suited to a robot, ending up with a small car-like form that had no moving parts other than its wheels. The first section explained why robots look the way they do, but it did not explain how they were manufactured.

In the first section, I used a variety of simple materials to make up the robot forms. These materials really weren't optimal as they took a long time to harden and were not very mechanically strong. In this section, I would like to investigate the different materials that robots are built from, along with some of the science behind robot structural design. When you have finished this section, you will have the basis for some of the robots I work through in the later experiments as well as a basic understanding of which materials and products are best suited for your own robots.

To avoid getting into a chicken-and-egg argument, I want to look at an already existing structure we can use to model our robots. The obvious one is the human body. Although you might be reluctant to consider the human body for this role because of the results from the first experiment, you should realize that smaller parts of the body (call them subsystems if you will) can be used as a guide for building a robot.

The parts of the human body that are of the most interest to robot designers (or roboticists) are the ones that can move. When looking at an anatomy book for information on the body's joints, you can see a basic finger joint is built like the one shown in Figure 2-1. I have simplified this diagram to show the most important pieces of the joint to more clearly illustrate what they are doing.

The purpose of the finger joint is to allow the muscles to change the angle at which the two bones of the finger meet. Muscles can only shorten (contract), so to change the bones' angle, one of them must contract, as shown in Figure 2-2. To straighten out the two bones, the muscle on the opposite side of the joint must then be commanded to contract and the other allowed to relax. To reproduce this action, you might be thinking of something like a solenoid, which I will discuss later.

The muscles are not connected directly to the bones. Instead, both the muscles and bones are con-

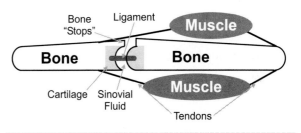

Figure 2-1   *The anatomy of a finger joint*

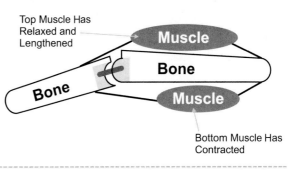

Figure 2-2   *Bottom muscle contracting to bend finger*

nected by a thin length of tissue known as a tendon. The tendon is very thin and strong, allowing the force of the muscle to pass over the bending joint without taking up a lot of space.

Moving the angle of the two bones relative to each other is only half the problem. The other half is coming up with a connection between the two bones that will allow the movement. In Figure 2-1, take a look at the joint section itself; it is made of four parts. The first part is the cartilage, which has a smooth, hard surface and is built into the bone to allow for the bones to rub together easily. The cartilage of the bones does not come into contact with one another, but they are separated by a thin film of liquid known as *sinovial fluid*. The ligaments pull the bones together and keep the cartilage of the two bones aligned. Finally, the "bone stops" are features that keep the bones from moving at an angle that the bone is not designed for. These different parts of the finger joint are analogous to a mechanical hinge.

The cartilage serves as a surface that bears the force applied to the joint, as does the hinge pin, and for this reason it is known as a *bearing*. To minimize the friction of the bearing, the finger joint is lubricated using sinovial fluid while in the hinge, and oil is used to minimize friction. The finger joint is held together by ligaments, and the curved metal pieces encasing the hinge pin provide the same function. Finally, the metal of the hinge is formed in such a way that it can only move so far in one direction, exactly analogous to the bone extensions built into the finger joint.

Using the hinge along with a couple pieces of wood, some fishing line, and a couple of devices that can contract, we could create a mechanical device that is the robot equivalent to a finger joint (see Figure 2-3). The joint's angle changes (or the joint moves) when one of the contracting devices (also known as *actuators* or *solenoids*) pull the fishing line connected to them.

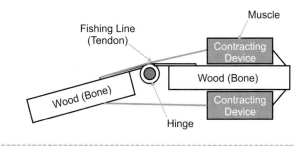

Figure 2-3 *Mechanical analog to finger joint*

This mechanical joint could also use a device that can push or pull, such as a hydraulic ram. In this joint, the actuator (muscle analog) is connected to the wood pieces (bones) by hinged rigid connections (called *push rods*) and can expand or contract, depending on what part of it has fluid forced into it. Creating a joint using a single actuator can be a bit simpler and cheaper to build than using two actuators.

After going through this, probably two issues immediately come to mind. The first is, in a real joint, a feedback mechanism (known as *reflexes*) is used to indicate the current position of the joint. When I introduce you to radio control servos, you will see how position feedback works.

The second question involves the results from the previous section. Instead of recommending a robot that has legs (like we use), I pushed the idea of robot that moved on wheels. The mechanical analogs I have given to the different parts of a finger joint probably do not seem like they would apply that well to a wheeled robot, but you can find strong similarities to the finger joint model presented here. Figure 2-4 indicates the motor is the actuator, and the axle bearing allows parts to move smoothly like the cartilage and sinovial fluid. The gears, wheels, and the mechanical bits holding everything together are the tendons.

Table 2-1 lists the different parts of the finger joint structure discussed in this introduction along with mechanical joint analogs and the wheel analog.

Table 2-1 Robot base materials

Finger Joint	Mechanical Analog	Wheel Analog
Bone	Connecting pieces	Base
Muscle	Actuators	Motor
Tendons	Fishing line/push	Rods drivetrain/wheels
Ligaments	Hinge body	Mounting pieces
Cartilage	Pin	Bearing
Sinovial fluid	Oil lubricant	Bearing lubricant

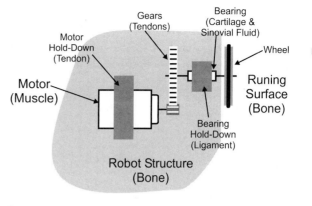

Figure 2-4 *Wheel/drivetrain system with the same part functions to the finger joint identified*

Experiment 5
Cutting Plywood

Parts Bin

12- by 12-inch sheet of 3/16-inch aircraft plywood

Tool Box
Saw for cutting plywood

When looking at different materials commonly used for robot framework, I came up with Table 2-2. The first two criteria, availability and cost, are self-explanatory. Strength is the material's relative strength and its resistance to being damaged. Cutting ease is how easy it is to cut down stock into a desired shape; the harder a material is (its strength), the harder it is to cut. Stability is how well the material keeps its shape (and precision) over time and use. "Vibration Resistance" is a measurement of the ability of the material to maintain its function when the robot is running. Will it crack or lose its shape as the robot runs over an uneven floor?

When paying a visit to a local hardware store, you will probably discover literally hundreds of different materials to choose from. Looking at Table 2-2, you'll see that for many of the different categories, I have given wood the widest possible range of ratings. The different ranges are due to the different types of wood available and the different properties that the various types can have depending on the cut type and where the piece was taken from the tree. Choosing the proper wood for a specific application requires a lot of knowledge and makes wood suboptimal, in my opinion, for use in a robot's framework.

Plywood consists of a number of thin sheets of wood arranged in different orientations and glued together under a press to create a strong and durable product, as shown in Figure 2-5. In the case of plywood, different types and cuts of wood are chosen for the application. The strongest and most durable type of plywood is known as aircraft plywood and can be found in small quantities at a hobby store.

G10FR4 is commonly used for making *printed circuit boards* (PCBs). It is built by compressing glass fibers together with an epoxy binding them together. G10 refers to the glass fibers and epoxy mixture, and maintains its dimensions despite changes in temperature. FR4 indicates that it has fire-resistant

Table 2-2 Material characteristics

Material	Availability	Cost	Strength	Cutting Ease	Stability	Vibration Resistance
Wood	Excellent	Good	Poor–excellent	Poor–excellent	Poor–excellent	Good
Plywood	Excellent	Fair	Excellent	Fair	Good–excellent	Excellent
Steel	Good	Good	Excellent	Poor–fair	Excellent	Good
Aluminum	Good	Fair	Good	Fair–good	Excellent	Fair
G10FR4	Fair	Poor	Excellent	Poor	Excellent	Excellent
Particle board	Excellent	Good	Fair–good	Fair–good	Poor–fair	Poor
Cardboard	Excellent	Excel	Poor–fair	Excellent	Poor	Poor
Foam board	Good	Good	Fair	Excellent	Poor	Excellent
Plexiglas	Good	Good	Fair	Poor–fair	Good	Poor
Polystyrene	Good	Fair	Poor–fair	Good–excellent	Good	Poor

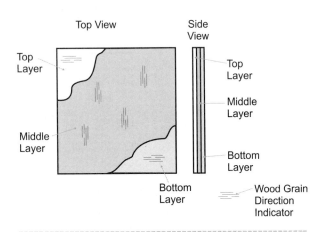

Figure 2-5 *Plywood construction*

formulation 4 built into it. Finding large blocks of G10FR4 is difficult and expensive, although many people use old PCBs for their robots, because they can be found for pennies at surplus stores.

You can consider a number of paper-based products, including particleboard (wood chips and fibers compressed and glued together), cardboard, and foam board (cardboard-cladded polystyrene foam). Of these, I would only recommend using foam board (available in art stores), as it survives well over time and is easy to form, but difficult to drill or machine precisely.

You can also consider plastics. Plexiglas (especially clear) can be used to make for an attractive robot.

Polystyrene usually comes in thin sheets and can be "vacuum formed" into interesting shapes and is excellent for making covers and bodies for robots. I tend to avoid plastics because they usually have poor vibration resistance. You will find that as you use your robot more and more, the likelihood of cracks appearing in the plastic increases.

Plywood is probably the best base material you can use when you are creating your own robot and it is what I will be using in this book. Before getting into the experiment, I wanted to cover the history of plywood along with some of the most interesting facts about it. After doing a bit of research, I discovered that plywood is about as boring as anything you could hope to find on this planet.

For this experiment, I would like you to cut a 12 by 12 piece of 3/16-inch aircraft plywood (available from a hobby shop for less than $5) into the 10 pieces shown in Figure 2-6. The three large rectangles will be used as the bases for the mobile robots presented in this book, whereas the four strips will be used to demonstrate the strength of plywood in the next experiment.

This experiment consists of figuring out the best way to cut the plywood into the shapes in Figure 2-6. The order of the cuts is shown in Figure 2-7. The first cut will be 4.75 inches from the edge of the plywood and will create a strip that is as wide as the length of the PCB included with the book.

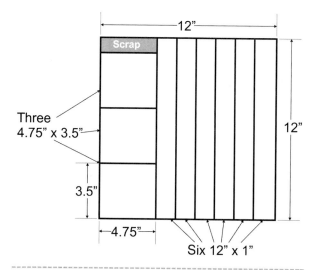

Figure 2-6 *Cuts to 3/16-inch plywood for experiments*

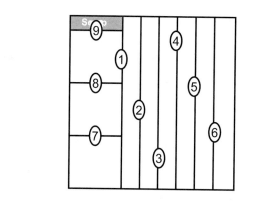

Figure 2-7 *Order of cuts to 3/16-inch plywood*

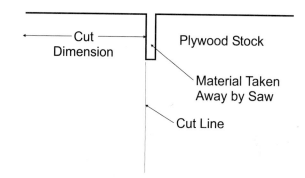

Figure 2-8 *Plan for lost material from previous cuts when marking your cut lines.*

Try different saws (such as a hacksaw, coping saw, jig saw, Dremel tool, sabre saw, table saw, band saw, circular saw, or a miter saw) and find which one works best for you. To cut plywood, I use a miter saw with a 12-inch circular blade. It is an expensive tool, but I can specify the cut angles that I want with accuracy. Depending on your means, you may only be able to use a handsaw, but make a note to yourself to try some of the other options listed here.

Note: If you are using a power tool, make sure you have been properly trained and are supervised when you are using it.

When you cut the plywood, you will find that some of the material is lost in the cuts, as shown in Figure 2-8. This means you will have to cut to the outside of the cut line to make sure the final piece is not smaller (by the width of the cutting blade). For this

reason, you should be measuring and marking your cuts after the previous one has been completed. If you mark the position of the cuts in the wood beforehand, you will discover that almost *all* your pieces, except for the first one, will be cut incorrectly. With some experience, you will learn to compensate for lost material when you are cutting, but for now, measure and mark your cuts just before doing them to make sure they are as accurate as possible.

Experiment 6
Strengthening Structures

Parts Bin

Four strips of 12- by 1- by 3/16-inch aircraft plywood cut in previous experiment

Tool Box

Carpenter's glue
Wood clamps

Now that we have decided upon using plywood for the robot structures in this book, I want to look at its physical properties and what we can take advantage of in the robots. In this experiment, I want to look at how strong plywood actually is and what can be done to strengthen it. This experiment is somewhat ironic, because although I am going to show you how you can strengthen a piece of plywood by two to three orders of magnitude (100 to 1,000 times), I recommend that you design your robots without having to do this.

Figure 2-9 shows the four different forces that can be applied to an object. Tension consists of trying to pull the piece apart. Trying to make a smaller piece of wood by pressing on it is known as compression. Twisting is called torsion. Finally, you can also try bending the strip in the middle. To look at the first three of these different properties, try these options:

- **Pulling the piece of plywood apart** You may want to get someone to help you. Each person should grab an end of the strip and pull.

- **Compressing the plywood** You and the helper from the previous step should try and push against the strip. This should be done without bending it. The idea is to just to test the strip with just one type of force.

- **Twist the plywood** This could be done either by two people or just yourself and something like a bench vise. Stop if you begin to hear cracking (although this will take a lot of force).

To test and quantify the strip of plywood's strength against bending, set up the apparatus shown in Figure 2-10. Place two bricks (or lengths of 2×4)

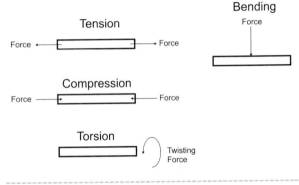

Figure 2-9   *Structural forces*

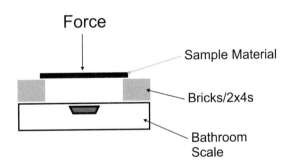

Figure 2-10   *Measuring bending force using a bathroom scale*

on a home scale with the strip of plywood between them. Record the weight of the bricks and the strip.

Next, slowly press down on the center of the strip with your foot, watching the indicated weight on the scale. I found that I had a measured weight of 5 pounds (2.3 kilograms) for 2 bricks and the plywood strip. As I slowly increased the force on the strip, I saw the scale go to 9 pounds or 4.1 kilograms before the strip started to make cracking sounds and break. Applying 4 pounds of force is surprisingly small and

Experiment 6 — Strengthening Structures

much less than would have been required for any of the forces to break the strip.

To summarize the results, we can say that the plywood strip is extremely resistant to tension and compression, resistant to torsion, and not very resistant to bending. With this information, let's create a different form for the plywood to see if we can increase its resistance to bending forces.

You might be confused by what is happening in this experiment. Thinking about it, you might be inclined to look at the bent strip of plywood like the cable shown in Figure 2-11. As a force pushes down on the cable, the different elements experience more tension. In reality, when the plywood is bent, the different fibers (or material particles) that make up the structure are having an asymmetric force applied to them. This asymmetric force breaks the fibers at the microscopic level by shearing them apart, resulting in the strip breaking. An important rule to remember is that the larger the bending movement the strip is allowed, the greater the shearing force applied to the individual fibers and the easier they will break.

The approach I used to find this better shape is to take advantage of the strengths of the plywood strip to overcome its weakness. In this case, I want to create a shape that will use the excellent tension and compression properties of plywood to overcome the poor bending quality.

The shape I decided to use should not be a surprise if you have ever seen a building put up. I decided on the I-beam shape (shown in Figure 2-12) because it will convert the incoming bending force to compression (on the top) and tension (on the bottom), as shown in Figure 2-13, rather than the shearing.

You can build a simple I beam to test using three strips of plywood and gluing them together in the

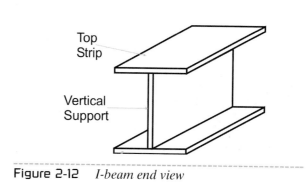

Figure 2-12   *I-beam end view*

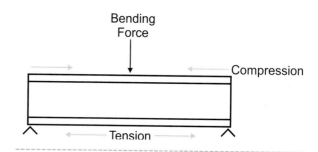

Figure 2-13   *Bending forces on the I beam*

I-beam shape using carpenter's glue. When I built my plywood I beam, I first glued the bottom piece to the vertical support and clamped the assembly together at the two ends and the middle. After waiting a day for the glue to harden, I then glued the top piece on and again clamped it together and let it harden overnight.

Repeating the experiment with the two bricks and the scale, I found that I could not break the I beam, even when I stood on it (and I am 200 pounds or 91 kilograms, which is 50 times more force that what was required to break a single piece of plywood). I was amazed at the strength of the I beam.

You might wonder what would happen if you simply glued three pieces of plywood together, flat side to flat side, essentially creating a thicker piece of plywood. You may expect to get a similar gain in bending resistance as the I beam created here. If you were to try it, you would discover that the bending resistance would be increased, but in a very linear fashion. You would find that two pieces of plywood would have essentially twice the bending resistance of a single piece, three pieces would have three times, and so on.

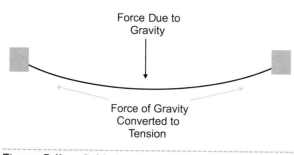

Figure 2-11   *Cable bending while suspended between two points*

Despite the phenomenal gains in bending strength from an I-beam shape and the small amount of effort needed to create one, I am not going to take advantage of it in any of the robots presented in this book simply because it is not needed. When you look at the robot structures I have created for this book, you'll see that all of them are quite short because the greater the movement of the wood, the greater the

shearing force on the fibers. By keeping the pieces of plywood relatively short (as well as wide so that the force is distributed over more area), you will find that the wood's resistance to bending forces increases significantly. By keeping the pieces short and wide, I don't have to worry about having to strengthen the pieces of plywood used for the robot's structures.

Experiment 7
Finishing Wood

Parts Bin

Two strips of 12- by 1- by 3/16-inch aircraft plywood cut in previous experiment

Three 4.75- by 3.5-inch pieces of 3/16-inch aircraft plywood cut in previous experiment

Tool Box

220-grit wet/dry sandpaper
Bottle caps
Old newspaper
Old cloth rag
Autobody primer aerosol
Acrylic marine aerosol

Nothing seems to be a bigger turn-off when it comes to home-built robots than bare plywood or other wood on the robot. I've been to several robot meets and seen well-designed robots that other people said looked as if they were just "thrown together." I have also seen robots that were obviously quite poor yet received many accolades on their appearance, and the only difference I could see was a nicely finished piece of wood that gave the appearance that considerable effort went into the design and construction of the robot. In this experiment, I will show you how to quickly and easily finish pieces of wood not only to make the robot look better, but also to provide a few advantages over using unpainted wood in your robot's structure.

The process will require a few minutes of time over the course of a few days. It will allow you to create finished pieces of plywood that will have the functional benefits of

- Eliminating the dust on the surface of the plywood, allowing for an effective surface for two-sided tape attachment and removal. Eliminating the fibers also allows for more effective glue and tape bonds.

- Smoothing the wood's surface and reducing the lifted fibers that appear when the wood gets moist or wet over time.

- Eliminating splinters and minimizing surface splintering when drilling into the wood.

- Allowing pencil and ink markings for corrections on the surface to be easily wiped off.

At the end of this experiment, you will have two painted plywood strips that will be used for the next experiment, along with three pieces of plywood that will be used for robot bases later in this book. Some of the materials listed at the beginning of this section probably seem to be quite obvious, whereas others will be quite surprising. I have tried to list everything that you will need so that the painting operations will go as efficiently as possible; this is why you will see items in the list like bottle caps.

I tend to use aerosol paint; when properly used, very little mess occurs and no brushes have to be cleaned up. From an autobody supply house, you should buy an aerosol can of primer (gray is always my first choice), and from a hardware store, buy an aerosol can of indoor/outdoor (or marine) acrylic paint in your favorite color (I use Krylon®). Personally, I like to use red, as it catches the eye and isn't overwhelming. If there is a blemish in the wood or your work, it hides it quite nicely.

Set up a painting area in a garage or another well-ventilated area by laying down newspaper both on the floor as well as vertically. Next, lay down the bottle caps to be used as supports for the materials you are going to be painting over the newspaper. You don't have to use bottle caps; scraps of wood or other detritus can be just as effective. Just make sure that when you are painting something that the supports are smaller than the perimeter of the object being finished. You will want to finish the ends of the plywood and don't want to end up with paint flowing between the plywood and the support.

Lightly sand the surfaces of the plywood you are going to paint. When finishing the two strips, I just painted 6 inches (15 centimeters) on one end. You may also want to sand the edges of the strips more aggressively to take off any loose wood that could become splinters. Once you have finished this, moisten the rag and wipe it over the surface you have sanded to pick up any loose dust.

Shake the can of primer following the instructions printed on the can. Usually, a small metal ball is inside the can, and you will be instructed to shake it until the ball rattles easily inside. Start with the two plywood strips and place one on the bottle caps supporting the surface to be painted. Spray about 6 inches of the strip, starting at the supported end. Most primers take 30 minutes or so to dry. Check the instructions on the can before going on to the next step of sanding and putting on new coats of paint or primer.

After the first application of primer, you will probably find that the surface of the wood is very rough. This is due to the cut fibers in the wood standing on end after being moistened from the primer. Repeat the sanding step (along with sanding the ends of the wood and wiping it down with the damp rag) before applying another coat of primer. After the second coat is put down, let it dry, sand very lightly, and wipe down again.

Now you are ready to apply the paint. Shake the can according to the instructions and spray the plywood strips, putting on a thin, even coat. You will probably find that the paint will seem to be sucked into the wood and the surface will not be that shiny. This is normal. Once the paint has dried, lightly sand again, wipe down with a wet cloth, and apply a thicker coat of paint.

When this coat has dried, you'll find that the surface of the plywood is smooth and shiny. Some of the grain of the wood is still visible, but it is not that noticeable. You do not have to sand the paint again; the plywood is now ready to be used in a robot.

Once you have done the two strips, you can finish the three rectangular pieces of plywood the same way. The difference will be that after each painting step, instead of going ahead and sanding, you should turn the piece over and use primer or paint on the other side. Sand both sides and then apply the next layer. When I am painting pieces like this, I usually spray twice a day (to make sure the primer or paint is thoroughly dry), which means it takes me four days to produce the finished wood. I suggest you plan ahead and finish as many pieces as I think are required at one time. When you are painting the rectangular pieces of plywood, remember to paint the edges as well. The acrylic paint will help bind the edges of the plywood together, minimizing the chance that you will get a splinter from the wood.

The two partially painted strips of wood will be used in the experiment that follows and the three fully painted rectangular pieces will be used as the mounts for the mobile robots presented later in the book.

Experiment 8
A Gaggle of Glues

Parts Bin

Two 12- by 1- by 3/16-inch piece of partially finished aircraft plywood

Tool Box

Small wood clamps

Weldbond

Solvents

Krazy Glue®/Loctite®

Carpenter's glue

Five-minute epoxy

Contact cement

Two-sided tape

Hot glue gun

What do you think is the number one problem I see when a robot is brought out for a competition? Most people would think of things like dead batteries or code that can't work in the actual environment (causing problems with light background noise or the running surface), but what I usually see is a robot that falls apart or breaks because the different elements are not fastened together very well. Part of the problem is the use of an unsuitable material for the structure (like one that breaks during use), but the overwhelming problem is the use of inappropriate adhesives and fasteners for the robots.

This is not surprising because a plethora of different glues and mechanical fasteners (discussed in the next experiment) can be chosen from. The list in the Parts Bin is just a small fraction of the total number of glues that are available. In this experiment, I would like to examine how a number of different types of glue work, using the two partially finished strips of plywood from the previous experiments. This experiment contains a pretty skimpy explanation of different glues and not much more about them other than which applications they are best suited for. If you are interested in finding out more about the different glues and why they behave the way they do, you can do research at a library or on the Internet.

I should discuss a few points about glues before discussing the different kinds and testing them out on the two strips of plywood:

- Glues work best when they are attaching two pieces of the same material together. In situations where dissimilar materials are attached,

use a mechanical fastener (like a nut and bolt) as described in the next experiment.

- Glues are chemicals and as such should be handled with care to ensure that no opportunity exists for you or others to be chemically burned or glued to different parts of the robot. Make sure you read and are familiar with the instructions and warning labels.

- Glue can also be known as "adhesive."

- The layer of glue between two objects is called the "joint."

- "Solvents" are used to thin glues, increase their curing time, or to wash glues from surfaces.

- Glues do not dry; they *cure* or *harden*.

- You cannot reactivate glues by soaking a joint in water or the glue's solvent.

- The best test of a glue is to use it to attach two representative objects, wait for the glue to cure, and then pull them apart. The glue is appropriate for the application if the joint doesn't break, but the attached material does.

The last point is a useful rule and states that the glue and the bond it forms with the material must be stronger than the material itself.

To help you understand which type of glue (listed in the Parts Bin) should be used for which application, I have created Table 2-3, which lists the different glues, the materials they are best suited for, and any comments I have about them. These glues are reason-

Table 2-3 Glues and their best materials

Glue	Materials	Comments
Weldbond	Wood/PCB	Excellent for tying down loose wires and insulating PCBs.
Solvents	Plastics	Melts plastics together.
Krazy Glue/Loctite	Metal/plastic	Best for locking nuts.
Carpenter's glue	Wood	Unfinished wood.
Five-minute epoxy	Everything	Very permanent.
Contact cement	Flat, porous materials	Good for bonding paper/laminate to wood.
Two-sided tape	Smooth surface	Good for holding components on the robot structure.
Hot glue gun	Everything	Not recommended.

ably inexpensive and can be found in virtually all hardware stores.

As stated in the table, I only use Krazy Glue to prevent nuts from loosening on bolts. Also, two-sided tape is excellent for mounting battery packs and servos on robots. For best results, I recommend that you use 3M Scotch Super Strength (5 pounds) Exterior

Mounting tape. I would caution you against using hot glue (dispensed from a hot gun) because it does not hold as well as the other glues. It also does not handle vibration very well and can leave long, sticky strings.

This experiment is quite simple: Using the two strips of partially finished plywood, test out and document the appropriateness of the different glues on both the finished and unfinished portions of the wood. Then document the results in a table. A successful glue is one in which the joint doesn't break, but the material being held together does.

The idea of gluing yourself (or somebody else) to something like a table might seem funny, but there is a real danger of causing an injury if a caustic glue is used or attempts to pull the glue off are taken without reading the warning label on the package or consulting a doctor first. Always remember that glues are chemicals and can cause serious problems. Follow the instructions on the package to make sure the glues are used correctly and any required solvents are on hand for cleanup.

After carrying out this experiment, I found that carpenter's glue was best for unfinished wood, and two-sided tape was most convenient for finished wood. You'll probably find that five-minute epoxy can glue anything to anything. You will be surprised at how ineffectual the other glues are when gluing strips of wood.

Experiment 9
Nuts and Bolts

Parts Bin

Deck of playing cards

Tool Box

Flat work surface

In this experiment, I would like to talk about using nuts and bolts as the primary removable fasteners you will use when you are creating your robots. Chances are you have used them in numerous products and toys, and you probably feel like not much

needs to be said, but you don't understand the principle upon which they are based.

When I use the term fastener, I am referring to a device that will hold two pieces of material together. In the previous experiment, you tried out a number of different glues on finished and unfinished wood to

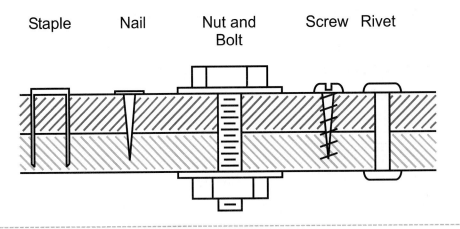

Figure 2-14 *Different mechanical fasteners*

better understand how the glues worked and which purposes they were best suited for. In this experiment, I would like to explain how a nut and bolt hold two pieces of material together but allow you to easily take them apart. You can consider other mechanical fasteners (including rivets, staples, screws, and nails, as shown in Figure 2-14), but for your robots I would like you to focus on the nut and bolt. The other fasteners shown in Figure 2-14 can only be used once (removal may damage the fastener or the material) and many of them will loosen over time due to vibration and stress.

Figure 2-15 shows a side view of a nut and bolt holding two pieces of material together. The bolt symbol has the horizontal bars that indicate the threads (the metal strip running around the shaft of the bolt screwed into the nut). The bolt's length should be chosen so that it extends through the material, through any washers, and past the nut for at least one and a half threads.

Washers are round pieces of metal used to spread the force applied by the nut and bolt on the material to lessen the chance the material will be damaged. I will further discuss washers at the end of this experiment and explain how they can be useful to you.

When you put on a nut and bolt, you have probably noticed that the nut slips on easily, but as you turn it down and the material is compressed, the nut becomes harder to turn. You should also notice that if you were to reverse the nut, the stiffness would also be present initially and then loosen as the nut is no longer in contact with the material being held together.

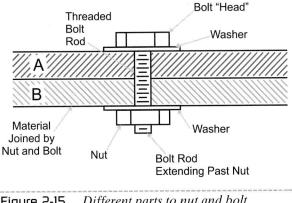

Figure 2-15 *Different parts to nut and bolt fastener*

The reason the nut becomes harder to turn is due to the material together, and placing tension on the shaft of the bolt. The technical term for this is called *pretension*, as shown in Figure 2-16 As the bolt undergoes pretension, the force it applies on the threads increases, causing an increase in the amount of *friction* the nut experiences.

The pretension force on the threads is in a different direction than the turning direction of the nut. As the pretension force on the threads is increased, the friction experienced turning the nut increases. This becomes a vicious cycle; the more the nut is turned, the greater the pretension, and the more pretension, the harder it is to turn the nut due to friction. This will continue until it is impossible to turn the nut any more. The force due to friction is defined by the following formula:

$$F_{friction} = k \times F_{pretension}$$

Experiment 9 — Nuts and Bolts

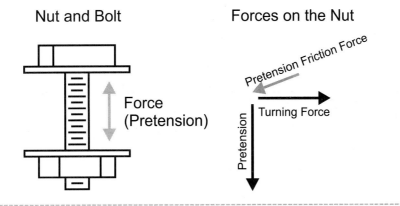

Nut and Bolt

Forces on the Nut

Figure 2-16   *Friction and force directions in a nut and bolt*

This equation says that $F_{pretension}$ (the pretension force) is multiplied by a constant to give you the nut's frictional force. Remember that this friction works in *both* directions, both when you tighten the nut as well as loosen it.

To demonstrate how this force works, I could have come up with an experiment that measures the pretension of the bolt as well as the amount of force needed to turn the nut, but I wanted to come up with something simpler that would be fun for you as well. To do this, build a house of cards (Figure 2-17).

The surface of a playing card is quite slippery. When you place a card on it at a very sharp angle (close to being perpendicular to the card), the friction of the card, which is a function of the force of gravity, will hold the card upright despite it trying to slide the bottom of the card. This is shown in Figure 2-18 along

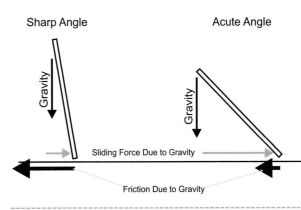

Figure 2-18   *Friction holding a card upright when the frictional force due to gravity downward is greater than the sideways force due to gravity*

with a drawing that shows what happens when the card is at a more acute angle. In the left drawing of Figure 2-18, the amount of sideways friction due to gravity is greater than the sideways force.

The increase in friction due to the increased force of gravity in the card on the left in Figure 2-18 is exactly the same as the increase of friction on the threads of the nut. For many applications, simply tightening the nut and bolt causes enough friction to prevent the nut from loosening.

Cases occur, however, when a bit of help is needed. Figure 2-19 shows the three most common washers you will work with. The flat washer is simply a ring of metal that is used to spread out the force of the nut and bolt as well as protect the material from being damaged by them (especially when the nut and bolt are being tightened).

Figure 2-17   *A house of cards showing how the force of gravity can increase frictional force*

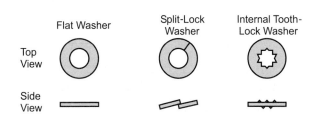

	Flat Washer	Split-Lock Washer	Internal Tooth-Lock Washer
Top View			
Side View			

Figure 2-19 *Different types of washers*

The other two washers shown in Figure 2-19 provide the same function as the flat washer but also help to hold the nut in place. For this reason, they are known as "lock" washers. The internal tooth-lock washer has a number of bent teeth inside it that will twist when being tightened and "bite" into the material, providing more friction. This type of washer is good to use on woods, plastics, and laminates (such as PCB materials).

The split-lock washer, acts like a small spring and pushes against the nut as it is tightened, adding to the pretension of the system and increasing the friction of the nut on the bolt. The split-lock washer should be used only on metal. If you do not want to use a lock washer, you can always use something like a drop of Krazy Glue on the threads to hold the nut in place (while still being removable).

In the experiments, I will specify some nuts and bolts, and in other cases let you decide. For most hobbyist robot applications, the type of nut, bolt, and washer used is not that important; the worst case is that the robot will break down and you will have to tighten the nuts up. If you are working on a powerful robot that is very heavy, you should consult with an expert or a machinist's text to find out what kind of nuts, bolts, and washers should be used.

Experiment 10
Soldering and Splicing Wires

Parts Bin

Wire sections

1/8-inch–diameter heat shrink tubing

Tool Box

Flat workstation

Soldering iron/soldering station

60/40 flux core solder

Matches

Clippers/wire strippers

One of the most important basic skills you will need if you want to be able to create your own robots is the ability to solder. Soldering (normally pronounced "soddering;" the "L" is silent) is the process of joining of two pieces of metal with an interposing material. This interposing material for electronic circuits is normally a tin-lead mixture called *solder* and is applied by applying heat. Soldering is different from welding or brazing in which the material being joined is brought to its melting point and then pressed together with or without interposing material. Welding or brazing results in a strong bond, and a solder joint has good electrical characteristics.

Figure 2-20 shows the cross-section of two joined pieces of copper. The edges of the two pieces of copper have been raised to the melting point of the solder, which flowed over them, forming an electrical and mechanical bond. If you want to break the connection between the two pieces of copper, just melt the solder. You should be aware of a few soldering technicalities before attempting to splice wire in this experiment.

Although the idea behind the soldering here is to create a bond by melting the solder between two pieces of copper, a mixing of the copper with the solder takes place and is known as the *intermetallic region* shown in Figure 2-20. This copper-solder mix is

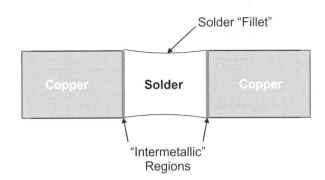

Figure 2-20 *Solder joint cross-sections*

actually an alloy (a mix of metals), often with greater strength and a higher melting point than the copper or solder. The goal of soldering is to apply heat for the minimum time possible to create a good joint and keep the intermetallic region as narrow as possible.

Tinning is accomplished by heating up a piece of copper and applying a thin layer of solder over the copper to inhibit corrosion. You will find that many wires and electronic components are pretinned to minimize the possibility that the copper will corrode, which will make soldering more difficult. You can tell if a wire or component is tinned because it will be a gray or silver color, rather than the expected copper color.

Solder is made up of a tin-lead mixture of varying percentages, and when solder is specified, the tin content is specified first (40/60 solder is made up of 40 percent tin and 60 percent lead). Some solders have silver added to them, but silver does not improve the joint's ability to pass electronic signals; it has been added to improve the joint's mechanical characteristics. Solder with silver added to it is not required for any applications presented in this book. Eutectic solder has a low melting point and is used for surface-mount electronics (not discussed in this book).

Lead-free solders are available and the electronics industry is changing toward having all electronic devices use lead-free solders by 2010. Currently, lead-free solders are difficult to find and use with conventional electronic parts. I recommend that you work with standard electronics 37/63 or 40/60 solders (with rosin flux cores) until you can be sure that the components you are working with are designed for lead-free solders. As I write this (mid-2003), very few components are designed for lead-free solders. The

lead in solders is not a health concern if you make sure

- You solder in a well-ventilated area.

- You do not smoke while soldering (cigarettes and lead fumes can create cyanide gas).

- You wash your hands after soldering.

The copper should be as clean as possible for solder to adhere to it. This should not be a big problem because new parts will have clean wires or, if you are connecting wires, they have been protected in a plastic sheath (called *insulation*). To further ensure that solder will stick to copper, most electronic solder comes with a weak, heat-activated acid called *flux* that cleans off copper oxides and debris. Many different types of flux are available, with rosin being the type of flux you should buy. Acid flux is used for plumbing applications, and no-clean fluxes (rosin flux is cleaned using water or isopropyl alcohol) should be avoided, as the residue from one joint can affect the solderability of another.

To apply solder, you will need a soldering iron, and a suitable one for digital electronics can be purchased for around $20. A a low-end soldering statio consists of a small lightweight soldering iron connected to a base that controls power to the iron as well as the heat on its tip. When you are soldering, the soldering iron's tip should be wiped on a wet sponge periodically to clean off any burned flux or excess solder.

Your soldering iron's tip will look something like Figure 2-21. An insulated grip contains a heating element and a removable tip that has been pretinned. If the tinning at the end of the removable tip is lost over time or if you can't get it clean, replace the tip. Do not try to file the tip down and then re-tin it yourself. When you file the tip, you will expose the copper of the tip and it will pollute the solder and result in the alloys I warned you about.

When you choose a soldering iron, make sure you get one designed for electronics assembly. It should be rated at 30 watts or so. A higher watt rating isn't better and can accidentally damage electronic circuits. A lower-watt rating may not result in satisfactory solder joints. A 30-watt iron or a solder station is ideal. Make sure a metal support for the iron keeps the tip away from the workbench surface.

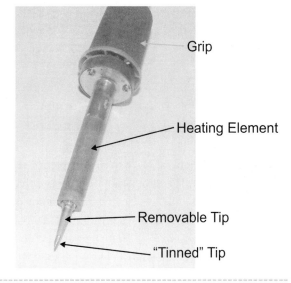

Figure 2-21   *Soldering iron tip*

Grip

Heating Element

Removable Tip

"Tinned" Tip

Soldering irons get hot, so make sure you are careful with them at all times, and put them down on a metal support to make sure they don't burn your table. If you aren't sure if the tip is hot, don't test it by using your finger or other parts of your body. See if it will melt a piece of solder or touch a solder sponge to hear it sizzle. Remember that although scars received in battle may say that you're heroic, scars received from soldering will just say that you are stupid.

With the theory of soldering behind us, let's experiment with trying to splice two wires together, as shown in Figure 2-22. This is an excellent, low-cost way of learning basic soldering skills and it will keep you from ruining the PCB that came with this book. In step 1 of Figure 2-22, the bared wires from two scraps of stranded wire are brought together. Baring or stripping wires is accomplished by either using a stripper tool or a sharp hobby knife. Strip a 1/4 -inch (6 millimeters) of insulation from the wires and try to merge the strands together as I have tried to do.

Next, wait for your soldering iron to heat up by testing it with a piece of solder. Wait for the solder to melt when it touches the iron. You may also see a puff of smoke when the solder is pressed to the iron. This is the flux vaporizing when heat is applied. When the iron is hot, hold it against the joined wires for a few seconds and apply the solder to it. The result should look something like step 2 of Figure 2-22 and the surface should be shiny, not dull. Once the wires are soldered together, you can protect and insulate the soldered splice using 1/8-inch–diameter (3 millimeter) heat-shrink tubing, as shown in steps 3 and 4 of Figure 2-22. As its name implies, heat-shrink tubing contracts when heat (like that from a match) is applied to it.

Step 1
Bringing Wires Together

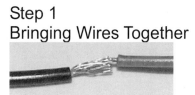

Step 2
Soldering Wires

Step 3
Fitting Heat Shrink

Step 4 - Finished Splice

Figure 2-22   *Splicing two wires using solder*

Experiment 11
Assembling the Included PCB

Parts Bin

Book PCB

Two 220 v 16-pin resistor Dual In-line Packages (DIPs)

Two 0.01 µF capacitors of any type

CKN9009 momentary push-button switch

Twenty-four-pin, 0.600-inch sockets

Single-row, 32-pin PCB mount socket

PCB mount, female 9-pin D-Shell connector

Keystone 1294 9-volt battery connector

Short (3.5-inch or 82-millimeter) breadboard

Two 1/4-inch (6.5-millimeter) 4-40 flathead screws and nuts

Tool Box

Soldering iron/soldering station

60/40 flux core solder

Matches

Clippers/wire strippers

Screwdriver for 4-40 screws

To make it easier for you to learn how to design your own robots, a PCB has been included with the book that, after soldering, can be used to explain and demonstrate basic electronic concepts. It also serves as a base for mounting your robot's control electronics. After assembly of the PCB (which means adding the electronic components listed in the Parts Bin at the start of the experiment), the PCB will provide you with

- A 9-volt battery connector to power your experiments.

- A breadboard that is used to temporarily wire circuits.

- A socket for a Parallax *BASIC Stamp 2* (BS2), the controller used for in the experiments presented in this book.

- A set of current-limiting resistors to protect the BS2's *input/output* (I/O) pins.

- A programming interface for the BS2.

Certain terms in the list of parts and features built into the PCB may be unfamiliar to you. Don't worry if something is confusing to you; as you work through the experiments, you will gain an understanding of these parts and features, including why they are built

into the PCB and how you would use them when you are designing your own robot. The Parts Bin description of the parts should be enough for you to go out and buy the parts, even if you've never worked with electronics before. If you are unsure, ask someone working at the store, but make sure you have the book and the PCB with you.

In the previous experiment, I introduced you to the basics of soldering and in this experiment I will expand on them and have you solder *pin through hole* (PTH) components to the PCB. As the name implies, a pin passes through a hole in a PCB (the hole is known as a *via*). The hole is plated with copper (or copper with a thin solder tinning), and after inserting the pin into the hole, the two are bonded together using a soldering iron and solder, as shown in Figure 2-23. When you have completed the solder joint and are about to cross section it, you should see something that looks like Figure 2-24. The solder more than fills the hole, but instead of forming a circular shape, it forms the conical fillet shown in the diagram, and it should be shiny. When you first start soldering PTH components, you will find that the shape may be more rounded than conical and the finish is dull and crinkled.

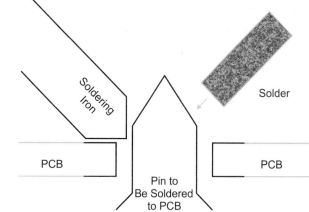

Figure 2-23   *Creating a PTH solder joint*

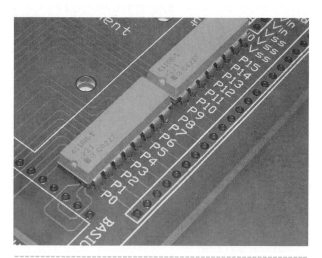

Figure 2-25   *Resistor DIPs inserted into the PCB*

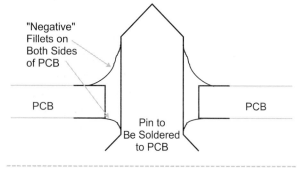

Figure 2-24   *Good solder joints*

When I am going to solder, I usually turn the soldering iron on and let it heat up for 15 minutes (like a watched pot never boiling, a watched soldering iron takes forever to heat up). To make a PTH solder joint, I apply the hot iron to the pin and PCB for about a second (you should see some residual solder flowing from the tip to the pin and the PCB) and then touch it with solder. Your first attempts might not look great, but you can fix them by touching the joint with the soldering iron for a few seconds without solder. I'm sure you will be a pro after getting through the first few joints in this experiment.

To start off, insert the two 16-pin resistor DIPs into the top of the PCB, as shown in Figure 2-25. The top of the PCB is the side that has the white markings on it. Component markings will be covered in more detail later in the book, but the two DIPs should have an indentation at one end and maybe a circle or square printed on the top by one of the corner pins. Push the DIPs into the PCB's RP1 220 and RP2 220 positions with the end indentation matching the

cutout on the PCB. You may find you have to bend the pins a bit narrower to get them to fit into the holes on both sides.

Once you have resistors inserted in the top of the PCB, turn everything over and start soldering the pins as described previously. You might want to first solder two corner pins to hold the DIPs rigidly. If the DIPs ride up, by pressing down on them while holding the soldering iron to the other side, you can push them down onto the PCB. Once the DIPs are in and the corners are soldered down, you can go ahead and solder the rest of the pins (Figure 2-26).

When you have finished with the resistor DIPs, you can solder in the 24-pin 0.600-inch socket and the momentary on switch. When soldering in the socket, remember to match any indentations or markings at the end of the socket with the marked indentation on the PCB. As for the switch, you should notice that one side is longer than the other and if you follow this, the switch will go in easily and be oriented in the right way.

Next, insert the two 0.01 μF capacitors into the PCB. The wires running from the capacitors should be 0.100 inches apart (2.54 millimeters) and should slide easily into the holes marked C11 and C12 on the PCB. After inserting the component wires into the PCB, turn it over, letting the components rest against the tabletop (they will be the same height as the resistor DIPs) and solder in the leads. After soldering the wires, clip them to the same length as the resistor DIP pins.

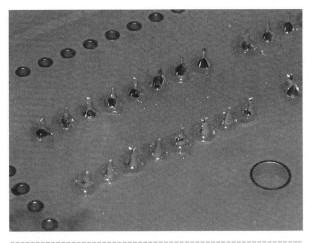

Figure 2-26   *Resistor DIPs soldered in*

If you have installed the components in the order I have given you here, you will discover that I have put all the short components together and now you will have to solder the higher components. Start with the single-row socket. First, clip off eight pins from one end and then solder the endpins to the PCB. While holding the PCB and connector, touch the iron to the soldered pins to make sure the socket is exactly perpendicular to the PCB. Once you are satisfied, you can then solder the rest of the pin.

It is pretty easy to solder in the final two components, the 9-pin female D-Shell connector and the 9-volt battery connector. The D-Shell connector should have metal tabs that fit through the large holes on either side of the J3 connector that lock it in. These tabs should be soldered to the PCB along with the connector pins.

The 9-volt battery connector has two tabs that are soldered into holes in the PCB, and the connector is held in place with two 4-40 screws and nuts. Ideally, the screws should be countersunk to allow a 9-volt battery to sit in the connector without being pushed out. Although I put in four holes for holding the battery connector, you really only have to use the two holes away from the battery contacts.

You're almost done! The last piece of work is to add the breadboard. It should have a piece of double-sided tape on its back so that you can attach it to the PCB, as shown in Figure 2-27. When I added the breadboard, I turned it so that the side with the red stripe on the outside is against the black single-row socket.

Looking at the PCB, you should notice two things. The first is that it is quite a bit heavier than you probably expected. You have added quite a bit of lead in the form of solder to the PCB, and even though it is only a small amount in each of the few holes (compared to many other PCBs), it still adds up. The second thing you should notice is that the bottom (the side that doesn't have the white markings) is actually quite sharp with a lot of little pins. So do not work with it on somebody's good table or even on your lap. The pins on the bottom of the PCB will cut the surface to shreds in very short order. When you are working on the circuits, I suggest you get a piece of antistatic matting to work on or place the PCB on one of the plywood robot bases that you cut and finished in this section, using 1-inch standoffs (available from electronics stores).

Figure 2-27   *Completed PCB with breadboard in place*

Section Three
Basic Electrical Theory

When I was a kid, a very popular theme for an episode of a TV show was to have two heroes (or the hero and a wisecracking terrified bystander) defuse a bomb. The process of amateurs defusing a bomb invariably follows after they discover an armed bomb where either a fake one was expected or a live bomb was expected somewhere else. After discovery of the bomb, the reluctant amateur *ordinance disposal experts* (ODEs) go through a complicated process of opening up the bomb (careful to avoid any booby traps) only to be confronted with two wires. These wires are used to connect the fuse (the part of the bomb with the timer, optional remote control receiver, and any booby trap sensors) to the detonator (the part of the bomb that causes the high explosive to blow up). The bomb usually looks like it is wired something like Figure 3-1.

With the booby traps behind them and the bomb's components exposed, the heroes are always faced with the dilemma of which of the two wires leading to the detonator to cut (adding to the tension, just before they make their decision and cut a wire, the show cuts to commercial). Somehow they know that by cutting the wrong wire, the bomb would explode, but by cutting the right wire, the bomb would be defused and safe.

For some reason, nobody (and this includes the special effects crew responsible for explosives) seems to have told any Hollywood screenwriters that they could cut either wire leading from the fuse to the detonator and the bomb would be safe. As I will show in this section, electricity must flow in a closed circuit—in the TV shows then, once one of the wires was cut, current couldn't flow from the fuse to the detonator and back to set it off.

The bomb's fuse is represented by the switch and the battery. Electricity is produced by a power source (usually a battery), and when the switch is closed, electricity flows through the wires to the "detonator" and then back to the battery.

The detonator can be thought of as a "load;" its purpose is to convert the electrical energy into something useful. In an actual detonator, electricity passing through it causes a wire to heat up and a small heat-activated charge in the detonator explodes. When the detonator's charge explodes, the shock of this explosion sets off the high explosive of the bomb. It is not widely known by most people, but high explosives do not go off when they are exposed to extreme heat—they may burn fiercely, but they will not explode. It is the shock of the exploding detonator charge that sets them off.

It should go without saying that throwing a stick of dynamite, or anything else you might find that is labeled "explosive" onto a fire to see what will happen is *not* a good idea. Many different kinds of explosives, as well as different products (such as aerosol hairspray), will explode and/or throw off burning materials if exposed to high heat. When I discuss explosives in this section, it is for your edification, not as an invitation for you to experiment with them.

In case you didn't get it the first time; do not place items labeled "explosive" (or that have an explosive warning symbol) onto a fire or other heat source.

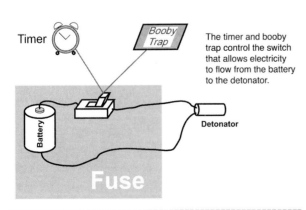

The timer and booby trap control the switch that allows electricity to flow from the battery to the detonator.

Figure 3-1   *Block diagram of a bomb circuit*

Benjamin Franklin said that electricity flows from the power source's positive connection to its negative when there is a path for it to follow. The black lines on a schematic diagram are used to represent the wires providing the path for the electricity through in the bomb's components. Looking at Figure 3-1, you will see that the different parts of the "bomb block diagram" are linked together in a closed loop—this is the closed circuit that is required to allow electricity to flow from the fuse's battery to the detonator and set off the bomb.

When a switch is said to be open, the contacts within the switch are not touching one another and electricity cannot flow through the circuit. This is the same as two wires held apart. When the contacts come together (as two wires held together), the switch and the circuit are described as being closed, and electricity can flow through them.

The need for a closed circuit for electricity to flow is the basic electricity rule that Hollywood screenwriters do not understand. In this section, I will expand on this basic rule and give you a better idea of what electricity is and how different values of it can be calculated and measured in a circuit.

In case you don't believe me about how bombs are treated in TV shows and movies, the following are some references in which the stars are given the task to defuse an explosive device they know nothing about:

> *Hogan's Heroes*—"A Klink, a Bomb and a Short Fuse"

M*A*S*H—"The Army-Navy Game"

Get Smart!—"Stakeout on Blue Mist Mountain"

Mission Impossible—"Time Bomb" (although to be fair, in *Mission Impossible*, the team seemed to have to defuse at least one bomb every three or four episodes)

Laverne and Shirley—"The Right to Light"

FBI—"Time Bomb"

Ironside—"Not with a Whimper, but a Bang"

Barney Miller—"Lady and the Bomb"

Remington Steele—"Premium Steele"

Lethal Weapon 3—Riggs and Murtaugh blow up a building by cutting the wrong wire in a bomb that they are trying to defuse.

At the start of this section, I noted that the bomb-defusing process on TV was very similar and generally followed the same lines for each of the different shows. What is amazing is that when you watch these shows, you'll see that in virtually every case, the wrong wire is cut, which results in a few moments of panic. Afterwards, everyone is okay and enjoys a good laugh because the bomb turns out to be a dud, or it was wired incorrectly, or it wasn't the type of bomb that was meant to destroy things.

Experiment 12
Electrical Circuits and Switches

Parts Bin

Assembled printed cir-
cuit board (PCB)

Resistor with brown,
black, and red bands

Light-emitting diode
(LED), any color

Single-Pole Double-Throw
(SPDT) switch toolbox

Tool Box

Wiring kit

Introducing this section, I presented, in a rather explosive (excuse the pun) manner, the concept of electrical circuits. Electricity must follow a closed circuit. For this experiment, the nature of an open and closed circuit will be demonstrated—a light will be turned on using electricity provided by a battery.

Every electrical circuit has three parts to it. Electricity is provided by a power source and passes through conductors (wires) to the load. The load converts the electrical energy into some other form and performs work with it. A load can be a simple light, a microcontroller (such as the BS2 presented later in the book), an electric motor, or a combination of parts.

In this experiment, I will use the PCB that came with the book, soldered together (*assembled*, as discussed in the previous section) with a battery, along with a few electronic devices to create a light that you can turn on and off. The circuit that you will build is shown in Figure 3-2.

In Figure 3-2, I have used the conventional diagram (the series of different-length parallel lines) to indicate the 9-volt radio battery power source placed in the clip built into the PCB. In my schematic dia-

grams, I usually mark the positive side of the power source with the + symbol to avoid confusion, but if you look at a circuit, remember that the end with the longer line is the positive connection to the battery. The positive connection of the battery comes out of the Vin connections and the other (negative) connection of the battery is wired to the Vss connections. In the later experiments, I will explain positive and negative as it relates to electronics, as well as explain what the connections Vin, Vdd, and Vss on the PCB mean.

The electricity coming from the battery is passed along wires from the connector to the components on the breadboard. From these wires, the electricity passes through the switch (Figure 3-3) to the resistor (Figure 3-4) and then to the LED (Figure 3-5) before returning to the battery. Wires are usually made from copper and may have another metal over them to resist corrosion. To prevent wires in different circuits (or different parts of the same circuits) from touching, the unused portions of the wire are wrapped (or *clad*) in a plastic sheath, known as insulation. The

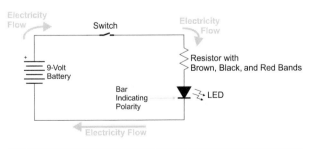

Figure 3-2 *First circuit showing switch control of electricity*

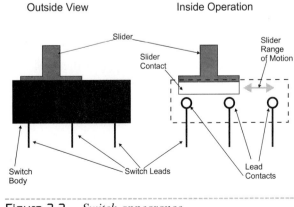

Figure 3-3 *Switch appearance*

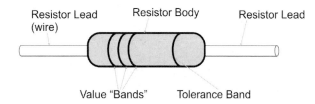

Resistor Lead (wire)   Resistor Body   Resistor Lead

Value "Bands"   Tolerance Band

Value Bands Should Be Brown, Black and Red, from the End of the Resistor to the Center

Figure 3-4 *Resistor appearance*

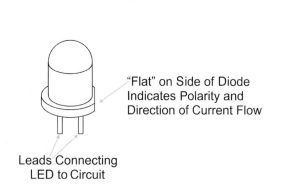

"Flat" on Side of Diode Indicates Polarity and Direction of Current Flow

Leads Connecting LED to Circuit

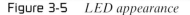

Figure 3-5 *LED appearance*

wires used in this application are part of a breadboard wiring kit that you will need to build the experiments in this book.

The switch is a device that brings two wires together to allow electricity to pass from one to another. Figure 3-3 shows a *Single-Pole Double-Throw* (SPDT) switch looks like. In the description, the *poles* are the number of circuits that can be switched and the *throw* is the number of connections that can be made in the circuit. I went with the SPDT switch simply out of convenience; it is a lot easier to find a breadboard-mountable SPDT switch than it is to find a breadboard-mountable SPST switch (the symbol for which is used in Figure 3-2). To use the SPDT switch as a SPST switch, connect the middle pins (called the *common terminal*) to one of the outside pins.

The resistor (Figure 3-4) is probably the most basic, true electronic device that you will work with. In this section, I go into detail explaining how resistors work, how their value is specified, and how they are used in an electrical circuit. For now, just select a resistor with a brown, black, and red stripe (or *band*) painted on it.

LED (Figure 3-5) is the acronym for light-emitting diode, and it can be pronounced either as the word rhyming with bed or as three individual letters. LEDs are semiconductor devices known as diodes that emit light when electricity passes through them in one direction. I discuss diodes in more detail later in the book. Right now, I just want to introduce them as a very inexpensive, reliable, and easy-to-use alternative to standard lightbulbs.

The breadboard is known as a prototyping system and allows you to quickly and easily wire circuits together. As you can see in Figure 3-6, a breadboard consists of a matrix of holes in which rows or columns of them are interconnected. To wire a circuit, you will have to place a component's pin or wire into one of the holes in the breadboard and then either put in a wire or another component into a hole that is connected with this one.

The different parts of the circuit are wired on the breadboard as I have shown in Figure 3-7. The polarity or orientation (the direction the component goes in) doesn't matter for the resistor or the switch, but it does for the LED. The battery's polarity also matters, but the PCB-mounted clip ensures the battery is wired correctly. The pin on the flat side of the LED is connected to the Vss holes in the PCB's connector. When you build your breadboard circuits, it is a good idea to keep the components' wires as short as possible and pressed against the breadboard. For simple circuits, such as the one in this experiment, this is not critical, but it will be for the more complicated experiments presented later in the book.

When you have put together your circuit, it should look something like Figure 3-8.

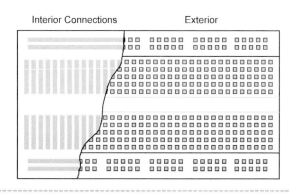

Interior Connections   Exterior

Figure 3-6 *Breadboard with interior connections shown*

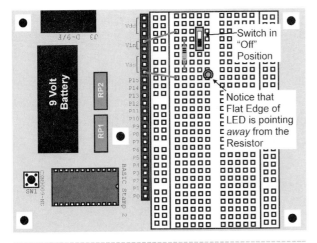

Figure 3-7   *First experiment wiring*

Figure 3-8   *What the first wiring experiment looks like on a breadboard*

When you move the switch's slider back and forth, you should see the LED light when the switch is in one position and not in the other. When the LED lights, the connection within the switch is made and the circuit is said to be closed with electricity flowing around it. When the switch is in the other position, the connection is broken and electricity cannot flow around the circuit. In this case, the circuit is described as being open.

If the LED does not light, then check your wiring (especially the orientation of the LED) and try another battery in the socket.

You can demonstrate the operation of the switch in a more concrete fashion by taking out the switch and using a length of wire that you can pull out of the breadboard and put back in to show what happens when the circuit is open and closed. Once you have the LED lighting when the switch is closed and turning off when the switch is open, you can go on to the next experiment. I will use this circuit (or its base) for the next few experiments.

Experiment 13
Electrical Circuits and Switches

Parts Bin

Assembled PCB with battery

Two SPDT breadboard-mount switches

Resistor with brown, black, and red bands

Three LEDs, any color

Tool Box

Wiring Kit

In the previous experiment, I demonstrated a simple electrical circuit that turns on and off a light (an LED actually) at the flip of a switch. Before going on and explaining what electricity is and how it works, I want to go back, examine how the switch used in the previous experiment works, and explain how a seemingly magical switch in your home works.

Chances are, in your home, probably in a hallway, you have a single light controlled by two switches. If you are at one end of the hall and you want to turn the light on, you change the position of the nearest switch. At the end of the hall, you don't need the light anymore so you turn it off by changing the position of the other switch. The two switches are very convenient to use and probably seem like they require quite a bit of complex circuitry to control.

In actuality, they just require two SPDT switches like the one that you used in the previous experiment. These switches are wired as I have shown in Figure 3-9, with the center contact connecting the switches to the rest of the circuit and the outside contacts connecting the two switches together.

As shown in Figure 3-9, electricity flows only if the switches close a circuit between one of the two connecting wires and the central connections. When the switches are closing the connections to different connecting wires, there is an open circuit and no electricity can flow through the circuit.

The movement of electricity in a circuit with two SPDT switches can be demonstrated by using the circuit shown in Figure 3-10. One of the two LEDs wired into the switch connection will light when the two switches allow electricity to flow through the wires that it is connected to. When one of these LEDs lights, the LED connected to the resistor will also light,

indicating that there is electricity passing through it. Figure 3-11 is the wiring diagram for this circuit.

You should spend some time flipping the switches back and forth in an effort to see how the switches

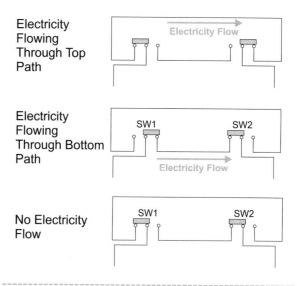

Figure 3-9   *SPDT light control operation*

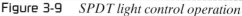

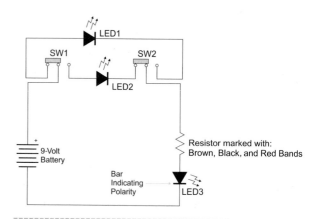

Figure 3-10   *SPDT control of two different paths for electricity*

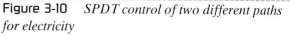

work (you might also want to compare the operation of a dual-switch light control in your home to convince yourself the circuit works in the same way).

It will probably be difficult for you to "see" Figure 3-10's circuit in Figure 3-11's wiring diagram. This is because I tried to come up with a way that the LEDs could be in line without requiring extra wires. When you design your own breadboard circuits, you will find that some methods will make the circuit easier to wire but obfuscate its operation. If you are confused by the wiring diagram, then I suggest that you trace it out with a highlighter to see the paths for electricity or try wiring it on your own and do not follow Figure 3-11.

Despite the simplicity of this circuit, it actually provides quite a sophisticated function; it allows control over a light from two physically different locations. This can be implemented a number of ways, but the method shown here is probably the most elegant.

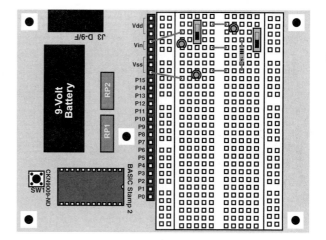

Figure 3-11 *Wiring diagram for SPDT switch being used to select different electricity paths*

Experiment 14
Voltage Measurement

Parts Bin

Assembled PCB with battery

Resistor with brown, black, and red bands

LED, any color

Tool Box

Wiring kit

Digital multimeter (DMM)

In the previous experiment, I said that the LED lights when "electricity" passes through it. It is not an accurate term, but to help you get through the first experiment in electricity, I wanted to keep the number of new concepts to a minimum. What you should have gotten from the previous experiment is that electricity must flow in a circuit for the circuit to work.

In this experiment, I would like to look a bit deeper into what electricity is. Electricity is a form of energy and is capable of performing "work." If you have taken introductory physics, then these two words should set off something in your head—energy and work imply a force acting on a mass. This is probably confusing to you because electricity doesn't have

any form (except as lightning), which means it doesn't have mass. Although you probably know it can be used to create a magnetic force, it doesn't seem to have any force that you can perceive or measure.

When you were in grade school, you probably read about Benjamin Franklin's experiment with flying a kite in a rainstorm—after the kite was hit by lightning, Franklin touched a metal key that had been tied to the line and got a shock. This shock was the same as to the static electricity shock that you get from shuffling your feet over a carpet and touching a door handle. At the time, touching things that were thought to have electricity was the accepted test to detect electricity: If you got a shock, it was there.

Along with "proving" lightning was made up of electricity, one of the most important theories Franklin postulated was that electricity flows like water in a closed system (if you dam up a river or block a pipe, you have effectively "opened" the system and water will stop flowing in it). As part of moving in a circuit, the electricity is *reused*; if you let all the water in a system leak away, the system will stop working. To move water you have to apply some force on it—because of the fluid properties of water, the force would have to be spread out over an area, so the force becomes *pressure* (which is defined as force over area). Using this model, it was realized that electricity would not move unless there was some kind of pressure applied to it.

Electrical pressure is called *voltage* and it is applied to an electrical circuit using a power source. This is analogous to using a pump to apply pressure to water in a swimming pool so that it can move upwards in a pipe (as shown in Figure 3-12). After rising to a level of the water in the pool, the water flows downward to the bottom basin where the pump draws the water again to continue the process.

As I have shown in Figure 3-12, the water at the top of the swimming pool is at the highest energy level. This energy is all potential, and it can perform work as it flows down to the bottom catch basin (the lowest energy level). If the catch basin has a crack in it so that water could leak out, then the system would not be able to perform any kind of work for very long. A crack in the catch basin is the same as an open in an electrical circuit.

We cannot measure the energy of the water at different points—but we can measure the pressure of the water at different points. In Figure 3-12, I have

put pressure indicators at different depths of the pool. To take these measurements, I am comparing the pressure of the water at different points to the pressure of the water at the bottom of the pool. At the top of the pool, the water pressure is quite a bit different from that at the bottom of the pool, and as you descend lower into the pool, the pressure approaches the pressure at the bottom of the pool.

The swimming pool is actually very similar to the electrical circuit you built in the previous experiment. To measure the electrical pressure (or voltage), a device called a *voltmeter* is used. Although simple voltmeters are available, I am going to ask you to get a *digital multimeter* (DMM) like the one in Figure 3-13. You can buy simple DMMs anywhere for $5 to more than $500. I'm going to suggest that you buy one for about $20.

For $20, you will get a DMM with reasonable accuracy that should be able to measure current and resistance. In DMMs less than $20, you may find that you have accuracy problems, and a more expensive unit will provide you with features that you cannot take advantage of. You should just be looking for the ability to measure voltage, current, and resistance. Features such as the ability to test diodes, measure transistor betas, and count the frequency of a signal are nice to have but are not required for any of the experiments in this book.

For this experiment, you will have to build the simple circuit shown in Figure 3-14, consisting of a resistor and LED on the breadboard. When you have done this and the LED is lit, set your DMM to measure voltages in the "20-volt DC" range (the instrument's instructions will explain how this is done). Hold the DMM's black probe connected to the cir-

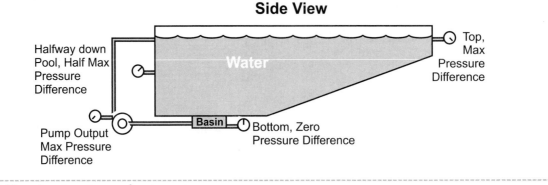

Side View

Halfway down Pool, Half Max Pressure Difference

Top, Max Pressure Difference

Pump Output Max Pressure Difference

Basin

Bottom, Zero Pressure Difference

Figure 3-12 *A swimming pool water system*

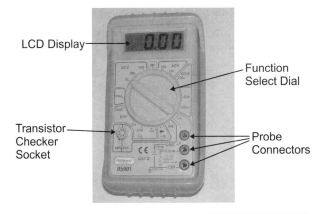

LCD Display

Function Select Dial

Transistor Checker Socket

Probe Connectors

Figure 3-13 *An inexpensive DMM's parts labeled*

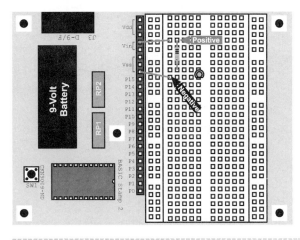

Figure 3-15 *Measuring the battery's voltage*

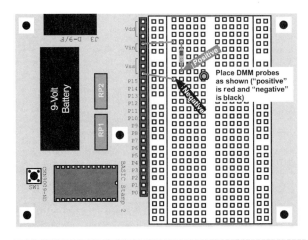

Place DMM probes as shown ("positive" is red and "negative" is black)

Figure 3-14 *Measuring the voltage across the LED*

Table 3-1 Voltage measurements for the LED and battery in this circuit

Point	Voltage Measured	Comments
Battery	9.25	The same at either the battery or across both the resistor and LED
LED	2.01	

cuit's negative connection (Vss shown in Figure 3-14); then measure the voltage at the two different points in the circuit as shown in Figure 3-14 and Figure 3-15 using the DMM.

In Table 3-1, I have listed the voltages that I measured in a prototype circuit.

These results are similar to what would be seen when measuring the water pressure relative to the basin at the bottom of the pool in the water example. When measuring water pressure or voltage, there must be some bottom reference that can be used to compare the values. In an electrical circuit, the negative output of the power source is normally used as this bottom reference. Instead of the term bottom reference, the word ground (abbreviated to Gnd or sometimes referred to as Vss as on the PCB) is used for electricity. The *ground* is the point in the circuit where there is no electrical pressure (it has a voltage of "0") and, if something was connected to the ground, it would be unable to do any kind of work.

Experiment 15
Resistors and Voltage Drops

Parts Bin

Assembled PCB with breadboard

Three resistors with brown, black, and orange bands

Tool Box

Wiring kit

DMM

Using the water analogy to continue to explain how electricity works, you should realize that water does not travel in a pipe effortlessly. The pipe subtly impedes the passage of the water running through it. If you were to measure the pressure at the inlet of a pipe and the exit, you would find that the restriction of the motion of water would cause a pressure drop.

This pressure drop is caused by the friction that the water experiences moving through the pipe. *Friction* is a force that resists motion, converting the energy of the moving water into heat. The first law of thermodynamics states that energy can neither be created nor destroyed; if some is lost in the pipe, then it must be converted into some other form. When energy is lost, it is almost always converted into heat.

When we are dealing with electricity, we assume that conductors are perfect, that they offer no friction (called *resistance*) to electricity moving through them. This is actually not true; except for materials called superconductors, everything will resist the movement of electricity through them. Conductors typically have a very low level of resistance, which is why the approximation of them being perfect is made.

In many electrical circuits, electrical energy has to be changed to be more suitable. The most common component for doing this is called a *resistor*, and different parts are available, offering different amounts of resistance to electricity. The units of resistance are ohms and are given the Ω symbol. Resistors themselves are quite small physically, so their values are indicated on them by a series of colored bands.

The bands specify the resistance using the following formula and are defined in Table 3-2.

Table 3-2 Resistor band color coding

Color	Band Color Value	Tolerance
Black	0	N/A
Brown	1	1%
Red	2	2%
Orange	3	N/A
Yellow	4	N/A
Green	5	0.5%
Blue	6	0.25%
Violet	7	0.1%
Gray	8	0.05%
White	9	N/A
Gold	N/A	5%
Silver	N/A	10%

$$Resistance = ((Band\ 1\ Color\ Value \times 10) + (Band\ 2\ Color\ Value)) \times 10^{Band\ 3\ Color\ Value}\ Ohms$$

Using the formula and the color chart, you can calculate the resistors used in this experiment to be

$$Resistance = ((Brown \times 10) + Black) \times 10^{Orange}\ Ohms$$

$$= (10 + 0) \times 10^3\ Ohms$$

$$= 10,000\ Ohms$$

Most resistors offer 5 percent tolerance and this is more than acceptable for the circuits presented in this book and the ones that you will work with. In practical terms, you will find that most resistors have a tolerance of 1 percent or less—they are specified as being 5 percent as the absolute worst case by the manufacturer.

To demonstrate the operation in an electrical circuit and how it affects the electrical pressure or voltage, build the circuit shown in Figure 3-16 and measure the voltage across the resistors. In your first test, set your DMM to "Voltage" (the 0- to 20- volt range) and place the black probe at the negative or Vss voltage and place the red probe at the four points noted in Figure 3-16 and record the voltages.

When you have recorded the voltages, you would have found that the voltages changed evenly from zero to the applied battery voltage at each resistor step. This is analogous to measuring the pressure of water as it travels down a pipe. Each resistor behaves similarly to a length of pipe in which the water pressure drops. Just as the pressure reduction in a length of pipe is called a *drop*, the voltage reduction through each resistor is called a *voltage drop*. You should notice that none of the voltages measured in Figure 3-16 is greater than the applied battery voltage; they are either less than or equal to the battery's voltage.

The voltages being less than or equal to the applied voltage should not be a surprise—especially considering the comments I made at the start of this experiment. If the voltage increased, then the energy of the electricity would have increased, which should be impossible due to the first law of thermodynamics.

For the energy to increase, an energy source, such as a battery, would have to be inserted into the circuit.

To further investigate the behavior of voltage in a circuit with resistors, measure the voltage across each resistor using the DMM probes as shown in Figure 3-17. You should find that the voltage across each resistor is approximately the same and is one-third the voltage applied by the battery. In doing this, you are measuring the voltage across each resistor.

Finally, measure the voltage across two resistors as I have shown in Figure 3-18. This voltage should be two-thirds of the total voltage applied to the circuit. As I will explain later in this section, the voltage across a resistance is proportional to its fraction in the total circuit. These two resistors have two-thirds of the resistance in the circuit, so it should make sense that they have two-thirds of the voltage drop in the circuit.

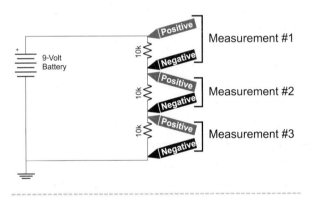

Figure 3-17   *Measuring voltage drops across individual resistors*

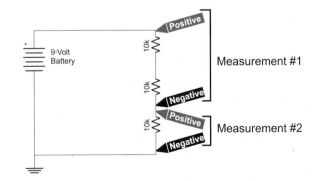

Figure 3-18   *Measuring voltage drops across multiple resistors*

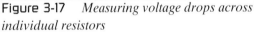

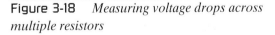

Figure 3-16   *Measuring voltages in a circuit relative to ground*

Parts Bin

Assembled PCB with breadboard

1,000 Ω resistor (brown, black, red bands)

10,000 Ω resistor (brown, black, orange bands)

Tool Box

Wiring kit

DMM

You should now be familiar with electrical pressure (voltage) and how it corresponds to pressure applied to water. As part of this, you should be familiar with the different voltage drops experienced by the different devices in the circuit. When the water examples were discussed, the force was applied to the molecules of water; now it is time to look at what is moved by electric force (voltage) to produce electricity.

As you are probably aware, electrons, which are negatively charged subatomic particles, make up electric current. When we talk about the size of atoms, it is hard to understand just how small they are and how many are in an object. The basic unit of charge is the *coulomb*, and it consists of 1.60219×10^{19} electrons. This exponent may not seem very large (your scientific calculator can probably do exponents to the base 10 of 99 or more). To give you an idea of how big it actually is, consider that there are about 2 trillion (2×10^{15}) planets in the Milky Way galaxy. One coulomb has about 8,000 times more electrons than there are planets in our galaxy.

To move the free electrons in a metal, negative electrical pressure (called voltage) is exerted on them and they will move away from the force toward a more positive location. This movement of electrons is known as *negative electrical current*.

When electricity was first being understood and defined, the nature of the atom was not understood at all. In fact, the prevailing theory of electricity was that electricity was a fluid that existed in certain materials and it could be moved by the application of some kind of electrical pressure (which we call voltage). To help define how electricity behaved, Benjamin Franklin postulated that this fluid moved from the positive terminal of a power source to the nega-

tive (the terms "positive" and "negative" being completely arbitrary). Unfortunately, this was wrong. Actual electrical current moves from the negative terminals to the positive terminals, but due to the acceptance of Franklin's theory of electricity travelling from positive to negative, we are stuck with this as the definition of electrical current. Working with this convention isn't too onerous because electrons are simply too small for us to see and move too fast for us to follow.

The most positive location in an electrical circuit is the positive terminal of the power source. If the positive terminal of the power source is not present (an open circuit exists), then the electrons will bunch up in the conductor until their collective voltage is equal to the voltage applied to the circuit.

The movement of electrical current can be observed and measured using your DMM, wired in a circuit as shown in Figure 3-19. First measure the voltage across the 1,000 Ω resistor and record it (I measured 8.89 volts). Then, setting your DMM to measure 0 to 20 milliamperes (mA) of current (explained in your DMM's manual), break the circuit and measure the current passing through it. I measured 8.93 mA.

Next, repeat the experiment using a 10,000 Ω resistor in place of the 1,000 Ω resistor. My voltage measurement was 8.84 volts and the current through the resistor was 0.90 mA.

Looking at these results, you should notice that the current through the 10,000 Ω resistor circuit was one-tenth the current through the 1,000 Ω resistor circuit. This implies that a voltage/current/resistance relationship exists that is similar to the following formula:

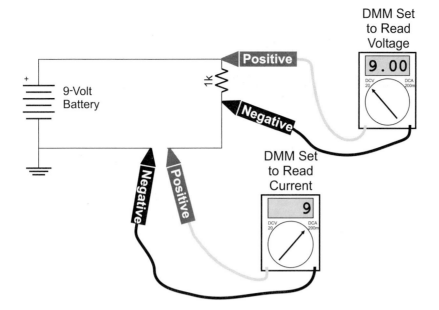

Figure 3-19   *Measuring voltage across a current as well as through it*

```
Current = Constant x Voltage/
          Resistance
```

You can test this by placing a different battery (with a different voltage output) into the circuit and repeating the measurement. No matter how many different ways you test this (with different resistors and batteries), you will find that the previous formula is true.

This is a basic rule of the universe known as *Ohm's law* and can be stated as "the current through an electrical circuit is proportional to the voltage applied to it and inversely proportional to the resistance within it." George Ohm first discovered this relationship between voltage current and resistance in 1826, and to recognize this accomplishment, the unit of resistance was named the ohm in his honor.

To simplify the application of the law, the *Systeme International* (SI) units for voltage, resistance, and current were chosen so the formula does not need a constant. Replacing voltage with the symbol "V," resistance with the symbol "R," and current with the symbol "i," Ohm's law can be written out simply as

```
V = i x R
```

Understanding and remembering this formula is critical if you are going to work with electronic circuits. You might want to remember it using a mnemonic like

> *Twinkle, twinkle, little star,*
> *Voltage equals eye times are.*

Another way of remember Ohm's law is to use the "Ohm's law triangle." This tool will return the formula for any of the three parts when you place your finger over one of the three symbols. In Figure 3-20, I show that by placing my finger over the "I" in the triangle, I can see that current is equal to voltage divided by resistance.

Find: I = ?

From Ohm's Law Triangle:

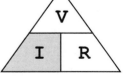

Result: I = $\dfrac{V}{R}$

Figure 3-20   *Ohm's law triangle example*

So far, I have been referring to resistors in units of ohms. You will find that most resistors that you work with are either in the thousands or millions of ohms, and it gets tedious writing them out as 1,000 Ω or 1,000,000 Ω. To simplify the writing down of large resistances, the symbol "k" is used for thousands of ohms, and "M" is used for millions of ohms. This means that 4,700 Ω is written out as 4.7k or 2,200,000 Ω is written as 2.2M. In some books and schematic diagrams, you will see these values written out as "4k7" and "2M2." They mean the same thing, but the space needed for the decimal point is replaced by the symbol for thousands or millions of ohms.

Experiment 17
Kirchoff's Voltage Law and Series Loads

Parts Bin
Assembled PCB with 9-volt battery

1k resistor

2.2k resistor

Tool Box
Wiring kit

DMM

With a basic understanding of Ohm's law, I think you can visualize what happens when electrical current is flowing in a circuit. It should seem obvious that as you increase the voltage (electrical pressure), the current increases. Similarly, if you increase the resistance in the circuit, you will decrease the amount of current that flows through it. If these statements do not make sense to you and you cannot visualize them, then you should go back and work through the experiments again to make sure that these concepts are clear in your mind.

I am pushing you to understand this concept because you must be absolutely clear on it when I start exploring what happens when multiple loads are in the circuit. As I will show in this and the later experiments, the behavior of an electrical circuit under loads is very predictable, but only if you have a very clear understanding of voltage, electrical current, resistances (loads), and Ohm's law.

You can add electrical loads to a circuit in two ways (Figure 3-21). The first serially connects the loads end to end. The second is in parallel, in which the loads are connected to the same power connections. Electrical circuits consist of some combination of these two circuit types; fortunately, when we are working with simple components and circuits, as we

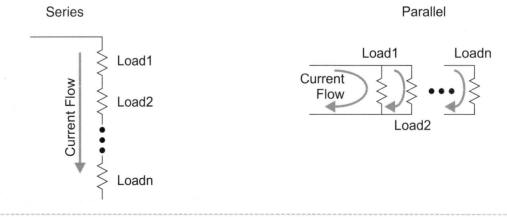

Figure 3-21   *Different ways of adding loads to an electrical circuit*

are in this section, we can look at how each type of circuit works without worrying about complications brought by other components.

In this experiment I will be investigating what happens when you put resistors in place serially. The analogy of adding resistors in series is similar to the situation shown in Figure 3-22 in which a pump is driving water through a pipe to work a water wheel.

Increasing the length of pipe is one way of adding resistance to a system; adding resistance will result in less water flow in the entire system and a drop in output water pressure (at the water wheel load). Adding more pipe (resistance) to the circuit is the same as adding another resistor to a circuit that is driving a load like the one I show in Figure 3-22. The longer the pipe, the larger the resistance and the less voltage and current is available for the load.

You should be thinking that adding more resistance to the circuit will reduce the amount of electrical energy that is available in the system by an amount proportional to the resistance. This should be intuitive to you—if you add a small resistance, then the impact to the circuit will be small. Adding a large resistance will have a large impact on the circuit.

It should also be obvious to you that the "equivalent" resistance of the circuit leading to the load is equal to the original resistance of the circuit plus the added resistance:

$$R_{equivalent} = R_{original} + R_{added}$$

Therefore, the equivalent resistance will always be larger than any of the single resistances summed together. This is an important fact to remember when you are calculating series resistances (such as on a test).

The impact of each series resistance is proportional to its value relative to the total resistance of the circuit. In Figure 3-23, I show a circuit in which there are three resistors—to the right of each resistor I present the formula for the voltage drop across it. The voltage drop is the fraction of the resistance to the total resistance in the circuit. The current is the same at every point in the circuit because there is nowhere, other than the circuit, to flow. Using the voltages across each resistor in Figure 3-23, the total voltage across the resistors is:

$$V_{total} = V_{R1} + V_{R2} + V_{R3}$$

$$= V \times R1/(R1 + R2 + R3) + V \times R2/(R1 + R2 + R3) + V \times R3/(R1 + R2 + R3)$$

$$= V \times (R1 + R2 + R3)/(R1 + R2 + R3)$$

$$= V$$

Simply put, you could say that the sum of the voltages across each resistive load is equal to the voltage applied to the circuit. This statement is known as *Kirchoff's voltage law*, and in this experiment, we will test this law using the circuit shown in Figure 3-24, which consists of two resistors in a series.

For this experiment, I would like you to fill out a table like Table 3-3. Remember that when you are measuring the voltage across a resistor, you are going to attach each test lead to a resistor wire. When measuring the current through the system, you will have to break the connection and set your DMM to 0 to 20 milliamperes.

Looking at these results, I can use Ohm's law to confirm that the total load resistance is 3.2k (1k added to 2.2k).

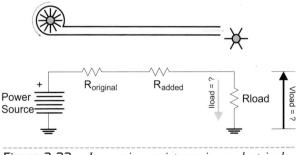

Figure 3-22   *Increasing resistance in an electrical circuit*

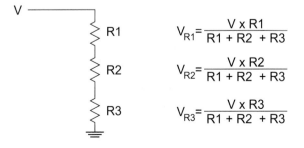

Figure 3-23   *Voltage drops across different series resistors*

Table 3-3 Circuit measurements

Measurement	Results
Battery voltage	8.77 V
1k resistor voltage	2.74 V
2.2k resistor voltage	6.02 V
Current	2.73 mA

$$R_{Load} = V/I$$
$$= 8.77 \ V/27.4 \ mA$$
$$= 3,201 \ \Omega \sim 3.2k$$

Checking the voltage drops across the two resistors, you should see that the voltage ratios of the two

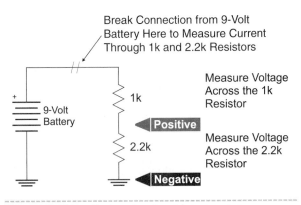

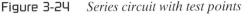

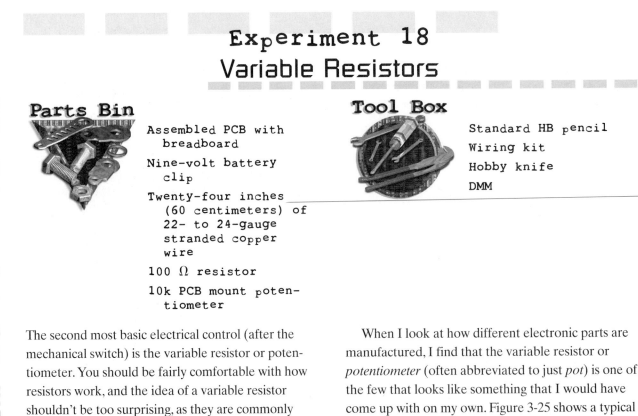

Figure 3-24 *Series circuit with test points*

resistors in this circuit can be expressed as the applied voltage multiplied by their value and divided by the total resistance.

Experiment 18
Variable Resistors

Parts Bin

Assembled PCB with breadboard

Nine-volt battery clip

Twenty-four inches (60 centimeters) of 22- to 24-gauge stranded copper wire

100 Ω resistor

10k PCB mount potentiometer

Tool Box

Standard HB pencil

Wiring kit

Hobby knife

DMM

The second most basic electrical control (after the mechanical switch) is the variable resistor or potentiometer. You should be fairly comfortable with how resistors work, and the idea of a variable resistor shouldn't be too surprising, as they are commonly used in different electronic devices. Whereas switches give you "on" and "off" (also called *binary*) control over a circuit, a potentiometer provides you with the ability to set a control to an arbitrary position between full "on" and completely "off." This type of control is usually referred to as an *analog* control.

When I look at how different electronic parts are manufactured, I find that the variable resistor or *potentiometer* (often abbreviated to just *pot*) is one of the few that looks like something that I would have come up with on my own. Figure 3-25 shows a typical rotary potentiometer; a copper "wiper" is moved across a resistive material under the control of an operator or some device. The schematic symbol of a potentiometer (Figure 3-26) is a fairly accurate representation of the potentiometer with the resistive material shown along with the wiper connected to it.

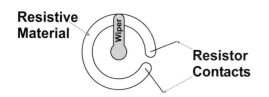

Resistive Material

Wiper

Resistor Contacts

Figure 3-25 *How a potentiometer is built*

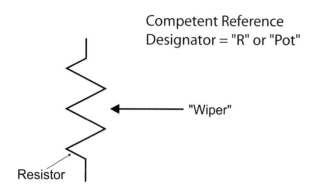

Competent Reference
Designator = "R" or "Pot"

"Wiper"

Resistor

Figure 3-26 *Potentiometer/variable resistor symbol*

Potentiometers are specified by the following three characteristics:

- Resistance of the material between the two contacts. The wiper resistance is assumed to be negligible and is used as a "tap" into the resistive material.

- Whether the resistive material in the potentiometer increases in resistances linearly (resistance changes evenly along its length) or logarithmically (the resistance changes according to a function like $10^{distance}$ or $2^{distance}$). Logarithmic pots are used in circuits in which orders of magnitude changes in resistance result in small changes in the circuit's operation.

- How much power the potentiometer can handle. As I will discuss later in this book and show why in this experiment, potentiometers are not well suited to control power being passed to loads. For this reason and because small potentiometers are much cheaper than high-power potentiometers, I only use potentiometers that can handle very miniscule amounts of power.

You can make your own potentiometer by removing the wood on one side of a simple HB pencil (not a colored pencil crayon) and exposing the lead inside, as I show in Figure 3-27. The lead of a pencil is made

from graphite (a form of carbon), the same material as is used in resistors and many types of potentiometers. You can demonstrate the operation of the variable resistor made from the pencil lead by placing your two DMM probes on the lead, setting the DMM to read resistances, and moving the probes back and forth across the lead. You should see the resistance increase as the two probes move farther apart (which is expected because there is more resistive material between the two probes).

If the resistance goes infinitely high on your pencil lead, then you have a crack in it. If this happens, you should try to expose the lead in another pencil. I found it took me three attempts to get the results shown in Figure 3-28.

The potentiometer can also be used to output changing voltages as can be demonstrated by wiring the PCB's 9-volt battery (with a loose battery connector) to the extreme ends of the exposed pencil lead. Next, set your DMM to measure voltage and place the black probe on the negative connection and the red probe onto the pencil lead (Figure 3-29). As

Pencil for Experimentation:

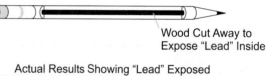

Wood Cut Away to Expose "Lead" Inside

Actual Results Showing "Lead" Exposed

Figure 3-27 *Potentiometer made from a pencil*

Pencil for Experimentation:

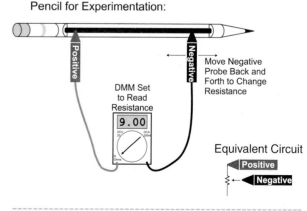

DMM Set to Read Resistance

Positive

Negative

Move Negative Probe Back and Forth to Change Resistance

Equivalent Circuit

Positive

Negative

Figure 3-28 *Measuring varying pencil lead resistance*

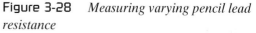

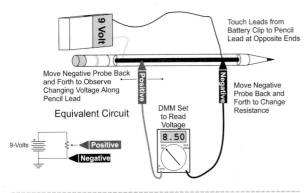

Figure 3-29 *Measuring varying voltage in pencil lead potentiometer*

you move the red probe back and forth on the pencil lead, you will see the voltage change from being equal to the applied battery voltage to zero (the negative battery connection).

The potentiometer in this configuration can be called a *voltage divider*, taking the form of Figure 3-30. Instead of being a single resistor with a copper wiper that "taps" into the resistive material, you can consider the voltage divider to be two resistors in a series with the voltage between the two brought out for use in the circuit.

You can test out the operation of a real potentiometer by creating the three circuits shown in Figure 3-31, Figure 3-32, and Figure 3-33, and wired using the PCB and breadboard. Figure 3-31 shows how the potentiometer can be configured as a variable resistor. Although the diagram on the right looks like the potentiometer has been changed into some kind of a voltage divider, it is equivalent because the wiper behaves as a dead short (no resistance). Current passes directly through the wiper and does not pass through the other contact.

The use of the potentiometer as a voltage divider (Figure 3-32) is exactly the same as the pencil lead voltage divider. The closer the wiper is physically to the contact connected to the positive power in, the

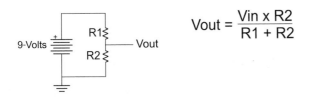

Figure 3-30 *Breaking the single resistor in a potentiometer into two to create a voltage divider*

$$Vout = \frac{Vin \times R2}{R1 + R2}$$

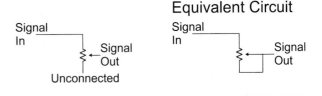

Figure 3-31 *Wiring a potentiometer as a variable resistor*

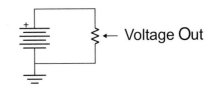

Figure 3-32 *Producing a varying voltage using a potentiometer*

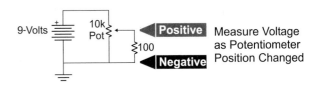

Figure 3-33 *Measuring varying voltage on a potentiometer under load*

higher the voltage will be. By turning the potentiometer, you should be able to display a literally infinite number of voltages between ground (the negative connection) and the applied voltage.

To finish this experiment, I wanted to show what would happen if a load was added between the wiper and ground. When you wire in the 100 Ω resistor to the pot's wiper and one of its contacts, you will see that the voltage at the wiper will no longer be linear and will either be very low or very high. It is much more difficult to set "intermediate" voltages when the 100 Ω resistor is in place than when it isn't. The 100 Ω resistor, similar to a load with an equivalent resistance of 100 Ω, becomes part of the circuit and part of the current path for the electricity passing through the potentiometer. So, instead of the circuit consisting of one current path through the potentiometer, it now consists of two, with the 100 Ω resistor being in parallel with the 10k pot, changing its capability to accurately change the voltage being output. For this type of application, a voltage regulator is actually required instead of a potentiometer-based (or two resistor-based) voltage divider.

Experiment 19
Kirchoff's Current Law and Parallel Loads

Parts Bin

Assembled PCB
1k resistor
2.2k resistor

Tool Box

Wiring kit
DMM

Looking at loads in a series, the addition of the resistances of the loads to make up a single, equivalent load should seem quite intuitive. Using the example of the lengthened pipe, the added resistance of the length of pipe is quite naturally added to any resistances already in the system and increases the total resistance of the system. With this in mind, you might think that the equivalent resistance of two resistors in parallel (Figure 3-34) is just as intuitive.

When you first try to figure out what is the equivalent resistance, you will probably hit a roadblock trying to visualize what is happening in the circuit. Going to the water analogies, it may help to think of the two resistors as two pipes in parallel passing water over a greater amount of area; this reduces the resistance, but the equivalent resistance is not intuitively obvious.

To calculate the equivalent resistance of the two resistors in Figure 3-34, knowing that the voltage drop is the same across each resistor, we can find the current flowing from the 10-volt power source using Ohm's law:

$$i_{5\ ohm} = 10\ volts/5\ ohms$$
$$= 2\ amps$$
$$i_{2\ ohm} = 10\ volts/2\ ohms$$
$$= 5\ amps$$
$$i_{total} = i_{5\ ohm} + i_{2\ ohm}$$
$$= 2 + 5\ amps$$
$$= 7\ amps$$

With the work done so far, looking at Figure 3-34, you should be comfortable with the concept that the sum of the currents passing through the branches of the parallel circuit is equivalent to the total current provided to the circuit.

Using Ohm's law, we can calculate the equivalent resistance:

$$R_{equivalent} = V/I_{total}$$
$$= 10\ volts/7\ amps$$
$$= 10/7\ ohms = 1.43\ ohms$$

Rather than jumping to the decimal value of the equivalent resistance, take a look at the fractional value for the resistance. The numerator of the fraction is the product of the two resistances and the denominator is the sum. With this information, you can generalize the equivalent of two parallel resistances (A and B) to

$$R_{equivalent} = (R_A \times R_B)/(R_A + R_B)$$

Looking at a circuit that has three or more resistors in parallel (such as Figure 3-35), and performs the same analysis (starting with the current through each resistor, adding them up, and calculating the equivalent resistance), you would discover that the general formula for the equivalent resistance is the following:

$$R_{equivalent} = 1/((1/R_1) + (1/R_2)$$
$$+ \ . \ . \ . \ (1/R_n))$$

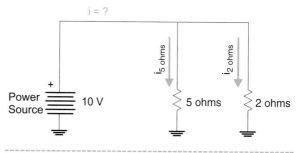

Figure 3-34   *Two-resistor test circuit*

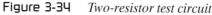

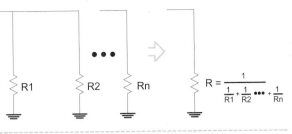

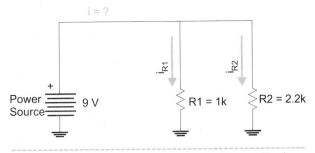

Figure 3-35   *Multiple resistor equivalency*

Figure 3-36   *Two-resistor test experiment circuit*

In this experiment, I would like to test this analysis using the circuit in Figure 3-36 and record the results in a table like Table 3-4. Using the results, calculate the equivalent resistance and see how closely it matches the calculated equivalent resistance.

The table contains the values I read from my circuit. As expected, the currents through the two resistors match the total current provided by the battery. This allows me to calculate an equivalent resistance of 701.3 ohms.

Using the formula I came up with earlier in this experiment, I can calculate the expected equivalent parallel resistance of the 1k and 2.2k resistors to be

$$R_{equivalent} = (R_A \times R_B)/(R_A + R_B)$$
$$= (1k \times 2.2k)/(1k + 2.2k)$$
$$= 2.2(10^6)/3.2(10^3) \text{ ohms} =$$
$$687.5 \text{ ohms}$$

Chances are that when you run the experiment, your calculated equivalent resistance will just be a percentage point or two off from the calculated value, just as I was. This difference is due to the actual values of the resistors that you are using as well as any accuracy errors in your DMM.

Table 3-4   Analyis results

Measurements

$V_{Battery}$	9.25 volts
$i_{Battery}$	13.19 mA
i_{R1}	9.02
i_{R2}	4.18 mA

An important observation is that the equivalent resistance of parallel resistors is *always* less than the smallest resistor in the parallel circuit. This observation is useful to remember when calculating the resistance of a parallel circuit to know if your answer is reasonable.

The statement that the sum of the currents passing through a parallel circuit is equivalent to the amount of current passing through the circuit is known as *Kirchoff's current law*. This law, like Kirchoff's voltage law and Ohm's law, is vitally important to remember and understand.

Experiment 20
Thevinin's Equivalency

Parts Bin

Assembled PCB

Four 1k resistors

Tool Box

Wiring kit

DMM

In cases where you have a combination of series and parallel resistances, you might be a bit overwhelmed by what you see. In basic electronics courses, you are often given a circuit like this one and asked to calculate the equivalent resistance of the circuit, the voltage across different resistors in the circuit, and the currents through the different parts. The problem really is not difficult when you use the rules for combining series and parallel resistances.

The process of simplifying a collection (or *network*) of resistances into a single resistance is known as *Thevinin's equivalency*, which states that all loads in a circuit can be reduced to a single equivalent one. The first step in reducing a circuit is to combine all series resistances as I have shown in Step 1 of Figure 3-37. Next (Step 2), the parallel resistances are combined. These two steps are repeated until you have the single resistance. I kept the result in a mixed fraction format when I solved the equivalent resistance rather than convert to a decimal number, to sim-

plify working with the values later, and to allow myself to mentally to check the results and make sure they made sense. For example, when calculating the value of "V1" (in Figure 3-38), I know that it will be less than half the power source voltage because the voltage drop at the 1k resistor across the top of the circuit schematic diagram will be greater than the drop across the other three resistors. The calculated value of 2/5 V is less than half the applied voltage, so intuitively, I know that my calculations are in the right direction.

With the equivalent resistance worked out, you can now work through different parameters in the circuit as in Figure 3-38. To carry out this analysis, you will have to use Ohm's law along with Kirchoff's voltage and current laws.

To test out the calculations that I made, I would like you to build the circuit that has been presented here. Once you have done this, power it up, create a table like Table 3-5, and record your results in the

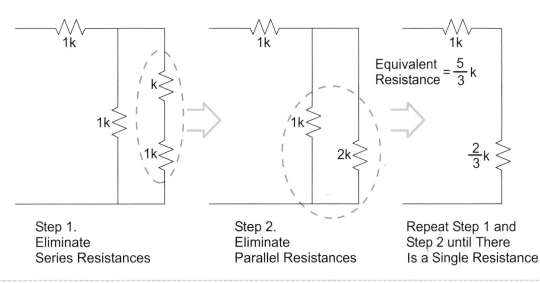

Step 1.
Eliminate
Series Resistances

Step 2.
Eliminate
Parallel Resistances

Repeat Step 1 and
Step 2 until There
Is a Single Resistance

Figure 3-37   *Reducing the complex electrical load to a single equivalent resistor*

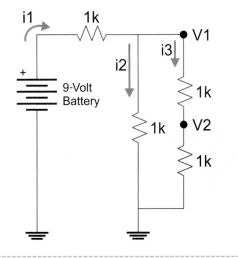

Figure 3-38   *Electrical circuit with complex load to test equivalent resistance*

Table 3-5 Comparing calculated to measured values in the Thevin equivalent circuit

Measurement	Value Formula	Expected Value	Actual Value
$V_{Battery}$	8.85 Volts		
i1	$V_{Battery} / (1k \times 5 / 3)$	5.31 mA	5.29 mA
V1	$2 / 5 \times V_{Battery}$	3.54 V	3.52 volts
i2	$2 / 5 \times V_{Battery} / 1k$	3.54 mA	3.52 mA

parameters listed in the table. Compare these values to the predicted values that I derived from Figure 3-38 by multiplying by the actual battery voltage (V) and 1k. You could expand the checks to make sure that each value (voltages across the different resistors and i3) was accurate to within the same fraction of a percentage point as mine was.

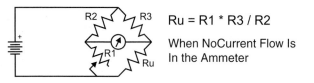

Figure 3-39   *Wheatstone bridge*

When I was first designing the circuit for this experiment, I created a "Wheatstone" bridge like the one shown in Figure 3-39.

The Wheatstone bridge is a very useful circuit when you are trying to measure minute changes in resistance. The resistors on each side of the circuit behave like a voltage divider and no current will flow when the ratios of the two voltage dividers are equal. In Figure 3-39, you can see that I put an ammeter between the two voltage dividers. The ammeter will read zero when the resistors in the two halves of the circuit are in balance.

If the ammeter does not read zero, then you will have to adjust R1 until the ammeter is displaying zero. Then take R1, R2, and R3 out of the circuit, measure their resistance, and use these values for the formula shown in Figure 3-39 to find out what the unknown resistance (Ru) is.

The Wheatstone bridge behaves as a resistance multiplier. If you had a strain gauge (which is glued to a structure and outputs the strain it's under by changing its resistance) that had a resistance that varied between 10 to 20 ohms, you could boost the resistance difference 50,000 times by using the Wheatstone bridge with R2 and R1 at a value of 1 MΩ. In this case, R3 would be 10 ohms. By adjusting R1 until no current flows through the ammeter, you can indirectly measure the resistance of the strain gauge.

Experiment 21
Power

Parts Bin

Assembled PCB with breadboard

100 Ω, 1/4-watt resistor

1k, 1/4-watt resistor

Tool Box

Wiring kit

Jogging shoes

Stopwatch

As you become more capable in electronic circuit design, you will take the power used by the application into account more and more. For most basic applications, you do not need to keep track of power, but as you begin to work with more and more complex applications (especially when you are working with robots), the importance of keeping track of power in the circuit becomes more important. Units in the electrical (as well as physical) arenas have been very cleverly specified in the SI measurement system to allow a simple conversion of units between the different fields to help us understand and relate power levels.

Power is the product of force times velocity and can be expressed quite simply as the following:

Power = F x v

where force is measured in *newtons* ($kg \times m/s^2$) and velocity is in m/s. Force is defined using the formula:

F = m x a

where "m" is the mass of an object and "a" is the acceleration the object is experiencing. If you are familiar with the English measurement system, you may refer to your weight as being some number of pounds. This is not correct, but the truth is somewhat confusing. Pound is a measurement of mass (how much matter an object has) and should have the abbreviation *lbm*. Where things become confusing is that a pound of mass exerts a *pound force* (*lbf*) when in a standard gravitational field (32 ft/s² or 9.807 m/s²). There is no kilogram of force in SI; the SI unit of force is called the newton (*N*). To convert a weight in pounds to a force in newtons, you should use the conversion

1 lbf = 4.45 N

When you have moved a mass some distance, you can be described as having performed some *work* on the object or added some energy to the object. The unit of SI work or energy is the *joule* and has the units *newton meters* (literally, force × distance).

The rate at which energy is being put into a system is called *power* and is given the label *watts* (*W*), which has the units $kg \times m^2/s^3$. I'm sure you've heard the term watts, but it is probably something that you have difficulty in visualizing what it actually is. James Watt, one of the inventors of the steam engine, defined the term horsepower as the standard amount of power output of horses pumping water out of a mine. Today one horsepower is defined as 746 watts.

To help you get a feeling for horsepower, try a simple experiment; measure a flight of stairs followed by the amount of time it takes you to run up it. After doing this, enter the units into one of the two formulas below to find out how much power was exerted running up the stairs:

Power = Weight (lbf) x Height (ft) x 1.356/Time up stairs (s)

= Weight (N) x Height (m)/Time up stairs (s)

My mass is around 200 lbm (90 kg, which becomes a weight of 890 N), and I was able to run up 10 feet (3 meters) of stairs in 7 seconds. Using these values in the formulas above, I found I expended about 378 watts or about 0.51 horsepower. This is surprisingly large and when you test yourself, I'm sure that you are able to output at least 0.25 horsepower, which also seems strangely large. The confusion lies in your vision of a horse—chances are you are visualizing a big strong beast. When Watt defined the horsepower, he used the scrawny work animals of his day. These horses were generally at the end of their lives or had been debilitated by an illness and could only put out a fraction of the power of a young, healthy, and well-fed horse.

In direct current electrical terms, power is defined using the following formula:

$$P = V \times i$$

Voltage (V) has the units joules/coulomb and current (i) has been given the units coulombs/s. Multiplying the two values together results in a quantity in joules/s or $kg \times m^2/s^3$, which is identical to the units for watts given above.

To test the ability of a battery as a circuit's power source, using the breadboard on the PCB that comes with the book, build the circuit shown in Figure 3-40 and wire it using first a 1k resistor and then a 100 Ω resistor. I recommend that you do not clip the leads of the resistors as short as possible so the resistors touch the breadboard—the reason why will become immediately apparent.

After wiring the resistor into the breadboard, wait about 30 seconds and then touch it with the *top* of your finger. You should not notice any difference with the 1k resistor, but the 100 Ω resistor will actually be quite warm. This is why I indicated that you touch the component with the top of your finger; the top is much more sensitive to heat than the tips or pads of your finger, and if you accidentally burn it, you won't have trouble holding a pen or pencil.

Actually, both resistors warm up because of the power being forced into them; the 1k resistor just isn't *dissipating* as much power as the 100 Ω resistor

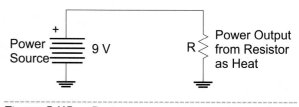

Figure 3-40 *Resistor power test experiment circuit*

is. When I say dissipating power, I am indicating the amount of power the resistor is turning into heat. Going back to the formula for power, Ohm's law can be used to express it in a number of different ways:

$$P = V \times I$$
$$= V^2/R$$
$$= I^2 \times R$$

We can calculate the power used by the two resistors by knowing the applied voltage; the 1k resistor is dissipating 0.081 watts and the 100 Ω resistor, 0.81 watts. The 100 Ω resistor is dissipating 10 times as much power as the 1k resistor, which is why it is noticeably hotter than the 1k resistor. Normally, in this situation, I would specify the use of a resistor rated for dissipating 1 watt of power instead of the standard 1/4-watt resistor that I use primarily throughout the book. A 1-watt resistor is physically much larger than a 1/4 -watt resistor, so that there is more surface area to dissipate the power over (and have a correspondingly lower surface temperature). Providing more surface area to a component that produces a significant amount of heat ("significant" generally being a 1/2 watt or more) is known as *heat sinking* and is commonly used for microprocessors, like the one in your PC.

Experiment 22
Batteries

Parts Bin

Assembled PCB with breadboard

100 Ω, 1-watt resistor

Nine-volt inexpensive carbon battery

Nine-volt alkaline battery

Nine-volt NiMH battery

Tool Box

Wiring it

DMM

Stopwatch

Power for robots almost always comes from onboard batteries. Some robots use photovoltaic cells, but these robots store power in capacitors or batteries before passing it to electronics or motors. Other robots are powered externally and have power cables running to the robot. I tend to discourage these types of robots because keeping the cord separate can be difficult, and every time I see a robot that has a wire running to it, I can't help but think of something from a *Far Side* comic strip. A very small number of robots are powered by fuel cells and internal combustion motors.

When choosing the method of powering your robot, you will have to make the decision on how to power the motors and any control/peripheral electronics. Many robots have two battery packs, one for the motors and one for electronics. The reason for using two battery packs is to minimize the power fluctuations experienced when the motors are turned on and off.

A single battery pack minimizes the cost and weight of the robot. You will find that as you minimize cost and weight in your robot, something amazing happens. A single battery pack means the overall weight of the robot is lowered, which means that smaller motors are required. Smaller motors are usually cheaper and require less current, which means a smaller battery. A smaller battery weighs and costs less than a larger one allowing you to use a smaller motor . . .

This loop of decreasing costs and weights is known as a *super-effect* and will help create the smallest,

lightest, and cheapest robot for a given set of requirements. I do want to caution you about one thing; it is very easy to design "down" your robot to the point where it does not fulfill the original requirements because of the attractiveness of using the smallest possible components.

When choosing batteries, the choice is between alkaline radio batteries, *nickel metal hydride* (NiMH)/*nickel-cadmium (*NiCad) rechargeable batteries, and lead-acid (motorcycle or car) rechargeable batteries. For all your robot projects, I recommend that you use NiMH batteries instead of NiCad batteries because they are less toxic for the environment. The correct choice of battery is important because it will affect the following:

- Size and weight of the robot
- Cell voltage
- Operational life of robot
- Speed of movement
- Cost
- Recharge time

Different battery types output different voltage levels per cell and discharge at different rates. Lithium cells provide relatively high voltage, but usually low current. For carbon- and alkaline-based batteries, you can expect 1.5 volts per cell or more. Rechargeable battery cells (such as NiCad and NiMH) output 1.2 volts per cell. As is shown in Figure 3-41, rechargeable batteries tend to output a constant voltage, whereas the output of single use cells (carbon and alkaline) decreases linearly as they are used.

When considering different batteries for use in a robot, it is important to remember that the critical parameter for choosing the battery to be used is the battery type's internal resistance. The higher the internal resistance, the less current available for motors and a disproportionate amount of power is lost within the batteries. Often when batteries heat up (due to power being dissipated within them), you will find that their ability to source current will drop. The higher internal resistance a battery has, the more the output is affected by voltage transients, resulting in increased needs for filtering the power being passed to the controller. Like the super-effects that appear when a single battery pack and the smallest possible motor is used, minimizing the internal resistance of the battery will result in a smaller, lighter, and cheaper robot.

Often people (describing themselves as "experts") will say that you should only buy cheap carbon batteries instead of expensive alkaline (or rechargeable) batteries because the A-H rating between the two is very similar. This is true, but I want to caution you about using these batteries in robots because inexpensive batteries tend to have high internal resistances. You can usually find out what a battery's internal resistance is by reading the datasheet for the

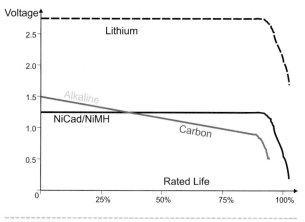

Figure 3-41   *Different battery operations*

battery available on the manufacturer's web page or what is specified in their industrial customer catalogs. The problem with very high internal battery resistances is illustrated in Figure 3-42. Most battery packs used in robots consist of multiple cells in a series, and although you generally visualize the circuit as being the "Idealized Circuit." the "Actual Circuit" is the diagram with each battery symbol and resistor representative of a single battery. In the "Effective Circuit" shown in Figure 3-42, I have lumped the internal resistances together to clearly show that as the current drawn from the battery increases, the voltage across the internal battery resistances will increase. According to Ohm's law, the voltage drop across a load is inversely proportional to the current

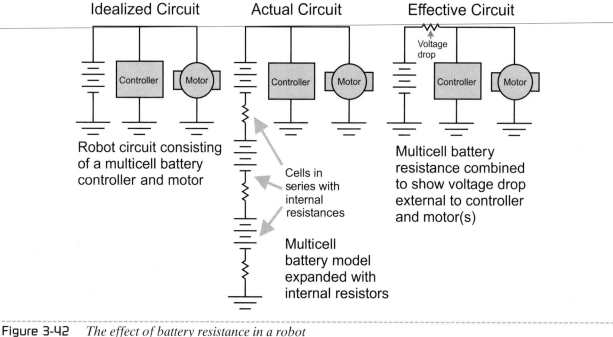

Figure 3-42   *The effect of battery resistance in a robot*

Experiment 22 — Batteries

drawn through it. This means that as more current is drawn through the battery's internal resistance increases, the effective voltage output drops.

When you look around, you will find that the lowest internal resistance batteries available are either premium alkaline radio battery cells or rechargeable batteries designed for high-current applications. As a potential source for robot applications, the 9.6-volt NiCad batteries used for remote-control electric racers have very low internal resistances and can be bought fairly inexpensively with chargers. The experiment that I would like you to perform is quite simple —to see if you can reproduce Figure 3-41 using an inexpensive carbon battery, a moderately expensive alkaline battery, as well as a rechargeable NiMH battery. These batteries will be used with the book's PCB in the circuit shown in Figure 3-43.

To test these assumptions, I recorded the life of a carbon, an alkaline, and a fully charged NiMH battery; then I recorded and plotted the results in Microsoft Excel (Figure 3-44). The results essentially match what I have stated here except for the carbon battery, which had a large initial voltage loss. After a bit of research, I found that this was due to a conversion of battery materials close to the electrodes within the battery, increasing the resistance within the battery. From Figure 3-44, it appears that the NiMH battery would be the best choice because the output

Figure 3-43   *Battery power life experiment circuit*

voltage stays closest to 9 volts for the longest period of time and falls off sharply instead of providing marginal power (which is actually an advantage in most applications).

For Consideration

The math that was used to calculate different values in this section should not have been new to you, nor should it have been very difficult. What will probably be a new experience for you is understanding what the values mean and whether or not they are reasonable. Toward the end off this section, I will explain more about the different values and what kind of ranges you should expect them to be in.

All electrical measurements are in SI units. SI is an agreed-to standard set of measurements for different quantities. Most countries around the world utilize SI

Time	NiMH	Alkaline	Carbon
0	9.02	9.08	8.24
5	8.69	8.44	7.02
10	8.62	8.32	6.57
15	8.61	8.22	6.37
20	8.62	8.12	6.3
25	8.61	8.04	6.28
30	8.61	7.96	6.25
35	8.6	7.9	6.23
40	8.59	7.83	6.19
45	8.55	7.78	6.16
50	8.53	7.73	6.09
55	8.5	7.67	6.05
60	8.46	7.62	5.98
65	8.4	7.58	5.93
70	8.34	7.52	5.86
75	8.27	7.46	5.79
80	8.18	7.43	5.73
85	8.05	7.39	5.67
90	7.86	7.36	5.61
95	7.63	7.33	5.56
100	7.22	7.3	5.49
105	6.3	7.28	5.43
110	3.3	7.25	5.37
115	1.88	7.22	5.31
120		7.19	5.25

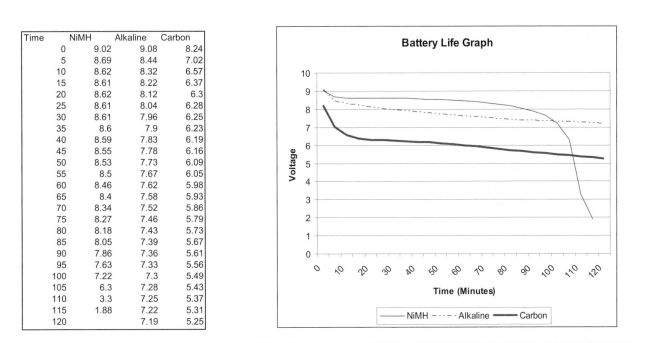

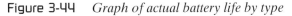

Figure 3-44   *Graph of actual battery life by type*

units for everything, but some countries (notably, the United States, Canada, Great Britain, and Australia) still rely on the imperial measurement system for units of length, volume, mass, and force. Despite this deviation, these countries use SI units for electrical quantities.

The way SI quantities are thought out allow for easy manipulation into different units. Being based on powers of 10 allows for simple manipulation of values. I remember as a kid learning that there were 12 inches to a foot, 3 feet to a yard, 22 yards to a chain, 10 chains to a furlong, and 8 furlongs to a mile. In terms of imperial volume, 2 pints make a quart, 4 quarts make a gallon, and 8 gallons make a bushel. Imperial weight measures are even stranger; 16 ounces makes a pound, 14 pounds to the stone, 8 stones to the hundredweight, and 20 hundredweight to the ton (an Imperial ton actually weighs 2,240 and not 2,200 pounds as most people assume). All this makes learning the wizard money used in the Harry Potter book series (29 knuts to a sickle and 17 sickles to the galleon) seem positively simple!

To convert SI units into a more convenient form, you merely have to multiply or divide by 10 or some exponent of 10 and add the appropriate prefix.

Using Table 3-6, if you have 1,000 metric somethings, you actually have 1 kilo-something (or "ksomething"). Most metric prefix tables include every power of 10 (for example, 10 to the power of 1 is "decca"), but I didn't place this information in the table because it is not typically used in for electronics.

A *second* is a SI unit that is defined as the microwave output frequency (9,192,631,770 oscillations in a second) of a cesium atom. Seconds are not normally listed with prefixes greater than one, but they are often listed in terms of thousandths of seconds (milliseconds or msecs) or millionths of seconds (microseconds or μsecs). For most projects presented in this book, you will see signal times in msecs and occasionally secs. If you are working on an experiment in this book and a calculation works out to billionths of a second (nsecs), then you have probably made a mistake.

The frequency of a signal is defined as the reciprocal of its period as shown in Figure 3-45. In this drawing, I have shown a repeating signal that takes "n"

Table 3-6 Converting SI units

Power Multiplier	Prefix	Symbol	Power Multiplier	Prefix	Symbol
10^3	kilo	k	10^{-3}	milli	m
10^6	mega	M	10^{-6}	micro	μ
10^9	giga	G	10^{-9}	nano	n
10^{12}	tera	T	10^{-12}	pico	p
10^{15}	peta	P	10^{-15}	femto	f

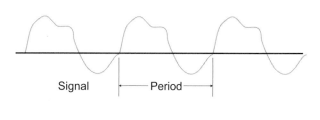

$$\text{Frequency} = \frac{1}{\text{Period}}$$

Figure 3-45 *Repeating signal showing period and frequency*

seconds to repeat. The length of time necessary for the signal to repeat is called its *period*.

So, if you had a signal with a period of 2 msecs (or 0.002 seconds), its frequency would be calculated as

```
Frequency = 1/Period

        = 1/2 msecs

        = 1/2(10⁻³) seconds

        = 500 1/seconds

        = 500 hertz
```

The SI unit for frequency is *hertz* (Hz) and has the units of 1 divided by seconds. Most audio signals are in the range of 100 Hz to 2,500 Hz (or 2.5 kHz). You can probably hear frequencies of 12 kHz or greater, but spoken words generally fall within the 100 Hz to 2.5 kHz range. In this book, you will see signal frequencies of several hundred to a million or so hertz (Hz, kHz, and MHz). You should never see fractions of a hertz (if you calculate a signal to have a frequency of millihertz, then you should know you have made a mistake).

Magnetic Devices

When we talk about something being difficult or non-intuitive (that is, not immediately obvious), we usually refer to it being "backwards" in some way. Colloquially, we use terms like "upside down," "inverted," and "180 degrees out of phase" to describe things that are difficult to learn, but we do not use terms to describe these things as being *sideways* (as in perpendicular or 90 degrees out of phase). I find this surprising because the most difficulty I had in visualizing something when I was a university student was magnetic fields and how they interacted with current-carrying wires.

Basic magnetism is quite easy to understand. In Figure 4-1, a magnetized piece of metal (called a permanent magnet or just a magnet) has lines of force running from the "north" pole (or end) to the "south" pole. The lines of force (also called the magnetic field) can be altered when a piece of a magnetic metal (normally iron) is placed into the lines of force.

Just as an electrical current travels through the path of least resistance, magnetic fields are most efficient when they travel in a magnetic material (usually iron) or have a minimal distance to travel when going

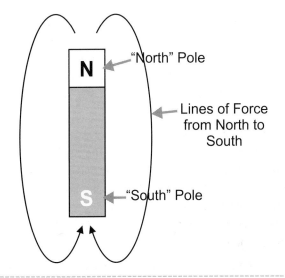

Figure 4-1   *Bar magnet showing line of force between the poles*

through nonmagnetic material. In the case of having a piece of iron close to a magnet, the iron provides a path for the magnetic fields to travel through. The magnetic field travels through the iron and draws it closer, so the magnetic field is smaller (and more efficient).

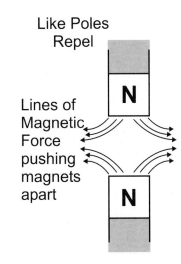

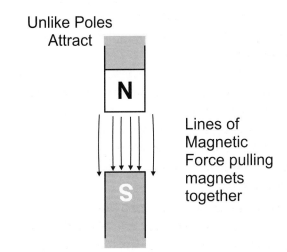

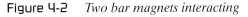

Figure 4-2   *Two bar magnets interacting*

This is why when a piece of iron is held away from a magnet, the force of the magnet can be detected but can be easily counteracted. As the iron is brought closer to the magnet, the force drawing it in grows stronger until the iron is in contact with the magnet and it takes a lot of effort to pull the iron off.

When you were younger, you might have taken advantage of the properties that like magnet poles repel and unlike poles attract (see Figure 4-2) to make yourself a simple compass. A simple compass can be made by rubbing a steel needle (or nail) with a magnet in the same direction several times and putting it on a cork floating in water held by a nonsteel pan.

Rubbing the needle with the magnet in the same direction several times will arrange atoms in the needle in the same direction, causing the needle to become magnetized. Many people will magnetize a screwdriver so that it will pick up and hold screws (so they will not fall) by rubbing it with a magnet. When the magnetized compass needle is placed on the cork, its North pole will be attracted to the earth's North Pole (and vice versa). Suspending a floating magnet in water (or alcohol) has been the compass' basic design for centuries.

Circular magnetic fields are produced when an electrical current passes through a conductor. Figure 4-5 shows the direction of the circular magnetic field that is produced when a current is passed through a wire. The direction of the current can be predicted by using your right hand. Curling your fingers, point the thumb away from your hand represents what happens when current flows through the wire. The thumb in the direction of the current and your curled fingers show the direction of the magnetic field.

When drawing current in a wire, the convention is to place a cross at the end of the wire where the current is going in (such as an arrow's feathers) and a dot at the end of the wire where current comes out (like an arrowhead). In Figure 4-3, I have drawn an X at the visible end of the wire where the current is passing through.

The magnetic field can be concentrated and increased by using multiple wires formed in a coil, as in Figure 4-4. The magnetic field can be predicted using the right-hand rule similar to the single-wire current-carrying conductor with the curled fingers

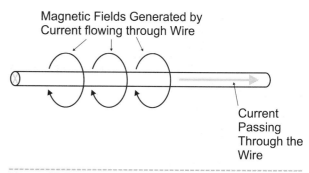

Figure 4-3 *Magnetic fields generated by a wire carrying a current*

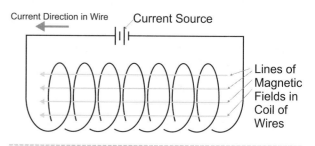

Figure 4-4 *Magnetic fields generated by a coil carrying a current*

being the direction of current in the coil and the thumb being the direction of the magnetic force.

By placing a bar of magnetic material (such as iron) in the coil, the strength of the magnetic field in the coil can be increased. This concentrates the magnetic field, resulting in a much stronger force on the magnetic objects. A coil wrapped around a metal bar is called an *electromagnet* and can produce electrical fields of considerable strength.

The operation of a permanent magnet and an electromagnet should seem quite straightforward. The difficulty in figuring out what happens becomes more complex when you mix the two. If you place a current-carrying wire in a magnetic field, the two magnetic fields will interact and cause a force to be applied on the wire. Figure 4-5 shows a current-carrying conductor between the two poles of a permanent magnet. The current flowing through the conductor is going *into* the page, as indicated by the cross on the small circle in the center of the page.

Figure 4-5 shows the lines of magnetism for both the permanent magnet and the conductor. The magnetic fields attempt to follow the same north-to-south connection, but the conductor causes the lines of the

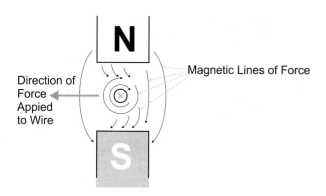

Figure 4-5 *Magnetic fields with a current-carrying wire between them*

magnetic fields to bend so that the lines coming from the north pole of the permanent magnet are drawn to the right side of the conductor. This movement of lines causes a force on the wire that doesn't draw it toward either permanent magnet pole but sideways between them.

The direction of force is surprising, but the reason should become obvious. The circular magnetic fields coming from the current-carrying conductor have both a north and a south pole that will be facing the poles of the permanent magnet. The like poles will repel, but the unlike poles will attract. The vertical

components of these forces will cancel each other out and the horizontal forces will become dominant. You can predict the direction of the force by using your left hand with the first two fingers and thumb at a right angle to each other. If the direction of the magnetic field is your first (or index) finger, and the direction of current in the conductor is your middle finger, the force applied to the wire is in the direction of your thumb. This is known as the "left hand rule."

You shouldn't be surprised to find out that the horizontal force exerted on the wire is very small. The amount of force can be increased by increasing the current passing through the conductor or by increasing the strength of the magnetic field (this can be accomplished by moving the poles closer together). Another way of increasing the force is to add multiple conductors.

For a traditional (rotary) electric motor, this sideways motion caused when a conductor is in a magnetic field cannot be easily taken advantage of because it would require one pole to be in the center of the motor with the other outside. A more practical implementation of this behavior is known as a *linear induction motor* and has been proposed many times in the past for levitating (frictionless) trains.

Experiment 23
Electromagnets

Parts Bin

Three-inch (8-centimeter) steel nail

Twenty feet (6 meters) of 22-gauge stranded copper wire

Elastic band

Tool Box

Two C batteries with clip

Miscellaneous nails, nuts, bolts, and washers

Miscellaneous coins

Probably the coolest thing ever done with an electromagnet was in the James Bond movie *Goldfinger*. The title character, Auric Goldfinger, has created a plan to rob the gold depository in Fort Knox, Kentucky, requiring the aid of organized criminals. After outlining his plot, one of the mobsters rejects the idea as being impossible and asks for his payoff and

leaves. While being driven to the airport, he is murdered by Goldfinger's bodyguard ("Oddjob"), and both him and the car he was murdered in (a Lincoln Continental Mark II with the gold payoff in the trunk) are crushed into a cube a yard (1 meter) across. The cube is picked up by an electromagnet and dropped into a pickup truck and returned to

Goldfinger so that the gold in the car could be reclaimed.

This is a great visual scene despite being illogical. I also consider the scene to be almost obscenely wasteful. I always loved the Continental Mark II and seeing a brand-new one destroyed (even though it's a 40-year-old movie) brings tears to my eyes.

What this scene demonstrates is the versatility of the electromagnet. After being crushed into a small cube, the car would have many sharp edges and would be difficult to pick up. The electromagnet, which can produce a magnetic field on command, will apply force to pick up *all* the metal in the crushed car without consideration of the shape of the final product or whether or not different parts will support the weight of the cube. The electromagnet is actually quite an elegant solution to the problem of picking up a crushed car.

To start off the experiment, wrap 20 feet (6 meters) of wire around a steel nail. Make sure the nail is normal steel using a permanent magnet. Although most nails are made from steel, you may find one that is made from stainless steel or brass, neither of which are magnetic materials and will not concentrate the magnet field the same way normal steel will.

I recommend using stranded wire rather than solid core because it will be easier to wrap around the nail. Don't worry about being neat or putting the loops of wire in any particular order. The only suggestion I have is to make sure the two wires coming from the nail come out at the same end. I secured the loose wires using a small elastic band (see Figures 4-6 and 4-7).

That's all there is to making an electromagnet. When power is applied to the wire wrapped around the nail, the current will cause a magnetic field around each loop on the nail that combines with the magnetic field of the other loops to produce a magnetic field that flows through the nail and allow it to pick up steel objects, as shown in Figure 4-8.

The magnetic field produced by the electromagnet will be surprisingly strong. Attach the two C cells to the electromagnet and test it out with different metal objects you find lying around. You should find that the electromagnet will easily pick up standard washers, nuts, and bolts but will not affect most coins. Nuts, bolts, and washers are commonly made from steel, whereas coins, made from brass, copper, and silver, are not affected by magnetic fields.

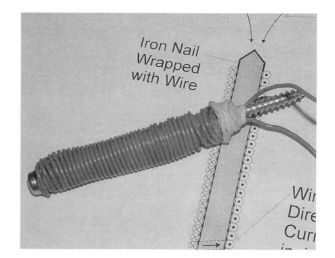

Figure 4-6   *The completed electromagnet*

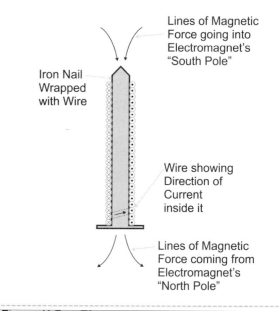

Figure 4-7   *Electromagnet cross-section*

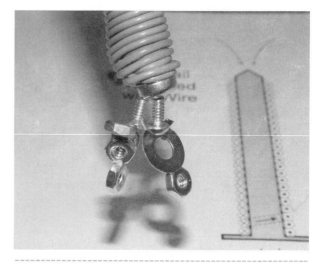

Figure 4-8   *Electromagnet in operation*

Experiment 24
Relays

Parts Bin

Two 3-inch (8-centimeter) steel nails

Twenty feet (6 meters) of 22-gauge stranded copper wire

Elastic band

Twelve inches of 24- to 28-gauge stranded copper wire

SPST or SPDT switch

Light-emitting diode (LED)

1k resistor

Two C batteries with clip

Tool Box

Nine-volt battery with clip

Soldering iron

Solder

Throughout the book, I emphasize that I do not think you should work with relays in your robot projects (or really any electronics projects). For now, I will discuss the operation of the relay and show you how to build one using the nail electromagnet from the previous experiment. Relays have two properties that will be important to understand as I explain more complex electronics.

The electronic symbol for the relay (see Figure 4-9) is a good representation of how it is built. The relay consists of an electromagnet that moves a "wiper," which has a piece of steel bonded to it. A wiper is a steel and copper connection that is attracted by the coil and provides a method of passing current between it and either the "Connection when Coil is Inactive" or "Connection when Coil Active." The wiper is normally pulled to the "Connection when Coil is Inactive" when no current is passing through its electromagnet. When current

passes through the electromagnet, the wiper is pulled down to make a connection to the "Connection when Coil Active." Although I have shown that only one wiper is in the relay depicted in Figure 4-9, I should point out that multiple wiper circuits' relays can control more than one device by the same relay.

The action of the wiper against the two contacts is exactly the same as the action of the switch wiper in the *single-pole double-throw* (SPDT) switch. It allows connections if the relay's electromagnet is active or not. When the electromagnet is off, the wiper is connected to one contact. When the electromagnet is on and active, the wiper is pulled away from the one contact and touches another.

By using the electromagnet built in the previous experiment, you can build the simple relay test circuit shown in Figure 4-10. This circuit consists of the electromagnet powered by two C batteries and controlled by a simple series switch.

The wiper for this circuit is actually another nail connected to an LED, a 1k resistor, and the positive connection of the 9-volt battery. To simplify building the circuit, the steel nail core of the electromagnet is the contact, which is made when the electromagnet is active and pulling in the wiper.

With a few modifications, this circuit could be used as quite an impressive science fair project. The

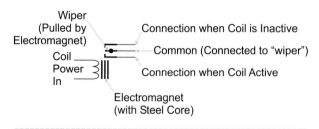

Figure 4-9   *Labeled relay symbol*

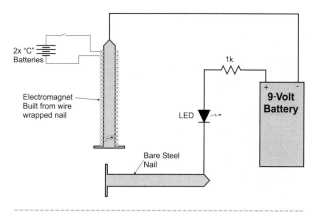

Figure 4-10   *Simple simulated relay circuit*

electromagnet can be mounted in a wooden frame (which is why I suggest you use a steel nail or bolt for the core) with the wiper held below. Instead of another nail, you could use a copper hinge with a steel bolt that would be attracted by the electromagnet. Also, use copper bolts for the contacts to the copper hinge wiper.

The two important features that relays bring to a circuit are (1) isolation and (2) the ability to control a high-voltage/current circuit from a low-voltage/current circuit. Isolation means that the driving circuit is completely separate from the driving circuit. This is important for applications controlling things such as household or other high-voltage AC circuits.

This isolation of circuits allows circuits with a low voltage and current to drive high-voltage/current circuits, which is a useful property when creating robots.

The digital logic (and analog) circuits presented later in the book run with much lower voltages and currents of even small hobby motors. Devices that allow power signals to control high-power devices are critical for being able to create robots.

Two other relay characteristics are rarely discussed, but you will appreciate them when you are designing your own solid-state electronic motor drivers: the very small voltage drop and the resistance of the relay's contacts. When I introduce transistors, you will see that they do have a definite voltage drop and a resistance that will affect the operation of motors.

Figure 4-11 is the transistor driver circuit required to control a relay from a simple logic circuit and is exactly the same as what would be used to control a motor from logic circuits. This circuit may be beyond your current level of knowledge, and if this is the case, you should make a mental note to come back to this circuit diagram for later reference.

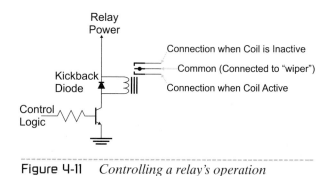

Figure 4-11   *Controlling a relay's operation*

Experiment 25
Measuring the Earth's Magnetic Field

Parts Bin

Assembled printed circuit board (PCB) with breadboard

Compass

Twelve to 20 inches of 22- to 26-gauge stranded wire

1k potentiometer

C battery

Tool Box

Wiring kit

Clippers

Knife

When you use your *digital multimeter* (DMM) to test the voltage, current, or resistance in a circuit, you must remember that this instrument is the result of literally millions of person-years of experimenting and theorizing the nature of electricity and how it is measured. In Ben Franklin's day, the test for electricity was touching a circuit and seeing if a spark was produced. This touch test for electricity only detected fairly high levels of electricity (to produce a spark, a potential of 1,500 volts or more is required). This test is obviously not very accurate and potentially very dangerous. An European scientist repeating Franklin's kite experiment was electrocuted when he touched the key.

Over time, different theories were postulated as to what was electricity, and to test these theories, different experiments were performed, with the apparatus used actually becoming tools to better understand what was happening. A very early example of this was the *electroscope* (see Figure 4-12), which consists of a copper plate connected to a circuit and to a piece of gold leaf, or a thin sheet of gold. When the copper plate would be connected to an electrical circuit, both it and the gold leaf would repel due to similar changes on the fixed and hinged gold leaf. The gold leaf, being very light, would be moved by the charge.

Gold leaf has a long and distinguished career in science (it being central to Rutherford's experiment on the nature of matter), because gold has some useful properties. First of all, it is an extremely good conductor. Secondly, it does not react with other

elements very easily; you may have heard that it is called the noble metal. This name does not come from it being used in royal crowns; it comes from the inability other elements have in combining with gold except under extreme conditions. The last property of gold that makes it ideal for this type of experiment is its malleability. Gold can be pounded into very thin sheets only a few atoms thick.

Electroscopes are difficult for the modern hobbyist to build and use. As you would expect, finding suitable gold leaf is difficult, but you would be surprised at how hard it is to find other materials that can be used in its place. Normal household aluminum foil cannot be used because it is too thick and heavy. I have seen some sample electroscopes built using the plastic foam "peanuts" used for protecting items during shipping. You could try to build your own electroscope using the foam peanuts, but remember that they tend to hold static electricity and are actually a considerable risk to the different electronic parts

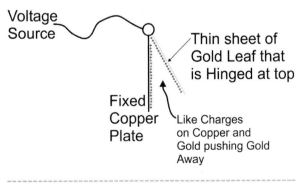

Figure 4-12 *Early electroscope*

used in this book's later experiments. After experimenting with the foam peanut electroscope, make sure the foam is disposed of and not a danger to electronic devices.

A much more useful historical electrical measurement device that you can experiment with is the compass-based ammeter. To build this device, use a stranded wire wrapped two or three times around a compass, as shown in Figure 4-13. Then orient the compass so that North is parallel to the direction of the wrapped wire, and connect a 9-volt battery to the bare ends of the wire.

You will find that the compass needle snaps, becoming perpendicular to the wires wrapped around the compass as it aligns itself with the magnetic field produced by the wires. I was surprised at how quickly the compass needle moved with the magnetic field to indicate the magnetic field produced by just two or three turns of wire. A C cell powered coil is much stronger than the earth's magnetic field.

Many early scientists used an ammeter to detect electricity in a circuit because of its simplicity and low cost. With a bit of work, after measuring the amount of movement of the needle to a known amount of current, the strength of the earth's magnetic field at a specific location can be found. Doing this takes a bit of work, but you can do it quite easily by adding a 1k potentiometer to the circuit, as shown in Figure 4-14. The potentiometer limits the total current passed to the coil of wire around the compass. I used the book's PCB for this experiment and found that I had to make sure that I kept the coil of wire and compass as far

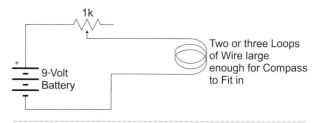

Figure 4-14 *Circuit to measure the earth's magnetic field*

away from the 9-volt battery as possible to ensure that any iron in the battery didn't throw the compass off.

The magnetic field produced by the coil (given the symbol "B" and is in units of "Teslas") is defined by the following formula:

$$B = \mu_0 \times N \times i/(r \times 11.18)$$

where μ_0 is the permeability of the vacuum and is $1.257(10^{-7})$ N/A^2. N is the number of turns of wire, i is the current (in amperes) through the wire, and r is the radius of the coil in meters. When arranging the compass, which is inside the coil of wire, so that North is in parallel with the coil of wire, put a 9-volt battery into its clip on the PCB and adjust the potentiometer until the compass is only deflected by 20 degrees. I then measured the voltage at the 9-volt battery (9.25 volts in my experiment) and measured the resistance of the potentiometer and the wire (12 Ω). Applying Ohm's law, I had a current draw of 770 mA. Knowing the current, that I used three turns of the wire, and that the coil was 2 inches (5.08 centimeters) in diameter, I found that the magnetic field inside the coil was $1.039(10^{-5})$ Tesla.

I started with North perpendicular to the direction of the coil's magnetic field so that I could use trigonometry to figure out how powerful the Earth's magnetic field is. In Figure 4-15, I have drawn out the Earth's magnetic field with the coils field and the resulting compass direction. Rearranging the formula for the tangent (and knowing the tangent of 20 degrees), I can calculate the Earth's magnetic field acting on the compass needle:

    Earth's magnetic field = Coil's magnetic
    field/tan(20)

       = 1.039(10⁻⁵) Tesla/0.364

       = 2.86(10⁻⁵) Tesla

Figure 4-13 *Ammeter made from a compass wrapped with wire*

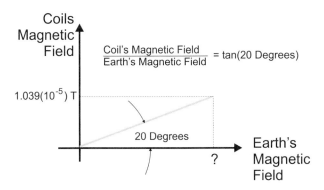

$$\frac{\text{Coil's Magnetic Field}}{\text{Earth's Magnetic Field}} = \tan(20 \text{ Degrees})$$

Coils Magnetic Field

$1.039(10^{-5})$ T

20 Degrees

?

Earth's Magnetic Field

Figure 4-15   *Trigonometry of the perpendicular magnetic fields*

The accepted value for the Earth's magnetic field is about $5(10^{-5})$ Tesla, so my measured and calculated value is a little more than half the value. This may seem like a large error, but I'm amazed I got as close as I did with the crude equipment I used. When I performed the experiment, I did it in the kitchen of my house; I might have gotten a more accurate result if I were to repeat it outside with no metal or power lines around. Of course, I might end up with a less accurate result because of my placement of the wire coil around the compass, the charge in the battery, or due to a variety of factors. It would be interesting to measure the Earth's magnetic field in different locations while keeping the apparatus as constant as possible to see what kinds of deviations are found.

Experiment 26
Direct Current (DC) Motor

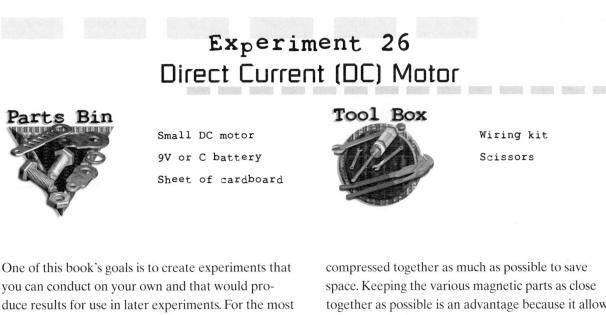

Parts Bin

Small DC motor

9V or C battery

Sheet of cardboard

Tool Box

Wiring kit

Scissors

One of this book's goals is to create experiments that you can conduct on your own and that would produce results for use in later experiments. For the most part, I think I have succeeded, except for one device, the electric motor. I have not been able to come up with a design for this machine that is both simple to build and that could be used in later experiments. This is surprising, especially considering that a *direct current* (DC) motor only consists of a few different parts (see Figure 4-16).

Figure 4-17 shows the different parts of a small electric motor. The armature consists of the motor's driveshaft and the parts that have been mounted on it. The different parts have been spread out so they can be easily identified, but when you look at an actual armature, as in Figure 4-17, its parts have been

compressed together as much as possible to save space. Keeping the various magnetic parts as close together as possible is an advantage because it allows the motor to run more efficiently.

An electric motor can be considered to consist of a number of electromagnets that can be switched on and off according to the position of the electromagnet rotor, as shown in Figure 4-18. I have drawn the motor as having three electromagnets (this is common for most DC motors, as I will explain below), and in the left drawing, electromagnet 1 is pointing upwards with electromagnet 2 producing a south pole and is drawn to the permanent magnet's north pole. Electromagnet 3 produces a north pole, is repelled from the permanent magnet's north pole, and is drawn to the permanent magnet's south pole.

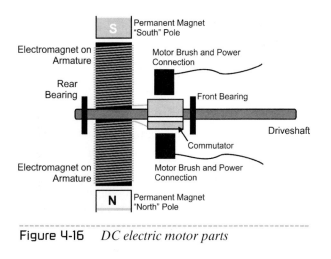

Figure 4-16 *DC electric motor parts*

In the drawing on the right of Figure 4-18, the motor's axle has turned 60 degrees and electromagnet 1 is turned on and producing a south pole. Electromagnet 2 is facing the permanent magnet's north pole and is turned off. Electromagnet 3 is still producing a north pole and continues to be drawn to the permanent magnet's south pole. As the motor's axle has turned 60 degrees in Figure 4-18, electromagnet 1 and 2 have changed operation to ensure that a force continues to draw the electromagnets to the permanent magnets and provide torque on the axle. The motion of the axle along with the torque created by the electromagnets is passed out of the motor and used to drive whatever it is connected to.

In Figure 4-18, I have marked the position of the commutator contacts on the axle as well as their rela-

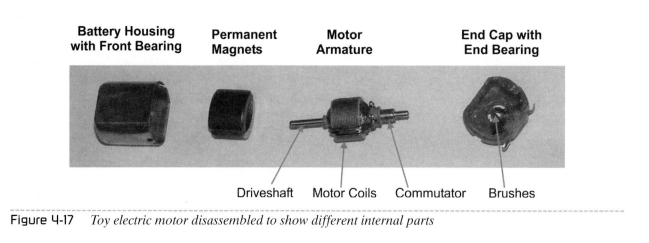

Figure 4-17 *Toy electric motor disassembled to show different internal parts*

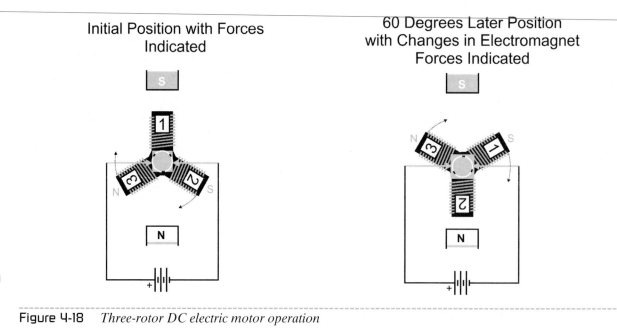

Figure 4-18 *Three-rotor DC electric motor operation*

tion to the motor's two brushes. Note that in the example only two of the commutator contacts are touching the brushes. This is not always the case. For the angles between the examples presented in Figure 4-18, all three contacts will be touching the brushes.

One thing that has not been discussed is the property of the motor that causes it to turn in only one direction according to the direction in which current is flowing. To indicate how the motor will run, by looking at the end cap of the motor you will see it is marked with both a + symbol by one wiring connection and by a flat side on the axle bearing (Figure 4-19).

Most small motors run anywhere from 2,000 to 4,000 RPM, and it is difficult to see the direction in which the motor is turning. To make the direction easier to observe, I used the fan disk shown in Figure 4-20 was made out of cardboard. When you make the fan disk, you should use a compass and protractor to make it as even as possible and make sure the center is accurately identified. Eight fan blades are bent from the cut cardboard. Using this fan disk, you can tell which direction the motor is turning from the direction in which air is being blown by the motor.

Before testing the motor with a battery, measure and record its resistance using a DMM (one of my motors had a resistance of 0.9 Ω). Once you have done this, press the fan disk onto the motor's axle and test it using the circuit shown in Figure 4-21. Measure the voltage across the motor, as well as the current through it, and calculate the resistance using Ohm's law. With 7.2 volts applied and 0.3 amperes being drawn, this works out to 24.3 Ω. This calculated value is much greater than the resistance of the stopped

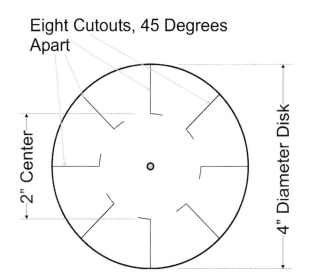

Eight Cutouts, 45 Degrees Apart

2" Center

4" Diameter Disk

Figure 4-20 *Cardboard fan disk to test motor operation*

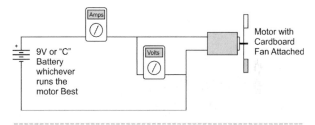

Amps

9V or "C" Battery whichever runs the motor Best

Volts

Motor with Cardboard Fan Attached

Figure 4-21 *Circuit to test the DC electric motor*

motor. The reason for the higher effective resistance is due to the "reluctance" of the switching electromagnets. When a coil is turned on or off, its effective resistance becomes very large until current is flowing smoothly in the motor.

As a final aside, a simple electric motor was first shown on the TV show *Beakman's World*. It is what I would call a single electromagnet motor because the electromagnet can only be energized one way. The motor is simple to build but is quite complicated to get running, and it usually only runs for a few seconds (in one direction) before jumping out of its cradle. If you are interested in building it, you can find the instructions at http://electronics.howstuffworks.com/framed.htm?parent=motor.htm&url=http://fly.hiwaay.net/~palmer/motor.html.

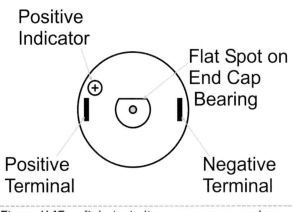

Positive Indicator

Flat Spot on End Cap Bearing

+

Positive Terminal

Negative Terminal

Figure 4-19 *Polarity indicators on motor end cap*

Section Five

Drivetrains

Electric motors are the most popular method of causing motion in robots, even though they are not very efficient and can be somewhat difficult to use in different situations, especially when they are compared to other devices such as

- Internal combustion motors (gas and diesel)

- External combustion (steam) motors

- Hydraulics

- Muscle wire

To allow electric motors to perform the same tasks as the devices listed above, they will have to have mechanical attachments to make them more useful in different situations. These mechanical attachments consist of devices such as axles, couplings, gears, and wheels, and rather than using the clumsy term *motors and mechanical attachments*, I am going to refer to the motor and hardware that drive the robot or perform some kind of action using the term *drivetrain*.

The simplest robot drivetrain that I can imagine is shown in Figure 5-1 and consists of motors with some kind of small sticky nub placed on the ends of their shafts. I always cringe when I seen a robot built with this configuration. Even though it will move the robot, it can be tricky to set up, and it does not handle any surface other than one that is perfectly flat and smooth. I have seen wheel nubs used for the robot design in Figure 5-1 made out of small washers that snugly fit over the motor's axles, or made from sections of hot glue gun sticks that have been melted onto the axles.

The reason why I cringe when I see a robot like this because it is so limited. Small electric motors tend to run at several thousand *revolutions per minute* (RPM) and running them continuously is impossible. This may not be immediately obvious, but consider the case where the motors are turning

at 2,000 RPM (somewhat slow for a small motor) and the nubs are 0.25 inches (6.35 millimeters) in diameter. The robot would move about 52 inches (133 centimeters) per second. This is quite a bit faster than an adult's walking speed and would allow the robot to cross a 10-foot (3-meter) room in just over 2 seconds. A small robot moving at this speed is very hard to control and will probably spend a great deal of time crashing into different things because it cannot detect objects far enough away to stop or turn away from them.

A question might be asked about how to slow down the motors electrically, and this is possible, but to do that, you will have to significantly cut down on the amount of current being passed to the motors. As you decrease the current to the motors, you do not get as much torque (rotational force) from the motors. In this type of robot where the nubs are small, you will find that a lot of force must be applied from the nubs to the surface the robot is running on to get the robot to move. If you decrease the amount of torque, you will find that the robot will move unpredictably (one side will turn while the other doesn't) or not move at all. Another solution to this problem is to "blip" the motors periodically, and this is the usual solution—the robot's controlling electronics determine the timing between each motor being turned on quickly. This isn't a terrible solution, but you have a robot that can only run on a very flat and level surface.

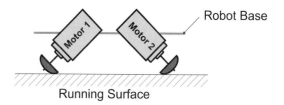

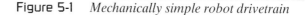

Figure 5-1   *Mechanically simple robot drivetrain*

The best solution to this problem is to reduce the speed the motor runs while increasing its torque. By doing this, larger wheels can be used with the robot, which will allow the robot to run over uneven surfaces and be controlled very simply. This change in speed and torque is accomplished using gears (Figure 5-2) that can both change the direction of rotation as well as change the speed and torque of the movement.

In Figure 5-2, I have presented an equation of the speed of rotation along with the number of teeth in the gears. This equation could be modified with the gear's radius, diameter, or circumference used instead —in either of these cases, the differences in sizes of the gears are taken into account. I want to point out that radius, diameter, or circumference can be used because this allows you to think in terms of two circles that are in contact and not only in terms of two gears. This is an important point and one that I will take advantage of in this section.

When I introduced the concept of power earlier in the book, I presented the concept that mechanical power was the product of force and velocity:

$$P = F \times V$$

This may seem to be a bit trickier when you are dealing with rotating components, but as I indicated, rotational force is torque, has the units of force multiplied by the radius at which the force is measured, and is specified in pounds-inches or Newton-meters. If rotation is measured in RPM, the power equation becomes the following:

$$P = Torque \times Rotational\ Speed$$

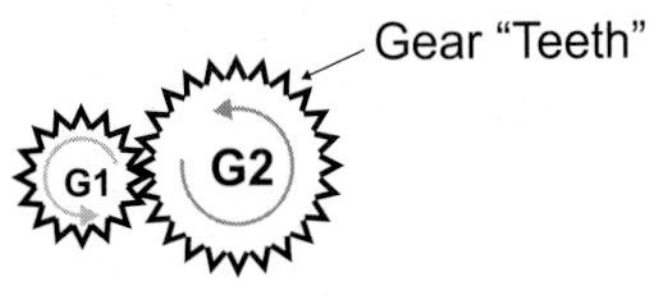

Speed$_{Large}$ x #Teeth$_{Large}$ = Speed$_{Small}$ x #Teeth$_{Small}$

Figure 5-2 *Gears meshing together and transforming rotation speed and torque*

It is measured in watts as in single direction motion or electrical power.

Going back to the original formula equating speed of rotation to gear size, if we have a small gear driving one that has three times the number of teeth, the speed of the larger gear can be calculated as the following:

$$Speed_{Large} = Speed_{Small} \times \#Teeth_{Small}\ /\ \#Teeth_{Large}$$

$$= Speed_{Small} \times 1/3$$

So the speed of the larger gear's output is one-third the speed of the smaller gear. Knowing that power is being driven into the small gear at some rate, and assuming that no power is lost in the system, we can equate the power of the two and find the torque at the larger gear:

$$Power_{Small} = Power_{Large}$$

$$Speed_{Small} \times Torque_{Small} = Speed_{Large} \times Torque_{Large}$$

$$Torque_{Large} = Speed_{Small} \times Torque_{Small}/Speed_{Large}$$

$$= Speed_{Small} \times Torque_{Small}\ /\ (1/3)\ Speed_{Small}$$

$$= 3 \times Torque_{Small}$$

The ability of gears to transform speed and torque is one that is critical for robots that work with small electric motors—they generally run at a very high speed while providing a small amount of torque.

Exeriment 27
Motor-Driven Crane

Parts Bin

Miscellaneous K'NEX Kits
   (see text)

Motor drive kit (see
   text)

Thread

Tool Box

Although I find it hard to agree that remote-control "robots" (such as BattleBots) fit the description of a true robot, I have to agree that the technology used to build them is applicable for autonomous robots. Saying that remote-control devices are true robots could be similar to accepting that motor-driven cranes (such as the construction cranes you see assisting in the creation of buildings) could also be called robots. I find cranes useful in explaining the operation of different robot parts because they cannot run off the table or bench you are working on and either smash into a million pieces or run under a couch.

A typical construction crane has three degrees of freedom (as shown in Figure 5-3). *Degree of freedom* is the term used to describe a movement that does not affect other movements. The crane can turn the boom left or right (one degree of freedom), move the hook up or down (second degree of freedom), or move the hook toward or away from the center of the crane (third degree of freedom). For comparison, you arm has seven degrees of freedom (two at the shoulder, one at the elbow, three at the wrist [twisting is a degree of freedom], and the last being the ability to open and close your hand). Most robot arms have three or more degrees of freedom, so the crane is not an unlikely device to consider as a robot analog.

The crane that I would like to play around with is the single-degree-of-freedom crane that I built out of miscellaneous K'NEX parts. It uses a motor taken from a K'NEX "Kart Racer" (Figure 5-4). If you are not familiar with K'NEX, I suggest that you buy a sample experimenter's kit—K'NEX is a building tool that consists of various rods and connectors that can be put together to create quite light, fairly strong, large structures. A nice feature of the different rods that make up the kit is that they are measured so triangular structures can be built along with square and rectangular ones very easily. Most of the experimenter's kits consist of instructions for building different creations (ranging from sculptures to moving

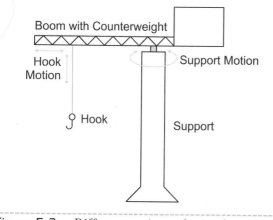

Figure 5-3   *Different motions of axis of movements possible for a construction crane*

Figure 5-4   *Crane built from K'NEX parts*

objects such as circus rides or simple vehicles), along with a plethora of parts (including gears and wheels) that will give you a great deal of flexibility in coming up with your own designs.

I used the Kart Racer gray motor as the crane's "winch" that pulls the sewing thread because it can drive the K'NEX shafts directly (without having to hack any parts to fit onto a motor driveshaft). This motor can be cut open to give you access to the motor wiring, and using the different gears and motors that are available in the different kits, you can very easily create a complex drivetrain of your very own.

Please do not feel that you have to use exactly the same parts that I did when building your crane; this design is by no means the most efficient, and depending on your skills and the parts you have on hand, you could probably come up with something better.

K'NEX is ideal for this type of experiment, better than LEGO or Mechano because it is well suited to building large, open structures such as a crane, and it has a good assortment of gears and wheels. Building structures is probably not as intuitive with K'NEX as it is with other building kits. For this reason, I suggest that before you create something new, look through instructions for something that is similar to what you want to build and start from there. The "crane" used in this experiment started life as a "windmill," and after removing the sails and modifying the base, I had most of the design for the crane completed.

Looking at the other different building kits available on the market, I categorize their strengths and weaknesses in Table 5-1. I have limited the list to just K'NEX, LEGO, and Mechano, although similar products are compatible to these and have similar characteristics. One thing you will notice is that these kits are fairly pricey, especially if you are looking for specific parts. You will find it a good idea to either buy a parts kit and try to work within it, or collect pieces (garage sales are excellent places to find them).

Table 5-1 Various building kits

Product	Description	Advantages	Disadvantages	Best Robot Applications
K'NEX	Rods and connectors	Large structures, lightweight. Good selection of gears and wheels that integrate easily with product.	Poor bending strength/ rigidity. Not intuitive to build without experience. Difficult to find motors and parts.	Differential drive robots (Good motor/gear/wheel integration). Prototype robot arms.
LEGO	Interlocking bricks	Strong small structures. Intuitive/fast to work with. "MindStorms" Robot Kit provides excellent robot base and sensor set. Widely available parts.	Large structures will be heavy/may not be very strong. Complex structures may be weak. Motors/gears generally designed for specific products.	Prototype mobile robots.
Mechano	Metal/plastic girders held together by small nuts and bolts	Very strong. Intuitive to work with. Parts can be easily adapted to work in other structures.	Fewer predefined kits than K'NEX and LEGO. Heavy. Nuts and bolts hurt when stepped on.	Poorly suited for complete robots, but individual parts can be integrated into robots very effectively and easily.

Experiment 28
Pulleys Added to Crane

Parts Bin

Crane from previous experiment

Four small tireless K'NEX wheels

Miscellaneous K'NEX Parts

Tool Box

Depending on the motor you used with your crane or the amount of charge left in the batteries, you will probably discover that the crane does not have a lot of lifting power. I found that the stock configuration of the crane would not lift two "C" cells—what was needed was some way of increasing the power of the crane. One of the obvious ways of doing this is to increase the gear ratio between the motor and the winch (what the thread winds itself around). This will work, but gears and the structure needed to support them are heavy and complex.

As I will do throughout this book, when I am confronted with a problem like this one, I will go back and look at how the problem was handled historically. Five thousand years ago, the Egyptians were faced with a similar problem: how to raise an obelisk only using human muscle power and a rope (Figure 5-5). Assuming the stone of the obelisk has a density of 5,000 kg/m³, is 20 meters high, and has an average cross-sectional area of 4 meter², the obelisk would have a volume of 80,000 m³ and weigh 400,000 kg (881,848 pounds). Assuming that one man could pull 50 kgf, then 8,000 men would be required to pull up the obelisk.

A density of 5,000 kg/m³ is five times that of water and should be a good estimation of the den-

sity of common stone (Earth has a density of 4,500 kg/m³, which was the basis I used for this value). Fifty kg of force ("kgf") is probably acceptable for a modern man, but for an Egyptian slave, it is probably optimistic. In any case, 8,000 men pulling up the obelisk is an unreasonable number and something has to be done to lower it to something more manageable.

The solution is to double up the ropes using a device called the pulley, which will effectively multiply the amount of force each slave could exert on the obelisk. The operation of the pulley is shown in Figure 5-6—a motor pulling a weight upwards on a single cable will have to exert enough force to lift 100 percent of the weight. By looping the cable around a pulley, the amount of force required is halved, and by looping the cable twice around two pulleys, the amount of force required from the motor is divided by four. The ancient Egyptians, by looping the rope that will pull up the obelisk around eight pulleys, will have reduced the number of men needed to lift the obelisk from 8,000 to a somewhat more manageable 1,000.

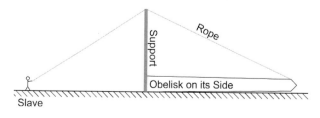

Figure 5-5 *Raising a large obelisk using muscle (slave) power*

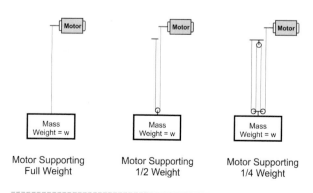

Figure 5-6 *Changing force to raise an object using a pulley*

Hopefully, you do not think that this force multiplication comes free—as with everything in life, when you change one thing, it affects something else. In this case, the length of rope or cable that is moved is increased with the number of times it is looped around the pulleys. The relationships between distance and force can be written out mathematically as the following:

```
Force (1 Pulley) = Force (n Pulleys)/n

Distance (1 Pulley) = n x Distance (n
Pulleys)
```

So, if it required 1 pound of force pulling for 1 foot using a rope looped around 3 pulleys, the force applied on the object would be 3 pounds and the object would travel $\frac{1}{3}$ of a foot.

In this experiment, you can demonstrate the increased power (and decreased speed) of adding a pulley to your crane. If the crane was built out of K'NEX, the top of the crane and the hook would be modified to look like Figure 5-7. In this case, I have looped the thread around two wheels at each end, making a four-pulley crane and increasing the

Figure 5-7   *K'NEX crane pulley detail*

amount of weight the crane can lift four times. After making this modification, you can test out this assertion by trying to lift heavier objects—if, like my crane, yours could not lift two "C" cells, you should find that with the pulley added to it, it can now comfortably lift six or seven.

Experiment 29
Switch DC Motor "H-Bridge"

Parts Bin

Assembled PCB with breadboard

Crane from previous experiments

Dual "C" battery clip

Two Single-Throw Double-Throw (SPDT), PCB-mountable switches

Tool Box

Wiring kit

Rotary tool

Soldering iron

Solder

Heat shrink tubing

Previously, I showed how the direction a DC motor turned in was controlled by the direction that current flowed through it. By changing the direction of the current flowing through the motor, the direction it turns changes. For most applications, motors only have to turn in one direction; even if they are required to reverse, their motion is often passed through a gear box that changes the output while

allowing the motor to turn in the same direction. For robot applications, the simplest way of changing the direction a shaft turns in is to change the direction of the current flowing through the motor.

If you were to think about the problem for a few minutes, you might come up with a circuit like the one shown in Figure 5-8. This circuit provides two battery packs, one of which is selected at any time

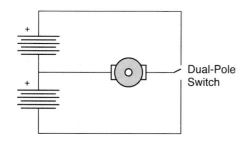

Figure 5-8 *Using two batteries to control motor operating direction*

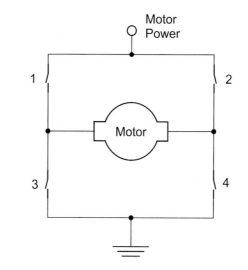

Figure 5-9 *H-Bridge motor driver*

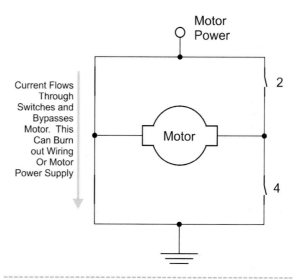

Figure 5-10 *H-Bridge motor driver operating incorrectly*

and allows current to either flow into the motor or out of it. This type of switch has been used in some robots but has a problem that should be quite obvious; the battery pack that powers the direction that the robot runs in most of the time (normally forward) will be run down much quicker than the other battery pack. The main advantage of the circuit in Figure 5-8 is that it only requires one switch. Another advantage is that there is only one switch voltage drop—this is not an issue when mechanical switches are used, but it can be a problem for electronic switches.

The motor control circuit that is most commonly used to control the direction of a DC motor is called the "H-Bridge" (Figure 5-9). By closing the switches catercorner from each other, you can control the direction of the motor. As I will discuss later in the book, turning the motor on or off, as well as how fast it is turned on and off, can be controlled easily using simple electronic devices. The H-Bridge allows a single power supply to be used to control the direction a DC motor turns in and can generally be built quite inexpensively.

The H-Bridge has two main concerns that you should be aware of. The first is that the motor current must always pass through two switches. When physical switches are used, this is not a problem, but there will be situations when electronic switches are used with low-voltage batteries where there may not be a sufficient amount of voltage for the motors to run properly.

The second concern is quite subtle and is one that you must be aware of at all times. The motors will turn when one switch on either side of the H-Bridge is closed. If both switches on the same side of the H-Bridge are closed (Figure 5-10), you will have a prob-

lem; by closing the two switches on the same side of the H-Bridge, you have created a short circuit. This short circuit will burn out the motor wiring, the H-Bridge switches, or the power supply. The best that you can hope for is that the battery's charge will be seriously depleted. Care must be taken in your motor control circuits and software to ensure that this condition can never happen.

If you have used the same K'NEX hardware as I have to build your crane, you might have figured out by now that the switch on the gray motor will command the motor to turn in one direction or the other and is probably wired internally as an H-Bridge. Rather than taking this on faith, for this experiment I

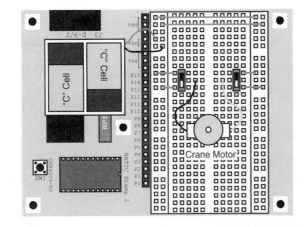

Figure 5-11 *Breadboard wiring for crane DC motor H-Bridge control*

would like you to disconnect the wires leading to the DC motor in the gray motor box by cutting it open (I used a rotary Dremel tool). After pulling out the switch and the small PCB that it is mounted on and exposing the DC motor wires, I soldered on some 24-gauge solid core wire to allow me to wire the motor to the book's PCB breadboard.

When this is done, you can wire the motor to the breadboard as part of the circuit according to Figure 5-11. For this circuit, I have wired two SPDT switches with the wiper connected to the DC Motor contacts —this will allow you to select either a positive or negative voltage for each side of the motor without there being a chance for both sides of the H-Bridge to be connected.

Once you have wired the circuit, you should discover that it works identically to the single switch of the original with one important exception: When both SPDT switches are set in the same direction, the motor stops. When both switches are in the same direction, no current can flow in the motors. So, if you were to continue opening up the gray box, you would probably see that in the center position (off), there is no actual connection to the motors, so no current can flow. I'm pointing this out because it is not often understood that if there is no voltage drop across a device (as in this case), there is no current flow and the device won't work.

Experiment 30
Differential Drive Robot Chassis

Parts Bin

Assembled PCB with breadboard

Four PCB mount SPDT switches

Switch with solder lugs

Four AA battery holder

Double-sided tape

Eight 4-40 0.5-inch (1-centimeter) screws

Four 1-inch (2.54-centimeter) standoffs

Two small toy motors

Two wheels (see text)

Two axles (see text)

Motor straps (see text)

Misc. nuts and bolts (see text)

Furniture glide, acorn nut, LED or clothes hook (see text)

Tool Box

Wiring kit

Soldering iron

Misc. screwdrivers/pliers

Glues/adhesives

When I present actual robots in the rest of this book, I will be primarily working with the differential drive platform that I introduced to you in the first section. The simplicity of this type of robot allows for simple building as well as easier circuit and software development, although you should be aware of a few things to make the assembly and operation of the robot as simple as possible. As I work through the chassis, I will cross-reference the design decisions to my 10 "Rules of Robotics."

I feel that the ideal layout for the differential drive robot is the one shown in Figure 5-12; the robot is as short as possible with the wheels centered and the center of mass at the center of the robot. In Figure 5-13, I have marked casters at the two ends of the robots; the castors are wheels or smooth plastic that allow the front or rear of the robot to slide on the running surface easily.

The center of mass is often called the center of gravity, but I prefer to refer to say mass rather than gravity when talking about robots because it reminds me that the inertia of the robot when it starts or stops will change the amount of force on the casters. By

keeping the wheels and center of mass as close to the center of the robot as possible, the amount of change in force on the casters is minimized, allowing the robot to run easily over different surfaces (see Figure 5-14). This is important when the robot is expected to run over carpets or from one surface to another. If a lot of force is on the caster, you may find that the robot will stall at the caster if too much force is being placed upon it and less weight is placed upon the wheels (Figure 5-15).

When a robot's wheels are placed at one end and the center of mass is located elsewhere, corrections may be required between the sensor/control input and the wheel commands. In the best possible situation, the sensors would be placed directly above the wheels (at the robot's center) so that the robot could turn and follow the sensor input without the requirement for any kind of correction caused by inadvertent body movement due to the wheels.

The purpose of this experiment is to add a battery connector, power switch, wheels, and motors to one

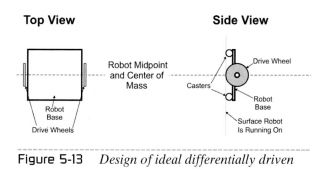

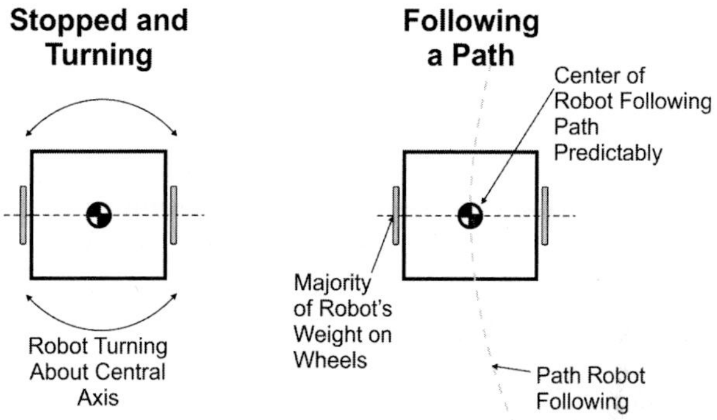

Figure 5-12   *What the differential drive robot with PCB will look like*

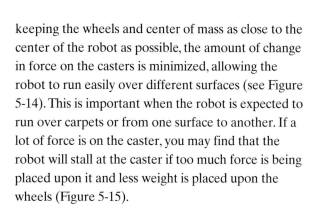

Figure 5-14   *Ideal differentially driven robot motion*

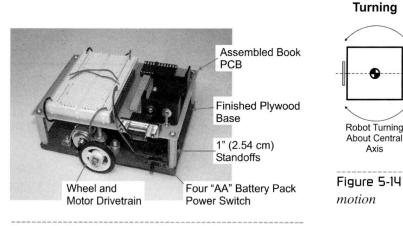

Figure 5-13   *Design of ideal differentially driven robot*

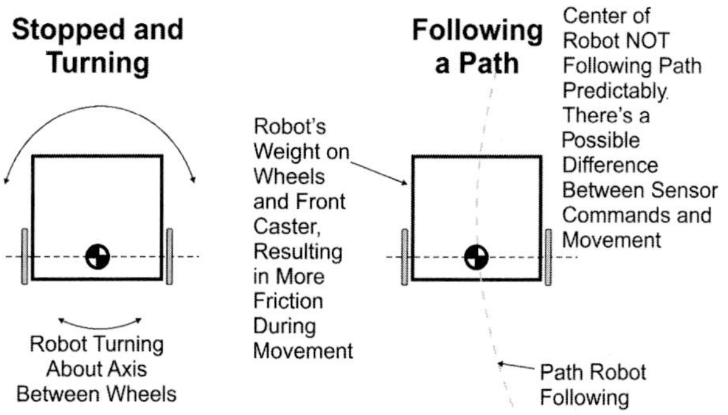

Figure 5-15   *Less than ideal differentially driven robot motion and some potential problems*

1.27 / 0.09 = 14.1

The Motor Speed is Divided 14.1 times, Resulting in a Slower Wheel Speed, but with 14.1 times more Torque than the Motor Provides

Figure 5-16   *Differential drive wheel dimensions and detail*

Figure 5-17   *Differential drive robot drivetrain details*

of the plywood bases that you finished earlier. Figure 5-16 is a side view showing how the motor axle is pressed against a wheel—this arrangement works similarly to two gears and reduces motor speed by 16 times or so. As you run the robot, you will discover that you will be cleaning the motor axle/wheel assembly a surprising amount (even if you think you are in a clean environment). The perfect situation would be a gear or pulley system sealed in a box—this will prevent hair, lint, and dust from fouling up the robot's drivetrain.

The wheel and axle support I used was taken from a Mechano set (shown in Figure 5-17). The axle support consists of an L-shaped piece that could be bolted down to the plywood using two 6-10 bolts while leaving enough space for the two toy motors to drive it. The motors are strapped down using Mechano metal pieces. Instead of using Mechano parts, you should be able to find LEGO pieces that work just as well or other toy kits that can be hacked together. If worse comes to worst, a wheel could be purchased from a craft store and be supported by a wood block that has been drilled out to accept a bolt axle and mounted to the plywood chassis.

Despite trying to offset the mass of the four AA battery holder with the motors and the 9-volt battery on the PCB, I found that my robot ended up heavy toward the AA battery pack. Because of this, I ended up only using one caster; I placed a wall-mountable plastic hook on the bottom of the robot. You could also use a Teflon-coated furniture slider, or even glue an LED to the bottom of the robot. The ultimate castor would be something like a free swiveling model airplane tail wheel, but these can be surprisingly expensive and take up a lot more space than is available on the bottom of the robot. The important thing about the castor is its ability to allow the robot to move and turn easily.

Once you have figured out a way to mount the wheels and the motors, place the AA battery holder on the plywood using two-sided tape. This battery pack will be used to power the robot's motors and it must have a power switch added in line to turn off power to the robot's motors. This is very important because there *will* be situations where your controlling hardware or software will send invalid commands to the motors and you will want to stop them. By powering the robot from a separate, switched battery pack, you will be able to stop the robot and observe what is happening without having the robot running over your table or causing other problems.

Now you are ready to test out the robot. After attaching the standoffs and bolting the robot together, follow Figure 5-18 to wire two H-Bridges that will allow you to manually control the robot and work through the switch settings to move the robot

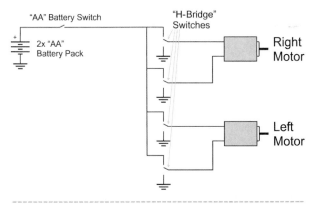

Figure 5-18   *Circuit to drive two robot wheels differentially with switch control*

forwards, backwards, and left and right. By wiring the four breadboard-mountable SPDT switches as I have shown, you will not be able to short-circuit the H-Bridge by allowing power to pass directly through the switches and bypass the motors. If you don't want a motor to turn, you simply have to set both motor switches to either battery Vcc or Gnd.

When the robot runs with just switch control, it should run somewhat faster than walking speed. This is desirable because transistor switches will drop the voltage available to the motors as well as limit the amount of current they will receive. Even with these losses you will still want the robot to move faster than you are comfortable with because it is much easier to slow it down than to speed up something you have already built and gotten running.

Experiment 31
Stepper Motors

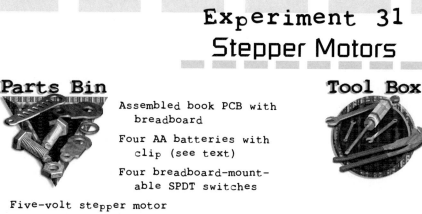

Parts Bin

Assembled book PCB with breadboard

Four AA batteries with clip (see text)

Four breadboard-mountable SPDT switches

Five-volt stepper motor

Four-pin breadboard to stepper motor connector (see text)

Paper

Tool Box

Wiring kit

Scissors

Krazy Glue

DMM

The "stepper motor" is another type of DC motor that is commonly used in robots, and chances are you are not familiar with them even though they are used in different devices. Stepper motors differ from standard DC motors because they lack the commutator of standard DC motors—stepper motors generally consist of an armature-mounted magnet with two perpendicular coils that can pull or push the magnet to different positions (Figure 5-19). The armature is generally geared down within the motor significantly, so that each time the armature moves (45 to 90 degrees), the output shaft only moves a few degrees —this gearing increases the torque output of the motor and allows for more precise movements.

To move the stepper motor, the coils are energized in a pattern something like the one listed in Table 5-2 for the stepper motor shown in Figure 5-19. When I listed the different coil polarities, I could have listed how just one coil is energized to move the armature 90 degrees at a time. This may be the simplest method of first implementing stepper motor control software.

To demonstrate the operation of the stepper motor, I would like you to build the circuit shown in Figure 5-20 and wire it like I did in Figure 5-21. The

Table 5-2 Coil energization sequence to move stepper motor

Step	Angle in Degrees	Coil A	Coil B
1	0	South	Off
2	45	South	North
3	90	Off	North
4	135	North	North
5	180	North	Off
6	225	North	South
7	270	Off	South
8	315	South	South
9	360/0	South	Off

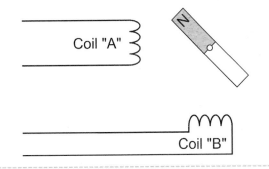

Figure 5-19   *Stepper motor*

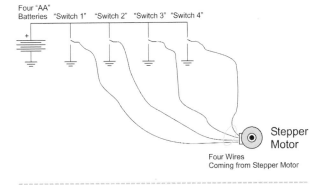

Figure 5-20   *Circuit used to test stepper motors*

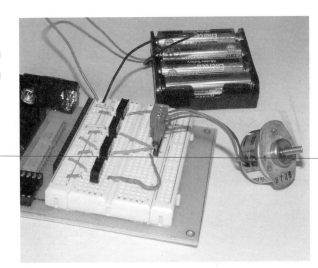

Figure 5-21   *Stepper motor switch control wiring on breadboard with stepper motor with paper pointer*

four SPDT switches are wired so that each coil in the stepper motor has its own H-Bridge. Before wiring the circuit, you should use the Ohmmeter function of your DMM to find out which wires are paired together to make the two coils in the stepper motor (this isn't as hard as it may seem—chances are the

pairs of wires for each coil will be side by side coming from the body of the stepper motor). When you have done this, I suggest that you place the two switches for each coil wire together to simplify the operation of the switch to move the motors. Rather than cut the connectors off of the motors that I had bought, I created the four-pin connector from two four-pin single row, breakable PCB mountable connectors.

When I built my test circuit, I cut out a simple paper arrow and glued it (using Krazy Glue) to the end of the stepper motor's output shaft. This allowed me to observe the operation of the motor easily and make sure that I could come up with a series of switch movements that would move the output shaft in a continuous direction. After you are finished with the experiment, the arrow can be pulled off the stepper motor's output shaft and any residual glue can be scraped off to return the motor to its original state. I used stepper motors that were specified for 5 volts of power. This made wiring them to four AA cells quite easy and avoided the need to come up with an alternative power supply. If you are unable to find 5-volt stepper motors, then you will have to come up with an alternate battery supply that meets your motor's requirements. For a 12-volt motor, you could use two four-AA-battery packs wired in series.

When you apply power to the stepper motor for the first time, don't be surprised if it jerks—the switches may be set in such a way that one or both of the coils are energized and the armature will move to this position. This is actually one of the drawbacks of using stepper motors in robots; as you have probably already discovered, the output shaft of the stepper motors can be turned easily when power is turned off from the robot. Although the motion of the motors is most likely to be small, this can be startling (and potentially dangerous) for users and observers. When stepper motors are used to control the position of a robot arm, it isn't unusual to have all the stepper motors move to a home position immediately on power up so that their positions are known while the robot is operational. To sense when the stepper motor is in the home position, a light sensor or a simple switch that is closed when an arm connected to the stepper motor touches it could be added.

With your circuit built, start working on the switch sequence that will move the pointer glued to the stepper motor in a constant direction. For my experimental setup, I created Table 5-3 to record the switch positions to move my stepper motor in a clockwise direction, and I recommend that you do the same for your motor. Note that the "Coil A" and "Coil B" polarities are simply arbitrary; I put them in as a check to make sure that the switch positions and motor response made sense. If you touch the motor, you will discover that it is quite warm—this is a characteristic of the stepper motor because current is always flowing through one or both coils.

In the table, I have highlighted the single switch setting change that is needed to move the stepper motor (the up and down positions are based on the orientation of the circuit in Figure 5-21). When you are wiring your circuit and working through switch sequencing, I suggest that you work toward being able to just change one switch at a time for each movement. This will make both the work to move the stepper motor manually easier, but also make it easier to program a controller to drive the motor.

Table 5-3 Actual stepper motor energization sequence

Step	Coil A	Switch 1	Switch 2	Coil B	Switch 3	Switch 4
1	South	Down	Up	Off	Up	Up
2	South	Down	Up	South	**Down**	Up
3	Off	Down	**Down**	South	Down	Up
4	North	**Up**	Down	South	Down	Up
5	North	Up	Down	Off	Down	**Down**
6	North	Up	Down	North	**Up**	Down
7	Off	Up	**Up**	North	Up	Down
8	South	**Down**	Up	North	Up	Down
9	South	Down	Up	Off	Up	**Up**

Experiment 32
Muscle Wire

Parts Bin

Assembled book PCB with breadboard

Breadboard-mountable SPDT switch

2.4-inch (60-millimeter) 0.004-inch (0.1-millimeter) diameter Flexinol muscle wire (see text)

39 Ω resistor (see text)

2.4-inch (60-millimeter) 5-millimeter-diameter piano wire

0.5-inch (12.5-millimeter) 10-millimeter aluminum tubing

SPDT switch, breadboard mountable

Tool Box

Wiring kit

Clippers

600-grit sandpaper

DMM

Medium-duty needle-nose pliers

In this book, I spend a fair number of pages discussing different DC motor-based robot drives. DC smotors can provide a fair amount of power in a small package, will run for a reasonably long time using standard batteries, and are easy to interface. They are, however, not the only type of robot actuator or drive that you can consider. One of the more interesting options that you can look at is muscle wire (Figure 5-22), which contracts when current is passed through it. The muscles in your body work in a similar manner, contracting to cause some kind of physical action.

Muscle wire, typically made from nickel-titanium, can be stretched at room temperature, and when heated, it returns to its original size with a surprising amount of force. The wire can be heated by an external force, but by passing a current through it, its natural resistance causes the wire to heat up and exert a contracting force. In Figure 5-23, I show how muscle wire will transition from its crystalline structure being in the *martensite* (stretched) state to the austenite state as it passes 70 C (158 F). When the muscle wire cools, it stretches back to its original size when it reaches 40 C (104 F). Obviously, muscle wire is best used at room temperature (20 C or 68 F).

For this experiment, I used 0.004 -inch-diameter (1-millimeter) Dynalloy *Flexinol*. This material can exert up to 150 grams (5.3 ounces) of force when 180

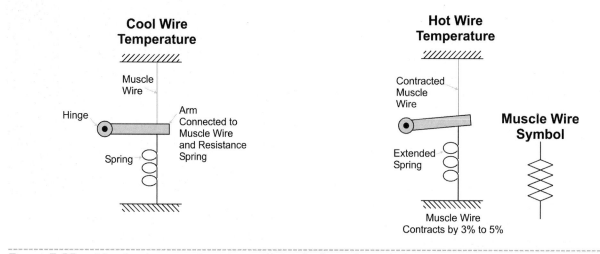

Figure 5-22 *Muscle wire temperature operation and schematic symbol*

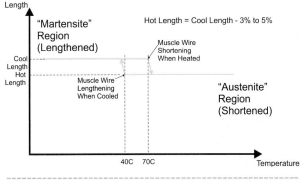

Figure 5-23 *Muscle wire length as related to temperature*

mA is passing through it. Flexinol comes pre-stretched, so when you are putting it into your application, it should be taught, but not under more than a gram (1/16th of an ounce) of tension. You can find more information about Flexinol (and muscle wire) at Dynalloy's web site at www.dynalloy.com.

You can buy it from "Mondo-Tronics, the Robot Store" at www.mondotronics.com or from the "Stiquito" Web page at www.stiquito.com/.

The Stiquito is a muscle wire-based robot (and book series) that you can experiment with. The Stiquito takes advantage of muscle wire's best features but avoids some of its shortcomings. The good part of muscle wire is its simplicity; as I will show in this experiment, muscle wire can be used to deform a metal rod for use as an actuator with very little work. When muscle wire is activated, it is silent, and as long as it is kept within its operating range, it will work almost literally forever. Muscle wire can be used to create simple insect robots quickly and with minimal effort and cost.

A disadvantage of muscle wire is its inability to support a large amount of weight safely. If muscle wire is stretched too far (around 8 percent longer than you started with), it will no longer be able to contract into a shorter length. Normally, muscle wire is prestretched to 3 to 5 percent of its original length. Most robots you will see that are based on muscle wire will have their batteries, controllers, and sensors external to the robot itself. Muscle wire's actuation is actually quite slow (taking up to a second to contract and then another second to expand), and it cannot be contracted a specified amount; it's either all or nothing. You will probably find that muscle wire is some-

what difficult to manipulate. Lastly, muscle wire uses more power than a comparable DC motor for doing the same amount of work.

Once you have bought some muscle wire and the other materials listed in the Parts Bin, you can see them in operation by wiring the circuit shown in Figure 5-24 that can be used to bend a piece of piano wire mounted to the breadboard as shown in Figure 5-25.

When I built my test circuit, I started by cutting and bending a piece of piano wire so that it had a straight edge 3 inches (7.6 centimeters) long with two right-angle bends a half-inch (12.5 millimeters) long at both ends. These half-inch-long ends were both sanded with 600-grit sandpaper until they were shiny. I then cut a 1-inch (2.5-centimeter) piece of piano wire and sanded it with 600-grit sandpaper. This sanding is to remove any oxidation and provide the best possible electrical connection. Next, I cut a 5-inch (12.7-centimeter) piece of Flexinol and sanded 1-inch (2.54 centimeters) of each end of the Flexinol *lightly*

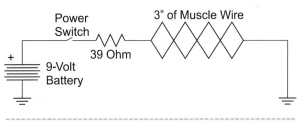

Figure 5-24 *Simple electrical circuit to test muscle wire's ability to shorten and lengthen*

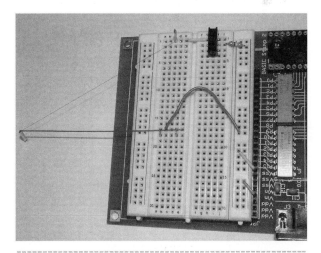

Figure 5-25 *Wire muscle circuit allowing current to flow through a piece of muscle wire to bend a piece of piano wire*

with 600-grit sandpaper. When sanding the Flexinol, take care to not stretch the Flexinol or nick it (which could cause it to break later). Along with the wire, cut two half-inch pieces (12.5 millimeters) of 10-millimeter aluminum tubing.

When you have the two pieces of piano wire cut, bent, and sanded, tie and wrap the Flexinol around the ends of each of the two pieces of piano wire. There should be 3 inches (7.6 centimeters) of Flexinol between the two pieces of piano wire. When this is done, gently push the pieces of aluminum tubing over the piano wire/Flexinol. When placing the tubing over the wire, you might want to twist it to make sure you do not push the Flexinol from the piano wire. This is somewhat of a delicate operation and it will take several tries to get it right. When you are satisfied with the fit of the aluminum tube, crimp it with a pair of medium-duty needle-nose pliers.

The specification for the 0.004-inch Flexinol is that it must have 180 mA passing through it. It can be connected directly to a battery, but if you do this, you will find that the Flexinol will heat up to the point where it will melt through the top surface of the breadboard. To avoid this, I added a 39 Ω current-limiting resistor. To determine the right value for the current-limiting resistor, I built the circuit and then measured its resistance at the PCB's battery terminals. In my case, it was 11.6 Ω, which is reasonable because the Flexinol that I used is rated at 3 Ohms per inch (2.54 centimeter). Knowing that I was going to use a 9-volt battery and I wanted 180 mA to pass through the wire, I needed a total circuit resistance of 50 Ω. Because I already had 11.6 Ω, due to the Flexinol and other wiring, a 39 Ω resistor was perfect for the experiment.

Semiconductors

A big science fiction theme in the 1950s and 1960s was how intelligence was defined and how it manifested itself in living organisms. The typical story (as in H. Beam Piper's *Little Fuzzy*) went along the lines of humans colonizing a planet, only to discover that some of the "animals" that already lived there show some remarkable capabilities that suggest they are "sentient." Other stories (such as Clarke and Kubrick's *2001*) explored how *humans* may have become intelligent. Different *Star Trek* episodes have explored this issue as well. These stories have died out over the past 30 years because you can't really write many different plots with this theme.

In all of the different stories, a simple test is used to determine intelligence. In *Little Fuzzy*, it was the ability to talk and use fire. In *2001*, it was the ability to use tools. The problem with many of these tests is that you will find animals on earth that will pass them. In terms of "talking" or communicating, many different species use different methods of communicating; for example bees use "dance" to pass along information on where food can be found, birds communicate danger by using different cries, and gorillas have been taught sign language and can "talk" to people. In terms of tools, the sea otter uses a variety of tools to gather and open food. Polar bears take advantage of fire to flush out food.

As I was creating the introductions to the different sections in this book, I realized something that humans do that separates us from animals, and that is the ability to change the properties of a substance. Now, before you start getting the picture of cavemen with test tubes in your head, try to think about what is the most basic change that can be made to something that appears naturally.

If you thought of cooking, then go to the head of the class. Early humans probably discovered that meat tasted better and was easier to digest if it had been heated first. One of the most dramatic transformations happens when eggs are heated; the molecules in the "white" are changed so that they can easily combine together, turning it from a clear liquid to a white solid.

Over time, we have discovered many ways in which materials can have their properties changed. In Table 6-1, I list some of the properties that we routinely change in different materials and a sample product that takes advantage of the property change.

Transistors and other *semiconductors* can change their ability to conduct electricity (and pass current) depending on some external condition. Semiconductors usually start as a pure crystal made out of elements like germanium, selenium, or silicon. Gallium and arsenic can be combined into a crystal (known as gallium arsenide) that is also a semiconductor. These crystals are usually very good insulators, but their conductivity is usually increased as the temperature of the crystal increases. By adding different atoms to the crystal (called *dopants*), the crystals become able to conduct electricity because the new atoms provide excess electrons that allow current to flow through the crystal.

Table 6-1

Property Change	Sample Product
Liquid to solid	Glue
Gas to liquid	Rocket fuel
Gas to solid	Dry ice
Increase tensile strength	Carbonized steel
Element properties	Plutonium (produced in nuclear reactor)
Electricity from chemicals	Batteries
Increase conductivity	Copper alloys
Controlled electrical conductivity	Transistors

As I have shown in the Figure 6-1, semiconductor crystals form structures in which dopants can be added very easily. The dopants replace the crystal atoms and, depending on the number of electrons in their outermost shells, provide incomplete bonds that leave excess electrons or holes that allow current flow. In Figure 6-1, I have shown a crystal atom with six electrons in the outermost shell and a dopant atom with five electrons. This leaves a hole in the crystal with an incomplete bond in one of the crystal atoms that can accept an electron easily.

Silicon is the most popular base crystal to use because of its wide availability, low cost (sand and glass are basically just silicon), ease of use, and low toxicity. Unlike the atoms used in the example diagram, which have six electrons in their outer shell, sil-

icon has four electrons—I used an example with six because a cube is much easier to visualize (and draw) than a three-dimensional structure that only has four vertices.

When an atom with fewer than four electrons is used to dope a silicon crystal, the resulting semiconductor is known as *P-type* and is used in the crystal in Figure 6-1. P-type semiconductors can accept electrons (and are known as *acceptors*), whereas *N-type* semiconductors (which use elements with an extra electron in the outer shell and are known as *donors*) can provide free electrons. Boron, which has three electrons in its outer shell, is a typical P-type semiconductor dopant, and phosphorus, which has five electrons in its outer shell, is used to make N-type semiconductors.

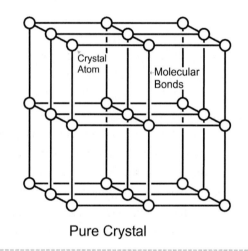

Pure Crystal

"Doped" Crystal

Figure 6-1   *Pure and doped semiconductor crystals*

Experiment 33
Diodes

Parts Bin

Assembled PCB

1N4148 or 1N914 silicon
  diode

1k resistor

Tool Box

DMM

Wiring kit

The most basic semiconductor application is the *diode*, which is a device that only allows current to pass in one direction. Using the water analogies that I used earlier in the book, the diode can be thought of as a one-way valve. When there is pressure in one direction, water will flow through the valve with just a small pressure drop. If the pressure is reversed, the valve closes and no water is passed. In this experiment, you will get a chance to see how a diode works in a circuit and learn about some of its characteristics.

The symbol for the diode, along with what the actual part looks like, is shown in Figure 6-2. The simple band around the diode is used to indicate the direction of current flow. In schematics, diodes are given the reference designator "CR" or "D."

One of the first applications for the diode was to convert (or rectify) *alternating current* (AC) into direct current. Alternating current consists of voltages that are both positive and negative, and you can see how the diode only passes positive voltages and currents in Figure 6-3.

In a diode, electrons fall from the high-potential N-type silicon semiconductor to the low-potential P-type silicon (Figure 6-4). As the electrons fall, they lose energy. This energy is converted to light energy

(photons) as I have shown in the diagram. For silicon diodes, these photons are in the very deep infrared range and not visible by the human eye. As I will discuss in the next section, by changing the diode's material, useful wavelengths of light can be produced.

It is important to note that in Figure 6-4, I show the direction of *electron flow* and not *current flow*. As I have stated earlier in the book, current flow is in the opposite direction of electron flow. When you wire your circuit, the N-type semiconductor is connected to the negative part of the circuit to pass current, not the positive as you might think from this drawing.

In this experiment, I would like you to wire a diode, 10k resistor, and wire into your PCB/breadboard combination as I have drawn in Figure 6-5. The voltages and currents within this circuit will be measured so you can understand the operation of a diode.

The 1N914 and its equivalent part, the 1N4148, are general purpose silicon diodes. The term *equivalent*

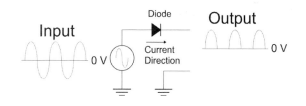
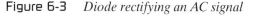

Figure 6-3   *Diode rectifying an AC signal*

Schematic Symbol

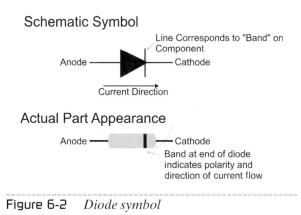

Actual Part Appearance

Figure 6-2   *Diode symbol*

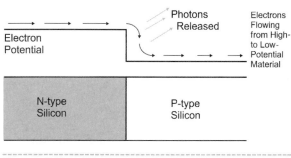
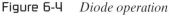

Figure 6-4   *Diode operation*

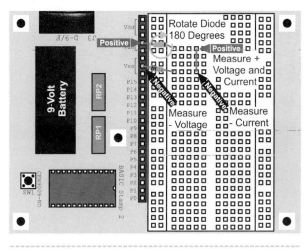

Figure 6-5   *Diode testing*

Table 6-2   Circuit 1

Battery voltage	7.46 volts
Diode voltage	0.59 volts
Resistor voltage	6.85 volts
Circuit current	6.69 mA

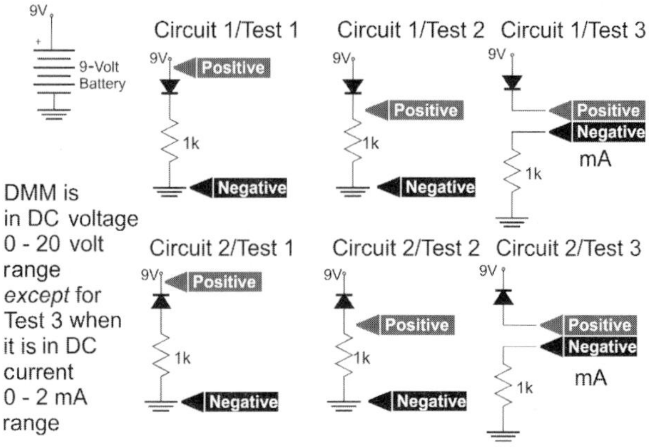

Figure 6-6   *Diode-testing circuits for testing diode operation*

when applied to electronics parts means that the parts behave identically, but different manufacturers have given them different part numbers. To avoid confusion, I normally don't specify parts that have common equivalents like the 1N914, but in this case it is a very useful, inexpensive, and easily found part and you will find that some suppliers will only stock it as the 1N4148 and will not recognize the 1N914.

Once you have built the circuit, make up a table like Table 6-2 and perform the measurements that I have outlined in the table and in Figure 6-5 for Circuit 1. I want to point out that I used a largely discharged NiMH battery when I created this experiment, in case the 9-volt battery voltage seems low. I have added an additional measurement, the voltage across the diode. This is obviously just the difference between the 9-volt battery source and the voltage across the resistor, but there are two points I would like to make with it.

Looking at Table 6-2, you should be able to make two conclusions that I alluded to earlier. The first is that voltage drop across the diode is about 0.6 volts—this is normal for silicon diodes and can be as high as 0.80 volts. The second conclusion that you should make is that Kirchoff's voltage law applies for the diode in this circuit—the voltage drop across it added to the voltage drop across the resistor is equal to the voltage applied to the two components.

Going further, you can see that the amount of current passing through the circuit is a result of the voltage across the resistor—the diode does not limit the current in any way. This observation will become important in the next experiment and shows how a resistor can be used to limit current.

Now reverse the diode and perform the measurements again.

When the diode has been reversed, all the voltage applied to the circuit is dropped across it and nothing across the resistor. Again, Kirchoff's voltage law is found to be valid with this semiconductor as the voltage across the drops is equal to the applied voltage. Applying Ohm's law, it shouldn't be surprising that the current through the circuit is equal to zero (because there is no voltage across the resistor). These results for the reverse-biased (backwards) diode should have been expected based on the introduction to the diode and its response to an AC input.

As a final point, if you were to use the laws I introduced you to before, you would discover that the diode is actually dissipating some heat. Looking at Circuit 1 in Figure 6-6, 6.69 mA is flowing through the diode with a voltage drop of 0.59 volts, and the power being dissipated is 4.01 mW. Although this is not much power in this example, in other circuits, where a large amount of current is being drawn, you will have to use a diode that is rated for the current passing through it as well as the power expected to be dissipated.

Experiment 34
Light-Emitting Diodes (LEDs)

Parts Bin

Assembled PCB

Light-emitting diode
(LED), any color

1k resistor

Tool Box

Assembled PCB

DMM

Wiring kit

In the previous experiment, I mentioned a by-product property of diodes that is actually quite useful. When I discussed the actual operation of the semiconductor parts of the diode, I showed that when the electrons fall from the high electron potential N-type silicon to the low electron potential P-type silicon, photons were released.

In the previous experiment, I went into some detail to show how the basic electronic laws are not violated by the introduction of semiconductors into circuits. The release of photons ensures that the diode does not violate the first law of thermodynamics—that energy cannot be created or destroyed. The energy lost by the electrons as they pass from the N-type to the P-type semiconductor is converted into the photons. As I noted, in a standard silicon diode, these photons are very low energy (long wavelength) and not really useful.

In Table 6-3, I have listed some of the different materials used for making LEDs and the different light that they output. Most of the materials used in the manufacture of LEDs are quite exotic, which is the reason why LEDs tend to be about 10 or more times the price of a simple silicon diode.

The LED itself comes in a cylindrical package that kind of looks like R2-D2, as I have shown in Figure 6-7. The schematic symbol for the LED is similar to that of the LED, but with "light rays" coming off it as in Figure 6-7. To indicate its polarity (and the direction of current flow), the LED package has a flat on one side of the circular base that indicates which side is the LED's cathode (negative connection). Like diodes, LEDs usually have the reference designators "CR" and "D," but in some cases, you will see the reference designator "LED" used instead.

Table 6-3   LED materials and light

Color	Diode Materials	Output Light Wavelength
Infrared	Gallium, arsenic	940–730 nm
Red	Gallium, aluminum, phosphorous	700–650 nm
Amber	Gallium, arsenic, phosphorous	610 nm
Yellow	Gallium, arsenic, phosphorous	590 nm
Green	Gallium, phosphorous	555 nm
Blue	Zinc, Selenium	480 nm

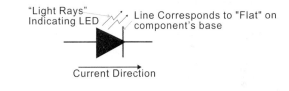

"Light Rays" Indicating LED   Line Corresponds to "Flat" on component's base

Current Direction

Actual Part Appearance

"Flat" on side of diode Indicates polarity and Direction of Current Flow

Figure 6-7   *LED symbol*

LEDs behave identically to regular diodes in a circuit with one important difference—they usually have a higher voltage drop than silicon (and other semiconductor) diodes. In this experiment, I am going to repeat the previous one but use an LED instead of a 1N914/1N4148 diode.

If you go to a reasonably well stocked electronic store, you will probably be amazed at the number of different LEDs available to choose from. Along with there being a number of different colors, there will be different packages and brightness levels. For this experiment (and the ones that follow), I recommend that you buy a bag of the cheapest ones you can find. These are usually red output and are in a 5-millimeter package, which has the flat at the base like the one shown in Figure 6-7.

This is the same circuit as was used in the previous experiment. Remember that the flat spot on the LED should be facing away from the Vin. Once you have the circuit wired, the LED should light up.

Once you have the LED lit, repeat the previous experiment's steps as listed in Figure 6-8 and fill out your results in a table like Table 6-4.

Looking over the data, you should see a big difference in the voltage drop across the LED versus the voltage drop across the silicon diode. Most LEDs have a voltage drop of around two volts, instead of the 0.6 to 0.8 volts of a silicon diode. Looking at the rest of the data, you'll see that Kirchoff's voltage law still holds and the current through the system is dependant on the voltage across the resistor and the resistor's value.

Most LEDs will light with 5 mA of current. Using the 1k resistor in this circuit provides just over 5 mA (as can be seen in Table 6-4). Adding more current will not make the LED brighter—although too much will burn it out. Decreasing the current will lessen the output, but it is hard to control the light output level.

Table 6-4 Electrical measurements for the LED circuit

Battery Voltage	Diode Voltage	Resistor Voltage	Circuit Current
7.38 volts	1.99 volts	5.37 volts	5.36 mA

The best method of controlling the LED light output is by rapidly turning the power to the LED on and off. This is known as *pulse width modulation* (PWM), and I will discuss how it works and how a pulse width modulated control signal is produced in more detail later in the book.

The 1k resistor is known as a current-limiting resistor and is required because the LED (or diode) has effectively no resistance in the circuit. If the resistor were left off in this circuit, you would find that the LED would probably still work (the 9-volt battery cannot source enough current to burn it out), but the battery's life would be quite short. By putting in the resistor, the amount of current passed to the LED is minimized to what is required to light the LED and no more. Using the current-limiting resistor will minimize the amount of current required by the application and maximize the application's battery life.

When working with a 5-volt logic circuit, I use a current-limiting resistor of 220 to 470 Ω. 470 Ω is optimal, but in situations such as when the BS2 is driving the LED being added to the circuit, the 220 Ω resistors built into the PCB eliminate the need for adding your own current-limiting resistors.

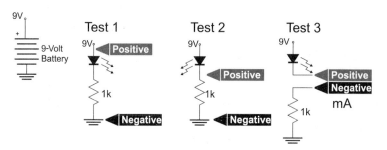

DMM is in DC voltage 0 - 20 volt range
except for Test 3 when it is in DC current 0 - 2 mA range

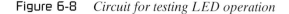

Figure 6-8 *Circuit for testing LED operation*

Experiment 35
NPN Transistor and Two-LED Lighting Control

Parts Bin

Assembled PCB

Two LEDs, any color

Three 1k resistors

Two ZTX649 NPN transistors

Tool Box

DMM

Wiring kit

So far in this book, when I have been introducing a new electronic component, I have been using a "water analog" to describe how the component works using a medium that you should be familiar with. Unfortunately, as I start working with more sophisticated semiconductor components, the applicability of using water as a demonstration tool becomes just about impossible. The operation of the diode can be modeled using a one-way valve as I mentioned earlier, but trying to come up with the transistor water model is quite difficult and some of the most important aspects of the transistor cannot be easily observed.

Figure 6-9 shows that the transistor consists of a simple valve that is controlled by the introduction of water into the control pipe. The water flow in the control turns a small turbine that pushes open a valve in the much larger pipe. The more water that passes through the control pipe, the faster the turbine turns and the more the valve opens. The more the valve opens, the more water passes through the large pipe. When no more water is flowing in the control pipe, the valve closes automatically.

This is actually a very accurate description of how an NPN transistor works, but it misses a few important points. The first is that the water that passes into the control pipe is dumped into the exit pipe. It is important to understand that the amount of water passing through the large pipe is dependant on the amount of water passing through the control pipe—the pressure of the water is not important. Finally, a multiplication factor exists between the amount of water passing through the control pipe and the large pipe. The amount of water that can pass through the large pipe is proportional to the amount of water flowing through the control pipe.

At the other end of the spectrum is the electrical model used by engineers and circuit designers to simulate the operation of a transistor in a circuit. In Figure 6-10, I have shown the simplified small-signal transistor model that is used by the *simulation program with integrated circuit emphasis* (SPICE). This circuit shows the parasitic resistances built into the transistor control or base as well as the coupling capacitances. The circle with an arrow is called a current source and it will allow a set amount of current, which is a multiple of the amount of current flowing through the transistor's control (or base) to the exit (or emitter). The current passing through the current source comes from the source or collector and is passed to the emitter. I didn't put in Figure 6-10 to

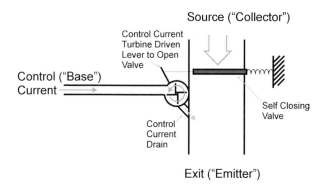

Figure 6-9   *Transistor water model*

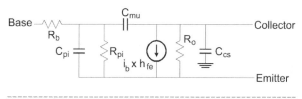

Figure 6-10   *Simplified transistor model*

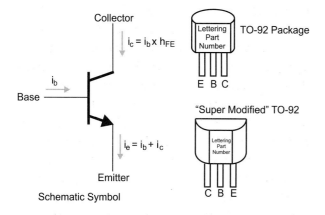

Collector

$i_c = i_b \times h_{FE}$

i_b

Base

$i_e = i_b + i_c$

Emitter

Schematic Symbol

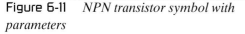

Lettering Part Number — TO-92 Package

E B C

"Super Modified" TO-92

Lettering Part Number

C B E

Figure 6-11   *NPN transistor symbol with parameters*

scare you. As you become more sophisticated in electronics, this model will be very important to you and the different parts will be easily recognizable and understandable.

Rather than trying to figure out the best model showing how an NPN transistor works, I am going to start off with no model at all and go straight to its schematic symbol and different transistor package pinouts with Figure 6-11.

In Figure 6-11, I showed the schematic symbol for the transistor with the different terminals labeled. The base corresponds to the control pipe in the previous water analog, whereas the collector and emitter correspond to the pipe with the valve. When you are looking at the flat side of the transistor with the leads

pointing downwards, the pins are, from left to right, always emitter-base-collector, or as I remember it, "Emitter Before Collector."

In the schematic symbol shown in Figure 6-11, I have labeled the current flows in the transistor. In simple terms, the amount of current that can pass through the collector to the emitter is the base current multiplied by h_{FE}. The multiplier, h_{FE}, is often called *Beta* (or β) and is specified in the data sheet for the transistor.

To demonstrate the operation of the NPN transistor, I have come up with the circuit that is shown in Figure 6-12. When there is no current flowing through the left transistor's base, current passes through the 1k resistor at its collector and goes to the right transistor's base. In this case, the right transistor is turned on and its emitter current turns on its LED.

When the left transistor's base has current passed to it, the current available at its collector is passed to its LED. In this case, there is no current for the right transistor's base, so it's turned off.

If you are familiar with transistors, you might be surprised at my choice of the Zetex ZTX649 NPN transistors. In many other basic electronics projects books, the 2N3904 is used because it is a very inexpensive and common general purpose transistor. The ZTX649 is somewhat more expensive than the 2N3904, but it can handle up to 2 amps of current, which makes it ideal for use as a motor driver in small robots. The transistor has a current amplifica-

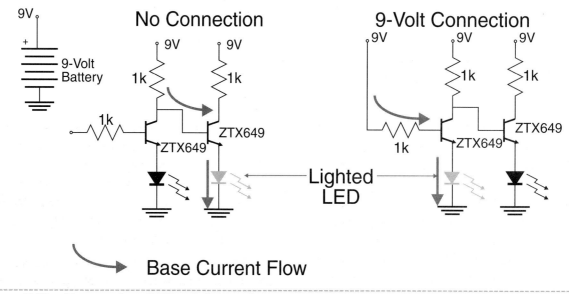

Figure 6-12   *Two-transistor LED switch operation*

tion factor (h_{FE}) of 300, which is about twice that of the 2N3904.

After building the circuit, spend some time watching the operation of the LEDs and then take out the connection between the left transistor's collector and the right transistor's base and measure the current. When the right transistor's LED is on, the current passing to its base is about 6 mA; when it is off, the base current is zero. This should not be surprising because the LED is off, indicating that there is no current flow.

You can also look at the voltage of the LEDs when they are turned on and off. In my circuits, I found a voltage of about 1.7 volts when the LED was off and 2.2 when the LED was on. You may wonder why the LED isn't "slightly" on when it has 1.7 volts across it, but you have to remember that there isn't any current flowing through it. For the LED to light, it must have both a voltage drop as well as current flow. I'm pointing this out because this illustrates an important fact that you will probably forget and have to rediscover—simply measuring voltages in semiconductor circuits is not always enough to explain what is happening. You must always be prepared to measure both the voltage drop and current through a component to be able to completely know what is happening.

Experiment 36
Driving a Motor with a Transistor

Parts Bin

Assembled PCB

ZTX649 NPN transistor

1N4148 or 1N914 silicon diode

Two C cells with battery clip

Any small toy motor capable of running with voltage inputs of 1.5 to 3 volts

110 Ω resistor

470 Ω resistor

1k resistor

10k resistor

Tool Box

DMM

Wiring kit

In the last experiment, I introduced you to the NPN transistor and discussed some models for its operation. I also gave you a simple formula showing the relationship between the base current and its collector current. I also showed how the NPN transistor could be used as a switch, turning one of two LEDs on. In this experiment, I go into more detail about the NPN transistor and show how transistors can be used to control high-current devices like electric motors.

So far, I have described the transistor as just an NPN transistor—the correct name for the transistor is the bipolar NPN transistor, and if you were to look at the side view of the transistor, you would see something like the bar-shaped device with an N-type semiconductor at each end and a thin P-type semiconductor in the middle as in Figure 6-13.

When the transistor is turned on, the base current draws electrons from the emitter N-type pole, creating a conduction region filled with electrons. To

Transistor "Off"

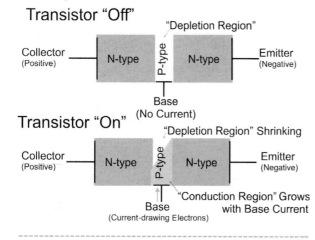

Transistor "On"

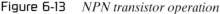

Figure 6-13   *NPN transistor operation*

understand how the transistor works, remember that electron flow is the reverse of current flow—as current is injected into the base, electrons are being drawn from it.

The P-type section of the transistor is very thin and the electrons pulled from the emitter jump to the collector; these electrons form the collector current and the amount of current is based on the amount of electrons being drawn from the base. The larger the base current flow, the larger the collector current flow that is possible. As the base current increases in the bipolar NPN transistor, the size of the conduction region increases and the collector current has more area in which to flow.

In this experiment, I would like to demonstrate the operation of the bipolar NPN transistor as a low-current-controlled, high-current switch. To do this, I will use the simple circuit shown in Figure 6-14. I used a 470 Ω base-current-limiting resistor, but when you test out the function of the circuit, I would like you to vary the base resistor using each of the resistors listed in the Toolbox.

When I ran my tests using an old toy motor, I found that the 470 Ω base-current limiting-resistor ran the motor most efficiently. When I put in the larger values for the base-current-limiting resistor, I found that the motor would run much slower (with less torque) or not at all. When I put in a 100-ohm base-current-limiting resistor, I could not detect any difference in motor operation with it or with the 470-ohm resistor. Thus, I found empirically that the 470-ohm resistor was the optimum choice for the motor that I used.

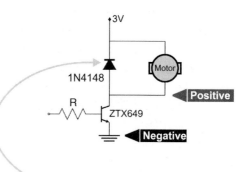

When measuring V_{ce} remove the diode while motor is running

Figure 6-14   *NPN transistor motor control*

Measuring the 9-volt battery voltage at 8.91 volts and the transistor base to emitter (or ground) voltage at 0.79 volts, I found a 8.12-volt drop across the 470-ohm base-current-limiting resistor and a 17 mA calculated (I measured an actual current of 17.1 mA) current being injected into the base. Assuming that the h_{FE} of the transistor is 300, the current flowing from the transistor's collector to emitter works out to be 5.18 amps. You should immediately recognize that the value of 5.18 amps is unreasonable. I indicated in the previous experiment, the ZTX649 can carry a maximum of 2 amps and, if you look at the datasheet for a C alkaline cell, you will see that it can nominally source about 350 mA. Measuring the current drawn by the motor, I found it to be about 190 mA.

The confusion comes from the range where the transistor is operating. When I have described the operation of the transistor, I have been doing so in the "linear small signal operating range" as I show in Figure 6-15. When a large current drain device (like

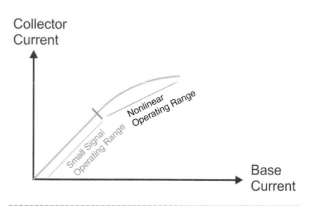

Figure 6-15   *NPN transistor base to collector current flow relationship*

the motor) is being controlled by the transistor, its operation has moved out of the small signal range and into a nonlinear range (or "saturation region") where the operation of the transistor cannot be predicted as easily. Most simulation tools have the ability to correctly model the operation of a transistor outside of the small signal operating range, but for driving small motors, I recommend following the path that I have used here—test out a number of different base-current-limiting resistors until you find the one that seems optimal for your application.

When the motor is turned on and off (or even when its armature turns and a new coil is energized), a large induced voltage (called *kickback*) causes noise in your circuit. If you go back to the original schematic diagram (Figure 6-14), you will see that I suggest that you measure the voltage between the transistor's collector and emitter. With the motor running, I suggest that you do this both with the diode in place and the diode removed. When I did this, I found that the collector/emitter voltage was a very constant 0.030 volts, with it varying by a millivolt. When I removed the diode, I found that the voltage stayed around 0.030 volts, but varied by as much as 10 millivolts. To try and confirm what I saw, I took a look with my oscilloscope at the collector voltage (the emitter voltage was "Ground" for the measurement) with the diode in place and with it removed. Figure 6-16 is an oscilloscope picture of what I saw.

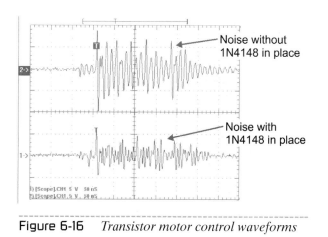

Noise without 1N4148 in place

Noise with 1N4148 in place

Figure 6-16   *Transistor motor control waveforms*

The upper oscilloscope trace is the collector voltage with no diode, showing up to $+/-10$ volts of kickback. The lower trace shows the same voltage, but with the diode in place—the noise output has been reduced to less than $+/-5$ volts. The diode provides this noise filtering by "breaking down" when a large voltage is applied to it. As I described the transistor behaving unexpectedly at extreme conditions, so does the diode when a high voltage is applied to it. At some voltage the diode's operation stops and it becomes like a short circuit, shunting current within it rather than passing it back to other parts in the application circuit. When you are building any magnetic device control application, you should always put in the kickback diodes to protect the other parts of the circuit.

Experiment 37
Bipolar PNP Transistor Motor Control

Parts Bin

Assembled PCB

ZTX749 PNP transistor

1N4148 or 1N914 silicon
diode

Two C cells with battery
clip

Any small toy motor cap-
able of running with
voltage inputs of 1.5
to 3 volts

110 Ω resistor   10k
resistor

470 Ω resistor   100k
resistor

1k resistor

Tool Box

DMM

Wiring kit

The bipolar NPN transistor is an excellent tool for many different electronic circuits, and it is the basis for basic transistor logic, as I will show you in a later section. The reasons for the popularity of the NPN transistor include its ease of manufacture on integrated circuits (which means low cost), high-speed operation, and good current-handling capability. Unfortunately, the NPN transistor cannot be used in all situations, particularly when a switch is needed to *source* (provide) current rather than *sink*.

The PNP transistor is used to complement the NPN transistor and has many of the same features as the NPN transistor as you can see in Figure 6-17. The symbol for the PNP transistor is similar to that of the NPN transistor, but in the PNP transistor, current flows are in the opposite direction as the NPN. The PNP transistor's collector current is calculated in exactly the same way as the NPN transistor's collector current. Finally, the PNP transistor's pins are labeled identically to the NPN transistor's pins, and they are orientated in exactly the same way as the NPN transistor.

When the PNP transistor is turned on, current is drawn from the base, which injects electrons into the N-type semiconductor of the transistor. These electrons are passed to the transistor's collector, with some jumping from the emitter's P-type semiconduc-

tor. The electrons coming from the emitter make up the current passed to the transistor's collector.

Like the NPN transistor, the PNP transistor is turned off if there is no current flow at the base and the current the collector is capable of passing is a multiple of the base current. Like the NPN transistor, the PNP transistor collector current multiplier is known as h_{FE} or Beta. When a complementary PNP transistor is manufactured for an NPN transistor, it normally has the same h_{FE} as the NPN transistor.

To demonstrate the operation of the PNP transistor, in this experiment I will create a similar circuit as

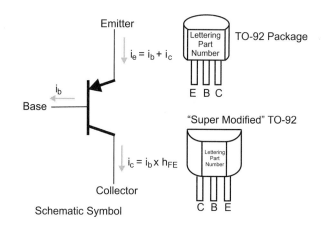

Figure 6-17   *PNP transistor symbol with parameters*

in the previous experiment, with a PNP transistor that is the complement to the NPN transistor used in the previous experiment. In the schematic diagram (Figure 6-18), you can see that I have reversed the position of the transistor within the circuit. To turn on the transistor, instead of connecting the base to a current source (the 9-volt battery in the original circuit), it has to be connected to a ground source.

You will probably notice that it is quite a bit different from the previous experiment, even though it looks similar. It is a good idea to take out all the components out of the breadboard from the previous application and start over rather than trying to modify the previous circuit. Although they are similar, some components are in different locations, and it can be a real pain trying to find the mistake.

Like in the previous experiment, you should try the different resistors until you find one that the motor will run most efficiently. For the motor that I used, the motor ran quite well with the 470 Ω resistor that I found was best with the ZTX649 NPN transistor. I did experiment with other resistors and found that the best speed and most torque were produced by a 220 Ω resistor, which may be a bit surprising because the ZTX749 is supposed to be the complement of the ZTX649.

PNP transistors are not as efficient as NPNs. This is due to the greater resistance of P-type silicon and the slower speed at which electrons pass through it.

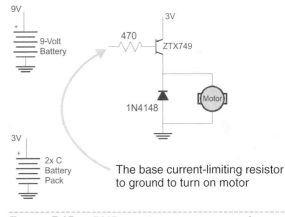

The base current-limiting resistor to ground to turn on motor

Figure 6-18   *PNP transistor motor control*

For this reason and the increased difficulty in manufacturing them on integrated circuits, PNP transistors are not as prevalent as NPN transistors in electronics. PNP transistors are very useful in some applications, and in the next experiment, I will be demonstrating how PNP and NPN transistors can be used together to create a bidirectional drive for the DC motors.

Before leaving this experiment, note that I have included a kickback diode in this circuit. The motor will produce the same large transients under PNP transistor control as it will for NPN transistor control. The difference is that I have connected the diode to ground rather than to the current supply, as I did in the previous experiment.

Parts Bin

Assembled PCB

Two ZTX749 PNP transistors

Four ZTX649 NPN transistors

Four 1N4148 or 1N914 silicon diodes

Two C cells with battery clip

Any small toy motor capable of running with voltage inputs of 1.5 to 3 volts

Two 100 Ω resistors

Two 1k resistors

Tool Box

DMM

Wiring kit

I have introduced transistors in this section to show how they can be used to control electric motors for robots. In the first section, I discussed the different formats that can be used to build robots, finally concluding that the differentially driven chassis was the most efficient because it avoided the need for any kind of steering gear—each motor would be used to drive and turn the robot. Creating a robot with a differential drive means that you must have a method of running the motors both forward and backward. I have seen a number of designs on the Internet that use relays to control a differentially driven robot's motors, but I believe that using transistors to drive the motors is a better solution as they cost less, can be part of the motor operating speed control, and are much more robust.

In the previous experiments in this section, I have introduced you to the NPN and PNP transistors along with showing you how they can be used to drive a motor. In these experiments, I have shown how the NPN transistor can be used as a switch, pulling a signal to ground and, by doing this, sinking current through a motor and letting it run. The PNP transistor can source current, and in this capacity, it is well suited for providing current to drive a motor. By combining the two types of transistors, you can create a motor driver that can make a DC motor turn either forward or backward.

I have already introduced you to the H-Bridge DC motor driver, in which four switches can be used to control the direction current flows through the motor. Creating the driver circuit for this is quite easy, although one thing to watch for is that if both switches on the same side are closed, then a direct path exists between the motor's power supply and ground. Ideally, the H-Bridge control circuitry and software that you come up with must ensure that both switches on the same side will never be closed.

If you have been thinking about how to implement the H-Bridge using transistors, you might think of something like Figure 6-19. This circuit uses two pairs of matched NPN and PNP transistors that provide the switching functions. The diodes in the circuit are used for filtering any kickback from the motor.

You might consider tying the bases of the transistors together, as in Figure 6-19, through a common current-limiting resistor. By only using two digital drivers, you can control the direction of the motor. The resulting circuit looks to be simple, and it appears that both transistors on one side of the H-Bridge driver circuit can never be driven because the two transistors turn on differently. The PNP transistor requires current to be drawn from its base, whereas the NPN transistor requires current to be injected into the base.

A very large problem with the circuit in Figure 6-19 is that current could be drawn from the PNP transistor's base and be passed to the NPN

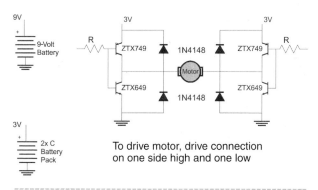

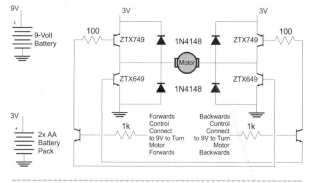

Figure 6-19 *Obvious H-Bridge transistor motor control*

Figure 6-20 *H-Bridge transistor motor control*

transistor's base. In this case, both of the transistors would be turned on. To make matters worse, this can be a self-perpetuating and amplifying problem; as more current passes through the transistors, more current is available for the bases, which increases the current again. This process can repeat until the transistors or the power supply are burned itself out.

This is not to say that the problem will always occur. In some situations and with some matched pairs of PNP and NPN transistors and motor load, it will never happen. The problem can be very hard to predict, and you may find that different transistor pairs, resistor values and wiring, battery levels, kickback diodes, and motors will have an effect on whether or not it will happen. I have been "bitten" by this problem once, and to make sure that it never happens again, when I design an H-Bridge, I use the

circuit shown in Figure 6-20. It is the one that I used for this experiment.

This circuit provides two terminals that are used to select in which direction the motor turns. Figure 6-21 shows current flowing in the circuit depending on which terminal has voltage applied to it. The only drawback to this H-Bridge is if both motor control terminals are driven or pulled high at the same time. If both terminals are energized, then the transistor switches on both sides will be turned on at the same time and you could burn out your power supply and/or the transistor switches. Software written for the interface should make sure that only one terminal has power applied to it at the same time.

Wiring the application is surprisingly easy; the only thing to watch for is the orientation of the transistors as the PNP transistors will be reversed relative to the NPN transistors.

Motor Running Forwards

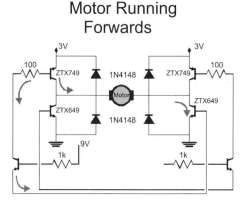

Left controlling NPS transistor turned on; current through motor travels from left to right

Motor Running Backwards

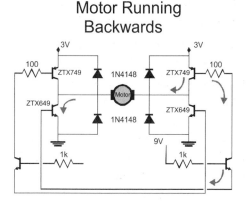

Right controlling NPN transistor turned on; current through motor travels from right to left reversing the direction the motor turns in

Figure 6-21 *Operation of the H-Bridge motor control*

This H-Bridge circuit is very reliable and will work for a wide variety of small DC motors. If you are going to use it with motors that require more than 300 mA, then you will have to choose different transistors and diodes and may have to change the resistor values. If you know the operating parameters of the motors that you are going to use, then this circuit can be very easily modeled using SPICE. This will let you optimize the circuit so the transistors do not pass more than the required current. It is important to make sure that there are not excessive current drains through the two controlling transistors' bases—a surprising amount of current can pass through a transistor's base, causing an unexpected power loss.

Section Seven

Our Friend, the 555 Chip

When I was a teenager, the most popular chip used by hobbyists (and I wouldn't be surprised if it was also the most popular in commercial products of the time) was the 555 timer integrated circuit or chip. The 555 is probably the most versatile nonprogrammable part I have ever seen. Hundreds of projects have used this chip in ways I'm sure the original designer never would have thought possible; the original function of the chip was to provide a regular train of pulses. In this section, I will introduce you to the 555 chip and work through some experiments to show how the chip is used in a circuit and how it can be used to create a simple robot.

In the previous sections, I have shown you the "pinout" of a number of different components—each one of them having a unique form factor. The 555 is usually built into a *dual in-line package* that is commonly used for chips. A dual in-line package is normally reduced to its acronym DIP and used to refer to chips that have leads that can be pushed into holes for mounting in a circuit. In Figure 7-1, I have put in an overhead view of the 555 along with a photograph of an actual 555 chip.

This overhead view of the 555 is the pinout of the chip and you will notice that I have labeled the pins starting at the top left pin, which is indicated by small circle on the chip. Along with this circle, many DIPs have a semicircle molded into the Pin 1 end of the chip. Depending on the manufacturer and the part, you may have the small circle or the semicircle or both. Once you have identified Pin 1 on the chip, the pins going in a counterclockwise manner are given increasing pin numbers as I have shown in the figure. This convention is used by all DIPs, regardless of the size, and you will see more of it as I introduce you to different parts in this book.

Looking at the labels for each of the pins, most of them do not make a lot of sense. What should jump out

at you is the Gnd (Ground) at Pin 1 and the Vcc (Positive Power) at Pin 8. These two pins are used to provide power for the part. When you work with chips, you will find that they do require power, and for chips that are built from bipolar transistors, such as the 555, you will always see Vcc and Gnd pins.

This is stepping a bit ahead, but when you look at chips that are built using MOSFET transistors, you will find that positive power is provided to the Vdd pin and Ground is the Vss pin. This is a convention that can be confusing; I tend to mentally convert Vdd to Vcc and Vss to Ground. Despite the difference in the names for power, the convention of numbering pins in increasing order counterclockwise from Pin 1 is still true for MOSFET transistor-based chips.

To try and get a better understanding of a chip, one of the first things I do is look at its block diagram. In Figure 7-2, I have drawn out the block diagram for the 555 timer.

Like the pinout diagram, I'm sure that at first glance, the 555 block diagram is quite intimidating (and maybe even a bit "scary"). There should be some things you recognize, but I'm sure many things don't make any sense to you at all. When I see a

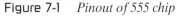

Figure 7-1   *Pinout of 555 chip*

107

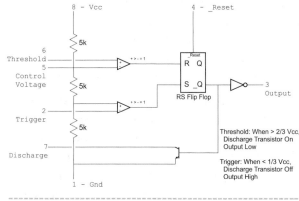

Figure 7-2   *555 block diagram*

chip for the first time, I feel the same way, but I try and figure out how the chip works from what I do know.

You should recognize two parts to the block diagram right off the bat. The first is the transistor at the bottom middle of the diagram. This transistor looks like it is wired similarly to the way it was wired when the motor was controlled in the previous section—the transistor is acting as a switch that will sink current to ground.

The next piece that you should recognize is the voltage divider running along the left side of the block diagram that I have separated out into Figure 7-3. If you were to work out the voltages at Vcontrol and Vtrig, you would discover that they are at 2/3 Vcc and 1/3 Vcc, respectively. This is actually an important clue as to how the chip works.

One apect of the 555's voltage divider circuit that you may find confusing is its connection to an outside pin called Control Voltage. This connection allows the circuit designer to change the voltage levels of the voltage divider circuit. Rather than Vcontrol being 2/3 Vcc, it can now be any value (less than Vcc) that the designer would like. Changing Vcontrol also changes Vtrig to 1/2 Vcontrol.

The voltages at the Vcontrol and Vtrig are passed to two triangular boxes with a "+" and "-" along with a funny-looking equation. These boxes are representations for comparators, and as I have shown in Figure 7-4, the comparators output a high voltage level when the voltage at the "+" input is greater than the voltage at the "-" input. The 555 uses the two comparators to continuously compare two external voltage levels to Vcontrol and Vtrig and passes the results to a box labeled RS Flip Flop.

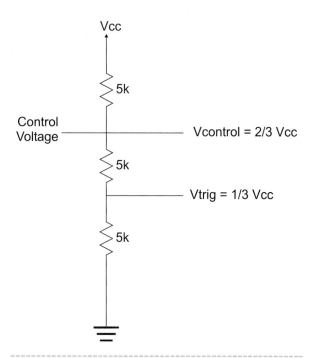

Figure 7-3   *555 voltage divider*

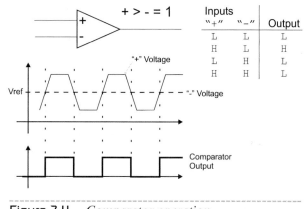

Figure 7-4   *Comparator operation*

I will explain how flip flops work in a later section of the book, but for now, I would like you to think of it as the two-coil relay (Figure 7-5). The device consists of two relay coils laid out horizontally with a wiper that will stay in the last position set by whichever coil was last energized.

The 555's RS Flip Flop performs the same function as this relay-based device. In the 555 it saves which comparator last passed a high voltage to it. If the comparator connected to the Threshold pin of the 555 and Vcontrol of the voltage divider output a high voltage, then the flip flop will output a high voltage at _Q, which turns on the transistor at the bottom of the

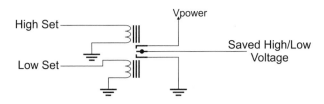

Figure 7-5 *Two-coil relay memory element used as a simple R-S flip flop*

block diagram. If the other comparator passes a high voltage to the RS Flip/Flop, then the voltage at _Q is driven low and the transistor is turned off.

The last component that will be new to you is the triangle with a ball at the end on the lower right-hand part of the block diagram (Figure 7-2). This component is known as an *inverting buffer* and converts a high input value to a low output and vice versa.

This is a fairly complete explanation of how the 555 works, and I'm sure that you are at least as confused as you were when I first showed you the block diagram of the chip. The individual parts are quite easy to understand, but I'm sure you're mystified as to how they work together. To fully understand how the 555 chip works, there is a new type of component that I will introduce you to in the next experiment.

Experiment 39
Blinking LEDs

Parts Bin

Assembled PCB

555 timer chip in 8-pin DIP package

LED, any color

470-Ω resistor

R1 = 33k resistor

R2 = 100k potentiometer

0.01 µF capacitor, any type

10 µF 35-volt electrolytic capacitor

Tool Box

Wiring kit

When I described the operation of the 555 chip, I neglected to take into account the components that would be wired to it. So far in the book, I have introduced you to resistors, diodes, and transistors, but I have not introduced you to any components that are designed to *store* energy. Resistors, diodes and transistors can all change the voltage and current of an electrical signal, but they cannot store any energy from it.

One of the two most basic electronic devices that can store energy is the capacitor. Capacitors consist of two metal plates that store energy as an electrical charge and have the schematic reference "C." The plates are represented in the capacitor's schematic

symbol (Figure 7-6). Figure 7-7 shows different capacitor packages and the indications that they are polarized.

Capacitors store electrical charge, which is measured in farads. One farad is an extremely large charge. It has only been in the last few years that capacitors have been created that can store a farad or more; most capacitors store charges in the ranges of millionths or trillionths of a farad. Capacitors that store charge in the range of millionths of a farad are rated in microfarads (µF) and charges of trillionths of farads are rated in picofarads (pF). Engineers and technicians often refer to microfarads as "mikes" and picofarads as "puffs."

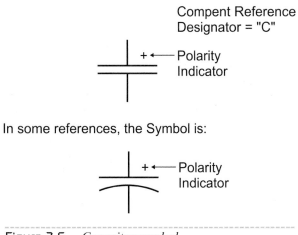

Compent Reference
Designator = "C"

+ ← — Polarity
Indicator

In some references, the Symbol is:

+ ← — Polarity
Indicator

Figure 7-6   *Capacitor symbol*

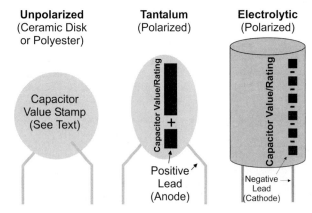

Unpolarized
(Ceramic Disk
or Polyester)

Tantalum
(Polarized)

Electrolytic
(Polarized)

Capacitor
Value Stamp
(See Text)

Capacitor Value/Rating

+

Positive
Lead
(Anode)

Capacitor Value/Rating

Negative →
Lead
(Cathode)

Figure 7-7   *Capacitor appearance and markings*

A number of different technologies are used to manufacture capacitors. All capacitors are built from two metal plates separated by a dielectric that enhances the amount of charge the plates can hold but does not let the two plates touch. In this book, the circuits are designed using either standard ceramic or electrolytic capacitors. Ceramic capacitors are not polarized and do not have any rated voltage. They are typically marked with a three-digit number to indicate their value. This value is similar to the value specified by a resistor's bands; the first two digits are the mantissa of the value and the third digit is the exponent of 10 with the base being in picofarads. For example, if you had a 330 pF capacitor, it would be marked with 331. Ceramic capacitors are typically available in the range of pFs to 0.1 μF. Electrolytic capacitors are polarized and use a liquid as the *dielectric* (the insulator between the two metal plates). They are usually built in metal cans with

their value stamped along with the negative lead (cathode) indicator and range from 1 mF to several farads. Basic capacitors use a ceramic material for the dielectric, more exotic devices use polyester, a tantalum solution, or an electrolytic solution for the dielectric. The more exotic the dielectric, the smaller the capacitor is, the more charge it can hold, and the greater its price.

A capacitor performs the same function as a water tower in a city water system. Normally, water is pumped into homes, but sometimes the demand exceeds the capacity of the system, or users don't use very much (such as when they are sleeping), and an excess is being pumped. To help the system, a water tower, like the one in Figure 7-8, is used. If it is hot and many people are watering their lawns, then water from the tower is added to the supply by gravity. At night, when the pump's capacity exceeds demand, water goes up the tower, saving it for later.

When used with a current-limiting resistor to make a *resistor-capacitor* (or RC) network, as I show in Figure 7-9, the voltage across the capacitor will change more slowly than if it weren't in place at all. The product of the value of the resistor and capacitor has a value of "seconds" and is known as the *RC time constant* and is given the Greek letter Tau (τ) as its

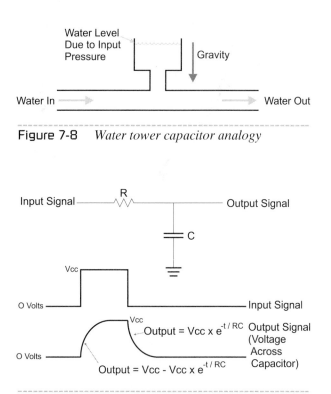

Water Level
Due to Input
Pressure

Gravity

Water In →   → Water Out

Figure 7-8   *Water tower capacitor analogy*

Input Signal —— R —— Output Signal

C

Vcc

O Volts

Input Signal

Vcc

O Volts

Output = Vcc x $e^{-t/RC}$ Output Signal
(Voltage
Across
Capacitor)

Output = Vcc - Vcc x $e^{-t/RC}$

Figure 7-9   *Operation of an RC network*

Experiment 39 — Blinking LEDs

symbol. The RC delay is used by the 555 chip (and its built-in comparators) to "time" an operation before proceeding.

To demonstrate how the RC network is used with the 555 timer chip to produce a repeating signal, I would like you to build the circuit shown in Figure 7-10.

When this circuit starts running, the 555 will be an "astable" oscillator with the LED flashing on and off at about once per second. You can change this rate by adjusting the 100k potentiometer that is wired as R2. As resistance decreases, the LED will flash faster and for a shorter period of time. The time the LED is on (555 output high) is found by the following equation:

$$T_{high} = 0.693 \times C \times (R1 + R2)$$

$$= 0.693 \times 10 \text{ µF} \times (33k + Rpot)$$

The time for the LED off is found by using the formula:

$$T_{low} = 0.693 \times C \times R2$$

$$= 0.693 \times 10 \text{ µF} \times Rpot$$

The 0.01 mF capacitor wired to the control voltage pin of the 555 is used as a filter for the internal voltages. This capacitor works very similarly to the "water tower;" if the input voltage changes, the capacitor will absorb or release charge to keep the voltage as even as possible.

To get a better idea of how the 555 timer works as an oscillator, the following values are labeled in Figure 7-10: the RC voltage (A), the RS Flip Flop output (B, which is in the inverted 555 output), the "Threshold" comparator voltage (C), and the "Trigger" comparator voltage (D). Figure 7-11 shows the waveforms for each of these parts marked in Figure 7-10,

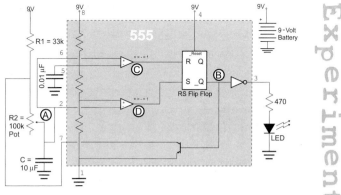

Figure 7-10   *555 oscillator circuit*

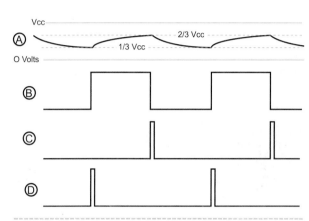

Figure 7-11   *Electrical signals within the 555 chip*

so you can see the changing RC waveform, the output from the two comparators, and the action of the RS Flip Flop.

I have covered a *lot* of material here; don't worry if you don't understand everything. Just accept that capacitors are used to filter out fluctuations in voltages or are used with resistors to delay the time it takes for a voltage output to reach the input level. How all this works will become clearer as you go on.

Parts Bin

Assembled PCB

555 timer chip

LED, any color

470-Ω resistor

R = 100k resistor

10k resistor

0.01 μF capacitor, any
   type

10 μF 35-volt elec-
   trolytic capacitor

Tool Box

Wiring kit

In the previous experiment, along with introducing you to the 555 astable oscillator, I have also shown you a capacitor for the first time. If I were to summarize the important points of the previous experiment, they would be the following:

- Capacitors store charge.

- Used with a resistor, capacitors can be used to delay electrical signals.

- The 555 can oscillate, creating a repeating signal with parameters defined by simple formulas.

What I did not note in the previous experiment is that the resistor and capacitor values used with the 555 should be within the following ranges for stable and reliable operation:

$$10k \leq R \leq 14M$$

$$100 \text{ pF} \leq C \leq 1,000 \text{ μF}$$

Another basic circuit the 555 is used for is the *monostable.* In the previous experiment, I introduced you to the astable oscillator, which will run forever; the monostable will only execute once and requires triggering. I'm willing to bet that after reading the previous sentence, you can think of applications where an astable oscillator is useful, but not the monostable. Actually, the monostable is very useful for a variety of different applications.

The typical circuit for a button and its operation is shown in Figure 7-12. In this configuration, the resis-

tor connected to the input pin of the 555 is known as a "pull-up" and it is connected to a button that, when pressed, will connect this line down to ground. The resistor limits the amount of current flow passing from the power source to ground and is called a *current-limiting resistor*, like a resistor used with an LED.

The button circuit is actually a small part of Figure 7-12; most of the diagram is taken up with an oscilloscope picture showing the voltage signal being passed to the input circuit. When a switch closes, the contacts within the switch do not simply touch and stay together; they actually bounce off each other a few times, resulting in the spiky contact. If this waveform is passed to a circuit, it probably would be registered as multiple button presses because each bounce would be treated as a unique button press.

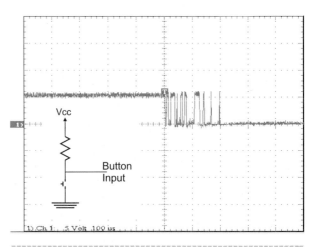

Figure 7-12   *Oscilloscope picture of a switch bounce*

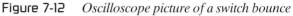

By using the 555 timer as a monostable, the "bounce" produced by the circuit could be ignored and a single button press would be registered in the application. The circuit shown in Figure 7-13 shows how the 555 timer will debounce a button input and turn on an LED for a second each time the button is pressed.

When the button switch is closed, the pulse output from the 555 timer is determined using the following formula:

$$T_{pulse} = 1.1 \times R \times C$$

$$= 1.1 \times 100k \times 10 \ \mu F = 1.1 \ seconds$$

I've "opened up" the 555 in Figure 7-13, as I did in the previous experiment, with a waveform (Figure 7-14) showing what happens on a button press.

The 555's RS Flip Flop is initially reset, and the transistor that passes capacitor charge to ground is turned on. When the button (A) is pressed and the "Trigger" input receives a low voltage input, its comparator signal (E) goes high, changing the state of the RS Flip Flop (C) (and turning on the LED). When the RS Flip Flop state changes, the transistor is turned off and the capacitor charges through the resistor. The capacitor charges according to the following formula until its voltage reaches 2/3 Vcc.

$$Output = Vcc - Vcc \times e^{-t/RC}$$

When it reaches 2/3 Vcc, the Threshold comparator (D) goes high and the RS Flip Flop changes state again, turning off the LED and turning on the transistor that shorts the capacitor to ground, returning the 555 and the circuit to its original state.

Note that in Figure 7-14, I have drawn the waveforms bouncing on the switch opening (line A going to a high voltage again). When a switch opens, it bounces just like when it closed. Secondly, I indicated that the switch is released before the capacitor charges to 2/3 Vcc; when you build the circuit, you will want to see what happens if you hold the button closed longer than the 1 second that the LED is on. In the figure, the capacitor does not discharge through the transistor instantly; it has the same exponential waveform (although much shorter) as the original charging waveform. I put in this waveform to indicate that the transistor has a very low resistance, and it behaves as if a resistor were connecting the capacitor to ground.

One question that you will have is how the button is implemented. For this experiment (and others that require a button input), I soldered a couple of 22-gauge solid core wires to a button and covered them in five-minute epoxy for strain relief as I show in Figure 7-15.

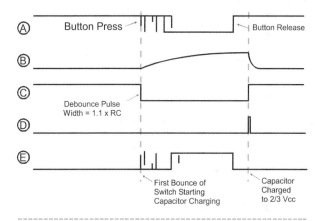

Figure 7-14 *555 button debounce operation waveforms*

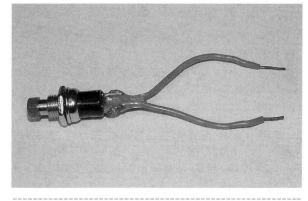

Figure 7-15 *Momentary push button with wires soldered to it for wiring to a breadboard*

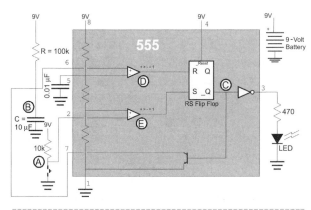

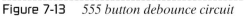

Figure 7-13 *555 button debounce circuit*

Trying out the circuit should not yield too many surprises until you hold the button down for longer than a second. Quickly pressing and releasing the button will turn on the LED for the second or so that is calculated for the resistor and capacitor chosen for the circuit. If you hold the button down for more than a second, you will find that the LED will stay on, but it will appear to "wink" off periodically. When the capacitor charges to 2/3 Vcc and the button is held down, both of the comparators will be driving a high voltage to the RS Flip Flop. This is an invalid condition for the RS Flip Flop and the output from the Flip Flop is indeterminate, resulting in the transistor tying the capacitor to ground periodically. To avoid this behavior, you should always make sure that the length of time for the pulse output from the 555 is longer than the expected input.

Experiment 41
R/C Servo Control

Parts Bin

Assembled PCB

556 timer chip in 14-pin DIP package

Four AA battery clip

Four AA batteries

2.7M resistor (made from 2.2M & 470k resistors)

Three 100k resistor

100k pot

Two 0.01 μF capacitor, any type

R/C servo

Servo connector (built as specified in text)

Tool Box

Wiring kit

If you have looked at a number of hobbyist-built robots, you will probably know that a large number of them use a *radio control* (R/C) servo for turning wheels or moving arms or actuating grippers. R/C servos (Figure 7-16) are excellent devices for using in robots; they are inexpensive and powerful actuators, and with a little bit of modification, they can be used as your robot's drivetrain. In this experiment, I will show how you can use the 555 to test and command R/C servos.

As you can see in Figure 7-16, the R/C servo consists of a small plastic box with a control/power cable running from it and a nylon arm that can be used to move the control surfaces of a model. Despite their small size (standard servos are about 1.5 inches [4 centimeters] long and about 0.8 inches [2 centime-

ters] deep), servos can provide 2 pounds (1 kilogram) or more of force. They come in all shapes and sizes for different applications. For the robots presented in

Figure 7-16   *Hobbyist radio control (R/C) servo*

this book, I will be using either standard, low-cost, general purpose R/C servos, which can be found for less than $10 at large hobby retailers, or "nano" servos, which cost about $20. R/C servos require a 4.5- to 6-volt power source and generally use between 150 to 300 mA of current when the motor is running.

The R/C servo uses a standard three-pin connector that is shown with an adapter you will have to make. The three pins are a control signal, Vcc, and Gnd, and if you look at most servos, the cable leading up to this connector consists of white (or yellow), red, and black wires, respectively, so that you can identify the purpose of the wires easily. The R/C servo adapter is made from two pieces of three conductor 0.100-inch breakaway header pins that are the "mate" to the sockets soldered to the PCB like you did with the earlier stepper motor experiment. To make the adapter, I soldered the short ends of two connector pieces together. When you are building the connectors, I recommend that you get your energy up and build as many of them as you can stand. These connectors are very useful and very easy to lose.

For the majority of nonrobot experiments, I have specified the 9-volt battery that is built into the PCB that came with the book. For this experiment and other ones that use an R/C servo or DC motor, I am going to specify that you use 4 AA cells.

The control signal used to specify the position of the R/C servo's control arm consists of a 1 to 2 msec pulse. When the control signal is a 1 msec pulse, the arm is at one extreme, a 2 msec pulse moves the arm to the other extreme, and pulses between 1 and 2 msecs move to a point in between. These pulses should be repeated once every 20 msecs, although if no pulses are passed to the servo, the arm will stay at its current position (although it will not offer any resistance if you were to try and move it). Figure 7-17 shows an R/C servo control waveform.

Figure 7-18 shows you what you get for less than $10. The servo itself consists of a gear-reduced motor that is driving the control arm. The control arm is mechanically connected to a potentiometer wired as a voltage divider. The voltage output from the potentiometer voltage divider is compared against a voltage proportional to the length of the incoming control signal pulse. If the position of the arm is different than that of the control signal, the compara-

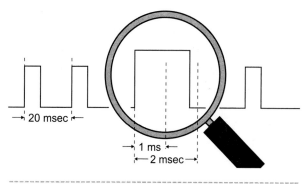
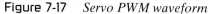

Figure 7-17   *Servo PWM waveform*

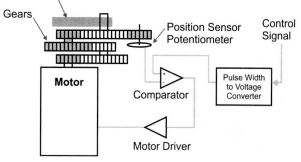

Figure 7-18   *R/C servo block diagram*

tor's output is amplified by the R/C servo's "Motor Driver" and the motor turns the arm in the appropriate direction. When the arm is in the same position as specified by the control signal pulses, then the comparator's output is zero and the motor doesn't move.

Looking at what I have written here and thinking back to the two previous experiments, you should be thinking that the 555 timer is ideally suited to driving an R/C servo. In the first experiment in this section, the 555 was used as an astable oscillator, driving a repeated negative pulse. In the second experiment, the 555 was used as a monostable to output a positive pulse from a negative input. It should be simple to wire two 555s, one as an astable oscillator and one as a monostable oscillator, to create a series of pulses for the 555. I would agree with you on all the points except one, instead of using *two* 555s, how about using one 556?

The 556 chip (Figure 7-19) consists of two 555 oscillators built together. This 14 pin chip provides the same function as two 555s (one on each side) and is ideally suited for this application. In Figure 7-20,

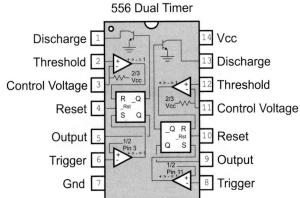

556 Dual Timer

Discharge	1	14	Vcc
Threshold	2	13	Discharge
Control Voltage	3	12	Threshold
Reset	4	11	Control Voltage
Output	5	10	Reset
Trigger	6	9	Output
Gnd	7	8	Trigger

Figure 7-19 *556 timer chip pinout*

is the same as holding the button down too long in the previous experiment). I "made" a 2.7M resistor out of a 2.2M resistor and a 470k resistor. For the monostable oscillator that provides the control signal for the R/C servo, I created a 1.1 msec to 2.2 msec delay circuit using the 100k resistor and 100k pot. The timing is slightly out of specification, but you will find that most servos will execute signals in this range without any problems. The oscilloscope picture in Figure 7-21 shows the 1 to 2 msec control pulse that repeats every 20 msecs.

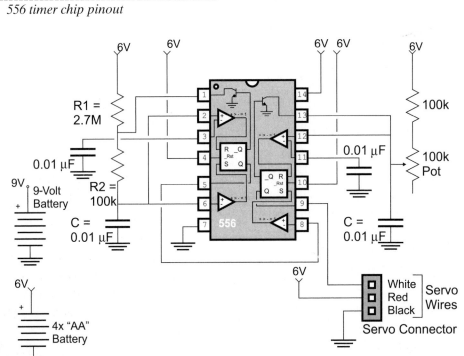

Figure 7-20 *556-based servo control/test circuit*

you can see that I have wired the left-hand circuit as an astable oscillator and the right as a monostable triggered by the left 555's astable oscillator.

For the astable oscillator portion of this circuit, I calculated resistor and capacitor values for a signal that is 19.4 msecs "high" and 700 msecs "low." I chose this timing to make sure that the "low" period would not be longer than the servo operating pulse (which

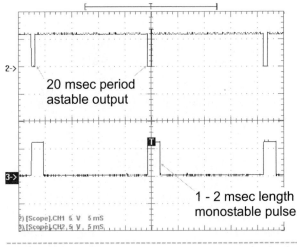

Figure 7-21 *Waveforms produced by 556 for R/C servo*

Experiment 42
Light-Seeking Robot

Parts Bin

Assembled book PCB

Plywood base with DC motors

556 dual 555 timer chip

Two ZTX749 PNP transistors

Two 0.01 μF capacitors, any type

Two 1,000 μF, 16-volt electrolytic capacitor

Four 100 Ω, 1/4-watt resistors

Two 10k, light-dependant resistors (CDS cells)

Tool Box

Wiring kit
Screwdriver

You now have all the information that you need to build a very simple robot. In this experiment, I will show you how to create a robot that will follow a light beam. This robot is very similar to the first light-seeking robot (the "turtle") created by Dr. Walter Grey in the early 1950s. The 555 timer provides the control signals for the robot's motors, replacing the vacuum tubes used by Dr. Grey.

This robot will use the PCB that comes with the book, combined with the DC motor base that you built earlier, and will look something like Figure 7-22. In Figure 7-22 I have labeled some of the most important parts of the robot. The wheels are pulling

the robot toward the light source found by the *light-dependent resistors* (LDRs).

The light sensors used in the robot are LDRs that are made out of cadmium sulfide and are often referred to as CDS cells. As the amount of light that falls on the LDRs increases, their resistance drops. For this experiment, I used LDRs that have a resistance of 10k and can fall to as little as 2k when they are exposed to a bright light.

The robot presented in this experiment uses this characteristic of LDRs to vary the resistance time-based signal used with 555 that controls the robot motors. In Figure 7-23, I have shown the block diagram of how the 555 is used as an astable oscillator to produce the time-based signals used by the robot.

The LDR, along with a fixed resistor and capacitor, is used to produce a series of low-voltage pulses that are used to turn on a PNP transistor to periodically provide current to a DC motor. When I am planning to use a 555 timer (or really any time-based signal), I like to draw out the most important parts of the circuit and the expected signals so that I can easily visualize what is supposed to happen in the circuit, and hopefully see problems before they become an issue.

In Figure 7-23, the downward-pointing arrow on the left-hand side of the drawing is used to indicate

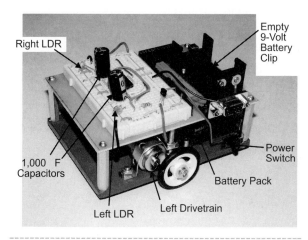

Right LDR

Empty 9-Volt Battery Clip

1,000 F Capacitors

Power Switch

Battery Pack

Left LDR

Left Drivetrain

Figure 7-22   *555 robot*

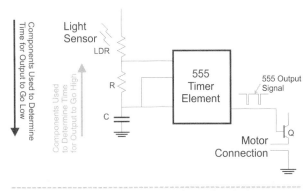

Figure 7-23 *Block diagram for robot light sensor/motor driver*

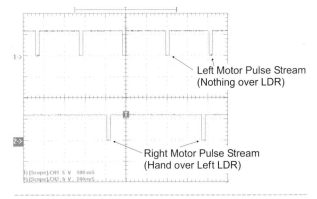

Figure 7-24 *556 outputs to robot motor driver transistors*

the resistances that are used to produce the high voltage portion of the signal. As I have pointed out earlier in this section, the two series resistors make up the RC network that provides the high to low delay. The upward-pointing arrow indicates that just the single, fixed resistor produces the low to high delay. This is a bit of a mnemonic that I use to remember how the 555 works, and it also reminds me that the time the signal is high is always longer than the time the signal is low because the total resistance of the downward arrow is larger than that of the upward arrow. As I showed in the earlier experiment, the larger the resistance in the 555's RC network, the longer the delay.

Using the formulas I presented earlier in this section, the time the output signal is low is going to be about 0.7 seconds, whereas the time the output signal is high will be anywhere from more than 10 seconds to about a second, depending on how much light falls on the LDR. The more light that falls on the LDR, the more low pulses (which turn on the motor) happen in a given period of time, moving the driven side of the robot faster.

To create the robot, it has two of these circuits, an LDR on each side of the robot used to control the motor on the other. As more light falls on one LDR, the motor on the opposite side is pulsed faster, turning the robot toward the LDR and the light source. Figure 7-24 is an oscilloscope picture of the outputs of the two 555 circuits showing how the pulses going to the left motor are happening more frequently than to the right motor, which had my hand over its LDR (which is on the right side of the robot). In this case, the robot would be turning to the left as it moves forward.

Although I could have used a 555 timer for this circuit, I decided to use a 556 and take advantage of the two 555s built into it. The schematic drawing for the circuit is shown in Figure 7-25.

This circuit is powered just by the four AA batteries in the clip attached to the plywood base. Remember to make sure you have an on/off switch for the AA batteries to make sure that the motors don't start working while you are wiring the circuit. Depending on the motors that you have used and how they are wired and connected to the drivetrain, you may find that their positive and negative wires are reversed from what you expect. The lighter (red) motor connection could go to the PNP providing the current, but in your actual robot, you may have to connect one or both of the black motor connections to the PNP transistors for the robot to run in the right direction.

Looking back to Figure 7-22, the robot should appear to be quite simple, but you should keep a few things in mind. The first is that the robot wheels pull the robot; they don't push it. The direction of movement for the robot is from the battery clip to the breadboard on the PCB. This will give the LDRs the best view of any lights in front of the robot. Second, before assembling the robot, remember to put in fresh AA batteries. Although it's not difficult to disassemble and reassemble the robot, save yourself a few minutes and make sure the batteries are good.

The 10k LDRs, 100 Ω resistors, and 1,000 μF capacitors with the motors that I have used provided me with a robot that moves at about one inch per second toward a light source, just about perfect for this application. When you build this robot, you might

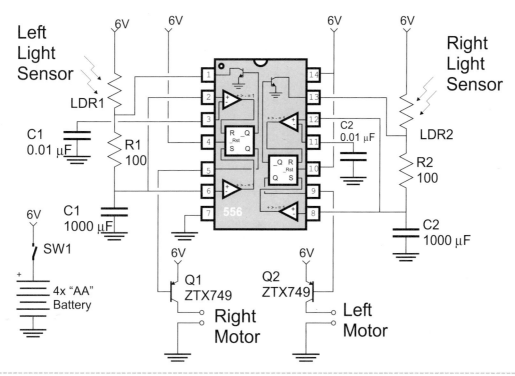

Figure 7-25 *555-based light-seeking robot circuit*

want to try other values for these components if your motors behave strangely or move too quickly and miss the light.

When you get the robot running, try it out in a normal room as well as a dark room with a flashlight in one corner. I will discuss the behavior of light-seeking robots in more detail later in the book, but for now, take a look at how the robot works and see if you can establish any rules for it. As a last experiment for you to try out, see if you can convert the robot into a light-*avoiding* robot (it's actually pretty simple; just reverse the LED connections to the 556).

Section Eight
Optoelectronics

One hundred years ago, the most important argument in science was trying to determine what is light. The field was divided into two camps—one that suggested that light was made up of particles, and another that was convinced that light was made up of waves. Confusing the matter were the discussions regarding what matter actually was. The final determination of what light was changed the course of human history.

In the early nineteenth century, matter was thought to consist of goo that was made up of atoms and electrons that were stuck together. No instruments were capable of discerning the physical characteristics of matter, and this idea seemed like a logical way to view how things were made (especially when you were confronted with a glass of water or a piece of metal). This model wasn't challenged until the discovery of materials that *fluoresce*, or give off light when they are struck by energetic particles.

The energetic particles that were used in these experiments were electrons generated in a vacuum in a device like the one shown in Figure 8-1. In this device, electrons were allowed to "boil" off a heated, negatively charged electrode (called a *cathode*) and allowed to travel towards a positively charged target (the cross in Figure 8-1). Some of the electrons would miss the target and hit the fluorescent material behind the target, causing it to emit light. The electrons that came from the electrode were known as *cathode rays*, which gave rise to the experiment's name: the cathode ray tube, which is the precursor to today's television set and computer displays. This result confused many researchers because light was produced by cathode rays, and *not* by heat, as was the accepted way light was produced; the fluorescent material gave off light while staying cool. One of the theories of the time was that light particles were part of the fluorescent material's atoms and when the cathode rays struck the fluorescent material's atoms, the light particles were knocked off.

Max Planck, in 1900, suggested that a unit of light (which he called a *quanta*) could only be ejected from an atom at a set energy level according to the following formula:

$$E = h\nu$$

Side View of CRT

End View of CRT

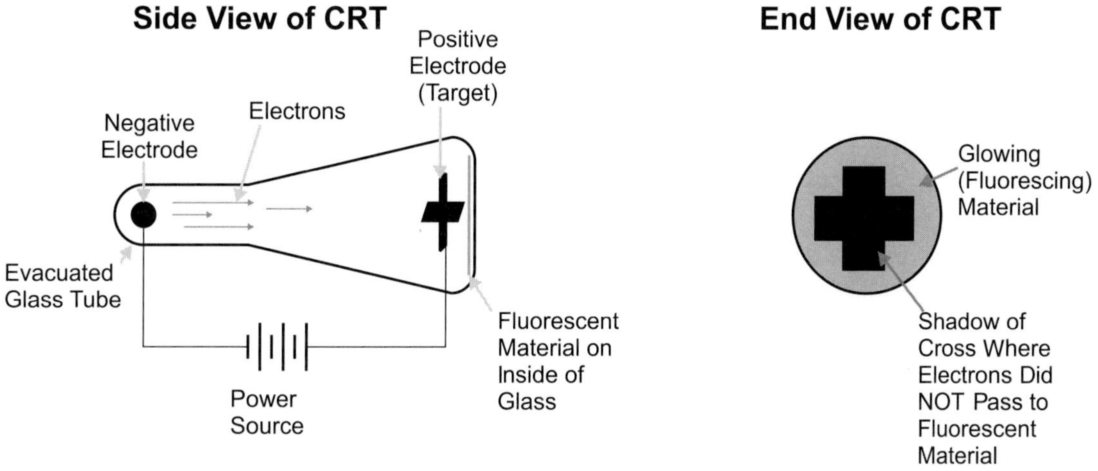

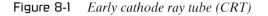

Figure 8-1   *Early cathode ray tube (CRT)*

"E" is the energy of the light, "v" is its frequency, and "h" is a constant, which we now call "Planck's Constant" and is 6.63 x 10⁻³⁴ Joule seconds (Js). Plank also discovered that for each element, the frequencies of light output were even multiples of h × v.

Making the determination of what light was became more difficult after the results of three other experiments. The first was when light was shone on a black target in a darkened area; a small but measurable amount of force seemed to be applied to the target. The results of this experiment, along with the Planck's theories, seemed to suggest that light was a particle.

The experiment shown in Figure 8-2 was another attempt to explain what light was. Light was passed through two narrow slits (called a *diffraction grating*) and then observed on a plate behind it. If light were made up of small particles, two bright spots would appear on the paper behind the material with the slits. The actual result is shown in Figure 8-2. Light and dark spots appeared, and the results were similar to that of a wave, such as a wave in water, when passed through a diffraction grating.

Further adding to the consternation were the results of Rutherford's experiment, in which he bombarded a piece of gold foil with alpha particles; conventional wisdom of the time stated that the alpha particles would blast their way through the gold. Today we would call "alpha particles" helium atoms, and they are one of the by-products of decaying uranium (and other radioactive materials). The purpose of Rutherford's experiment was that he was looking for evidence of damage to the gold leaf from the alpha particles. What he found was that not only was the gold leaf undamaged by the alpha particles, but occasionally particles were deflected by the gold leaf. Rutherford later said, "It was almost as incredible as if you fired a 15-inch shell at a piece of tissue paper and it came back and hit you." Rutherford postulated that matter must be made up of mostly empty space with the nucleus of atoms being quite small and relatively far apart so that when the alpha particle beam passed through the gold foil, only a few alpha particles hit any of the gold atoms.

In 1905, Albert Einstein proposed in a paper a model for the atom and light in which the results of these experiments would be explained. Einstein's

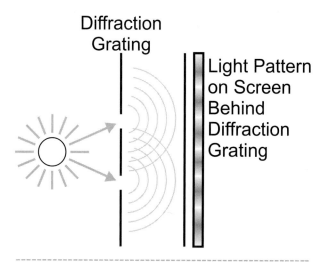

Figure 8-2   *Passing light through a diffraction grating with the actual results*

theory suggested that energy is passed to or from an atom via a photon (his term) that had Planck's quanta of energy. This energy was stored in the atom in the orbit of its electrons (as shown in Figure 8-3). When energy was added to the electron, it would absorb a photon and an electron would jump to a higher potential (or *excitation*) state. This energy was defined by Planck's formula mentioned earlier. If the photon did not have this level of energy, then it would not be absorbed; likewise, if it had more energy than what Plank's formula dictated, it would not be used to increase an electron's excitation state. When an atom released energy, an electron would fall to a lower excitation state, releasing a photon at the appropriate quantum of energy. This was called the *photovoltaic effect*.

The photovoltaic effect can be seen very clearly in the operation of LEDs that you have been working with for quite a few experiments. LEDs output light at a frequency that is dependent on the chemicals used to build them.

Optoelectronics are devices that either process incoming light or output it using the theories (known as quantum mechanics) presented here. Along with LEDs, which you were introduced to earlier, a number of different devices both react to light as well as output it.

Light is considered the part of the electromagnetic spectrum that has wavelengths from 0.01 millimeters to 100 nanometers. The human eye can see light in

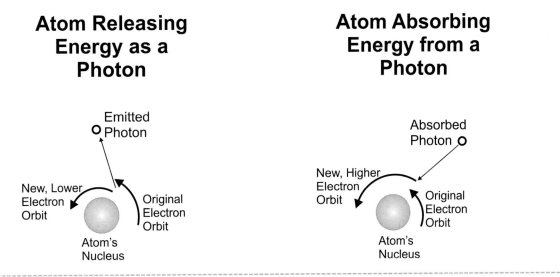

Atom Releasing Energy as a Photon

Emitted Photon

New, Lower Electron Orbit

Original Electron Orbit

Atom's Nucleus

Atom Absorbing Energy from a Photon

Absorbed Photon

New, Higher Electron Orbit

Original Electron Orbit

Atom's Nucleus

Figure 8-3   *Photovoltaic effect: changing an atom's energy level by photon absorption or release*

the range of 400 to 720 nanometers. Wavelengths greater than 720 nanometers are in the *infrared* range, and wavelengths less than 400 nanometers are in the *ultraviolet* range. The seven basic colors of the rainbow with their wavelengths are listed in Table 8-1.

When you are asked what the colors of the rainbow are, remember the name "ROY G. BIV," which is the acronym of the first letters of the visible colors.

Table 8-1   Colors and their wavelengths

Color	Wavelength
Infrared	720+ nm
Red	610–720 nm
Orange	580– 610 nm
Yellow	530–580 nm
Green	480–530 nm
Blue	430–480 nm
Indigo	410–430 nm
Violet	400–410 nm

Experiment 43
Different Color LEDs

Parts Bin

Assembled PCB with breadboard

1k resistor

Infrared LED

Red LED

Orange LED

Yellow LED

Green LED

Blue LED

White LED

Tool Box

Wiring kit

DMM

When I wrote out the different wavelengths of light for the different LEDs at the start of this section, I made the hypothesis to myself that the voltage drop across an LED was related to the wavelength of light output, with the shorter the wavelength (the higher the frequency), the greater the voltage drop. I did not make this hypothesis based on anything I consciously remembered, but on my understanding that higher frequencies of light require more energy to produce. Colored LEDs are different from light filaments as they do not emit a spectra of light due to heating; they emit light at a single wavelength of light caused by electrons falling from one energy level to another. This makes LEDs useful for calibrating light sensors because the wavelength is dependent on the material that was used to make them and cannot change due to environmental conditions (including variances in the amount of current passing through them).

There's one piece of the puzzle that you are probably not aware of, and that is how a light wavelength is related to energy. As you are probably aware, along with the photovoltaic effect, Einstein also postulated that the speed of light is the fastest speed at which anything can travel. This is part of the theory of relativity and to be totally accurate, it states that nothing relative to an observer can travel faster than light.

The question is, how do you add energy to light if you cannot make it go faster? The answer is that as you add more energy to light, its wavelength shortens. If you take energy away from light, rather than slowing down, its wavelength lengthens. This is probably hard to understand.

Astronomers measure the distance between objects in the universe by their red shift. The theory behind this is that after the Big Bang, galaxies and other objects were flung out from the center of the universe at different rates of speed. The galaxies that were thrown out at the fastest speed are at the rim of the expanding universe, whereas the Milky Way galaxy (where the Earth is located) has stayed relatively close to the center of the universe because it was not given as much energy from the Big Bang.

The problem comes about when you consider that the "object" sent from the other galaxy is light; the speed of light cannot be changed relative to its observer. So, when the light from the far-off galaxy reaches the Earth, it is still traveling at 186,000 miles per second (2.99792×10^8 m/s). Comparing the situations, it doesn't make sense that light from a faraway galaxy that is traveling away from the Earth is the same as when it left.

To solve this problem, don't limit yourself to thinking about speed; think about the energy of a physical object as well as the energy of the light as they pass between the galaxies. In the case of the physical object, the energy it has when it impacts the Earth is lessened because of the relative motion of the Earth away from the sending galaxy. This is also true for light; instead of slowing down, the energy of the photons decreases. As the energy of photons decreases, their wavelength lengthens. As wavelengths lengthen, light moves toward the red portion of the spectrum, which is why astronomers measure distances between objects by what astronomer's call the light's red shift.

With this background, I came up with the hypothesis that the shorter the wavelength of light produced by an LED, the greater the voltage drop across it because of the greater amount of energy passed from the electricity to the light. To test out this hypothesis, I used the simple circuit shown in Figure 8-4 and measured the voltage across a number of different LEDs. The results of the experiment are listed in Table 8-2.

Looking at the results, you can see that there is some correlation between wavelength and voltage drop: An infrared LED, which has the longest wavelength, has the smallest voltage drop; and the blue LED, which has one of the shortest wavelengths, has a large voltage drop. What doesn't make sense is the values for the "middle" colors (red, orange, yellow, and green).

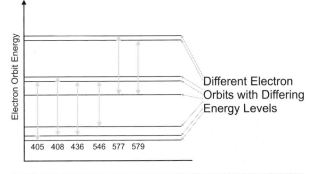

Figure 8-5 *Different wavelengths of light (measured in nm) released from or absorbed by tungsten atoms*

This was confusing to me until I considered the different materials that are used to make the different color LEDs. Each element emits light at different frequencies. For example, if you look at tungsten (Figure 8-5), you will see that it has eight electron orbits (each with their own energy levels), and a number of wavelengths of light can be produced, depending on which electron orbits are changed. From this information, I came to the conclusion that the voltage drop was more of a function of the doping material of the LED than the wavelength of the light produced.

When I tested the different LEDs, you'll see that I also tested a white LED. When white LEDs are manufactured, they start out as blue LEDs, but phosphorus is added to the LED before it is placed in the epoxy lens. This causes the LED to produce light across the entire visible spectrum, resulting in white light being output rather than just one frequency of light. I included the results for the white LED for reference, but I did not consider it as part of the hypothesis.

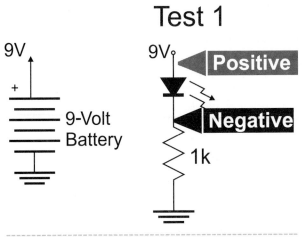

Figure 8-4 *Circuit for testing LED voltage drop*

Table 8-2 Results of the experiment

LED Color	Voltage
Infrared	1.12 V
Red	1.96 V
Orange	1.82 V
Yellow	1.86 V
Green	1.95 V
Blue	2.71 V
White	2.76 V

Experiment 44
Changing an LED's Brightness

Parts Bin

Assembled PCB

555 timer chip in 8-pin DIP package

LED, any color

1k-ohm resistor

Two 10k resistors

100k potentiometer

0.01 µF capacitor, any type

0.1 µF capacitor, any type

Tool Box

Wiring kit

The automatic solution that most people come up with when they are asked to vary the power level of a device is to lower the voltage being applied to it. This makes sense for a variety of different devices, ranging from light bulbs to motors. The problem with this method is that although it works for most devices (it doesn't for some), it can be very difficult to implement.

Implementing a variable voltage-level power supply could be accomplished using a PNP transistor and a voltage comparator in a circuit known as a *linear regulator*, and I will discuss it in more detail when I introduce the concept and operation of power supplies. Although the operation of the linear regulator probably seems quite simple to you, as you are familiar with the operation of the transistor as well as the comparator, it is actually very difficult to get it working correctly. One aspect makes it very undesirable for applications other than that of a power supply for control electronics.

The reason the linear regulator is undesirable for providing a variable voltage to a device is due to the amount of power that is lost in the PNP transistor. For example, if you wanted to drop 10 volts to 5 volts using this circuit, and the load drew 2 amps, the heat that would have to be dissipated through the PNP transistor would be 10 watts. This is a significant amount of heat that will have to be passed to the surrounding air. If the linear regulator is being used to control a robot's motor speeds, then the 10 watts of power that is being lost is a considerable drain on a robot's batteries.

A much better solution is to periodically turn power to the load on and off, resulting in the *average* power being what you want to control the device with. This switching on and off of the power is known as *pulse width modulation* (PWM), and if you were to look at it on an oscilloscope, it would look something like Figure 8-6.

You should be aware of two features of the PWM signal, and the first is the period of the signal. The period of the PWM signal should be outside the range of human visual perception; this means that it should be shorter than 20 ms (which results in 50 PWM cycles per second or more) if you are using the PWM with a device that emits light. When the device can make a sound, its PWM period should either be more than 20 ms (50 PWM cycles or less per second) or less than 66 µs (15,000 PWM cycles or more per second). The reason for specifying PWM speeds is to make sure that the throttling effect of the PWM is

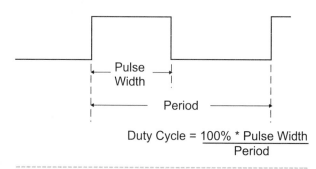

Figure 8-6 *Pulse wave modulated signal waveform*

observed, not the switching on and off of the PWM itself.

The *duty cycle* of the PWM signal is the percentage of time the signal is active during the PWM period. I have hedged a bit on the definition because the active portion of the signal can either be high (which is what most people consider it to be) or low (as I use in this experiment).

To demonstrate the operation of a device controlled by a PWM, I came up with the circuit shown in Figure 8-7. You should recognize this circuit and with a bit of studying figure it out as an astable 555 oscillator in which the time the output wave is high is variable. The purpose of this experiment is to demonstrate the operation of a PWM controlling the brightness of an LED, as well as show that a 555 may be poor device to consider as the basis for a PWM signal generator.

When you build and test the application, you will see that by varying the potentiometer, you can vary the brightness of the LED and you will not see any flashing of the LED being turned on and off very quickly. The turning on and off of the LED is happening so quickly that your eye "averages" out the time the LED is on and off and gives the appearance that it is on continuously, but at a lower brightness level than if it were on 100 percent of the time. If you were to look at the operation of the PWM using an oscilloscope, you would see that when the LED is quite dim, the signal would look like Figure 8-8. The LED is active when the output signal is low, and the proportion of time that the signal is low is quite short, about 17 percent of the PWM period. This 17 percent

is the duty cycle of the PWM signal when the LED is low. When the LED is brighter, the time in between the low periods of the repeating output of the 555 is reduced and looks something like Figure 8-9. The duty cycle for this signal is around 45 percent.

Looking at the two oscilloscope pictures in Figure 8-8 and Figure 8-9, you should see that something funny is happening. Instead of changing the PWM's duty cycle and keeping the PWM signal's period constant, I am actually decreasing the time the PWM signal is inactive (decreasing the PWM signal's period). If you were to look at the period, you would see that it ranges from about 10 msecs down to 2.2 msecs or from 100 Hz to around 450 Hz. If you have experimented with the 555's astable oscillator, you will know that regardless of how much you try to change the resistor and capacitor values used in the circuit, you will never get the PWM's duty cycle above 50 percent.

These are the two problems with using a 555 timer for a PWM. The first problem is that you *could* use

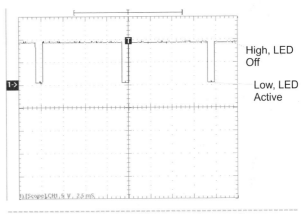

Figure 8-8   *PWM at 17 percent duty cycle*

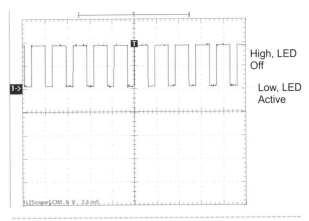

Figure 8-9   *PWM at 49 percent duty cycle*

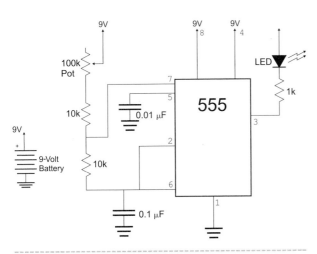

Figure 8-7   *555 LED PWM circuit*

the output of the 555 to light the LED when it's high, but you will find that the duty cycle ranges from 83 percent to 55 percent, which may not be good enough for some applications. The second problem is that by changing the PWM signal's period you could possibly be moving it into the range where a person could perceive the operation of the PWM. If you were to change the two 10k resistors with resistors of higher value, the turning on and off of the LED could become noticeable and the varying output could be impossible to observe.

A PWM can result in substantial power savings in an application. When the PWM signal's duty cycle is 50 percent, how much power is being used by the load? If you answered 50 percent, you would be wrong—it is actually 25 percent because in the power equation ($P = V \times i$), both voltage and current are active only half the time. This has some interesting implications for motor power. For example, motors running at 71 percent duty cycle are using half the power required to run the motors with a 100 percent duty cycle.

Experiment 45
Multisegment LEDs

Parts Bin

Assembled PCB with breadboard

Common anode seven-segment LED display

13 ZTX649 NPN transistors

Two LEDs, any color

Two breadboard-mountable SPDT switches

13 1k, 1/4-watt resistors

13 10k, 1/4-watt resistors

Tool Box

Wiring kit

One of the most common icons of our modern society is the seven-segment LED display (Figure 8-10). It first became popular in the 1970s with the advent of digital watches. In the 30 or more years since their introduction, they have become a ubiquitous part of modern civilization. Seven-segment LEDs can be found virtually everywhere, being used not only in digital watches, but also in kitchen appliances, cars, instruments, and, of course, in video cassette recorders (VCRs). The flashing "12:00" created using seven-segment LEDs on a clock or VCR is the symbol of a person's inability to handle the latest in technology.

Despite its commonality, the seven-segment LED display is not trivial to work with. A number of chips on the market make the component easier to work with in some applications, but when you are working

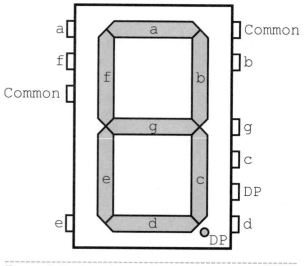

Figure 8-10   *Seven-segment LED display pinout*

with robots or your own projects, you will find that these "canned" functions don't quite do what you want them to do. You will find that you will have to come up with your own circuitry to decode data along with some way of handling multiple displays. In this experiment, I would like to introduce you to the seven-segment LED display and some of the circuitry needed to decode incoming numeric bit values before displaying.

In Figure 8-10, I have shown the appearance of the seven-segment LED display; it can be put in the same "footprint" as a 0.300-inch-wide 14-pin DIP package, but some of the pins (N/C for "no connect") are not present. The DP LED stands for the "decimal point."

The seven-segment LED display can be wired as either a common anode or common cathode. In this experiment we will be using a common anode, which is wired as shown in Figure 8-11. For this part, the two "common" pins are connected to all (and occasionally some) of the anodes of the eight LEDs built into the display. This simplifies the wiring you will have to do somewhat and makes working with multiple displays a bit easier, as I will show in a later experiment.

As you are probably aware, by turning on each of the different LEDs differently, you can create different digits. Figure 8-12 lists how different LEDs of seven segments can be used to display the 10 numeric characters. Along with the 10 numbers are a number of letters that can be displayed, although only a few of them look exactly like the characters they are supposed to represent. If you want to display letters as

well as numbers, then you will have to use an LED with more segments; these are available as either 16-segment displays or as matrixes of LEDs that display the character as a font, just as on your computer screen.

Each LED in the display can be wired conventionally to control whether or not it is turned on or off. Controlling individual LEDs in the displays is quite easy, it gets quite a bit more difficult when you have to control multiple LEDs and even more difficult when you want them to display something useful. When I first blocked out the book, I wanted to show how just a few logic chips could be used to display all the characters from "0" to "9" and "A" to "F" to demonstrate how the displays can be used for hexadecimal displays. As I worked through the logic for this, I found the number of logic chips required to do this to be prohibitive. I then worked down to wanting to display all the characters from "0" to "8" and also found that the complexity of the circuit was too much for the breadboard that is mounted on the PCB. I finally decided on the numbers "0" to "3" and used two lines for input.

To decide how to wire the circuit, I created Table 8-3, which lists which segments are active for the different four output digits, and then wrote out the "sum of products" statements (explained later in the book). The inputs were labeled "A0" and "A1" for the least significant and most significant bits used to select the numbers to be displayed, respectively. These two signals could be considered a two "bit" number. The bit numbering system as well as the AND and OR logic gates will also be described in more detail later in the book.

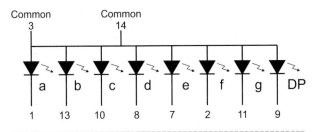

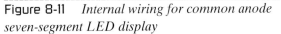

Figure 8-11 *Internal wiring for common anode seven-segment LED display*

Figure 8-12 *Seven-segment LED display values for the digits 0 through 9*

Table 8-3 Active segments for the output digits

	"0"	"1"	"2"	"3"	Terms	Comments
a	1		1	1	!A0 · !A1 + A1	Same as d
b	1	1	1	1	1	Always on
c	1	1		1	!A1 + A0 · A1	
d	1		1	1	!A0 · !A1 + A1	Same as a
e	1		1		!A0	
f	1				!A0 · !A1	Uses AND from a & d
g			1	1	A1	

When I created this table, I made sure that I noted any situations where I could simplify the circuitry. Essentially, six unique circuits will have to be designed along with the LED drivers. I could have done this with TTL logic chips, but I thought that I would use RTL (*Resistor-Transistor-Logic*) because it would make for a more interesting circuit. First, I created switch inputs (which are inverted) and two AND gates that are shown in Figure 8-13. Note that I have marked the outputs of each gate the same way.

For any of these four gates, current passed through a 1k resistor is passed to the transistor circuits downstream. The bases of the downstream circuits have 10k resistors to keep the actual current flows through the transistors essentially the 7 mA or so that comes through the 1k resistor. This is probably not absolutely required, but it is a good design rule to fol-

low to make sure that excessive currents do not build up in the transistors, and it also ensures an equal amount of current flows through each segment of the LEDs (making sure they are all equally bright).

The LED drivers consist of the various open collector transistors shown in Figure 8-14. The two OR gates required for the application are well suited to this type of logic and can be easily expanded if you wanted to create a circuit that could output more than the four digits of this one.

Wiring this experiment's circuit is a bit of a challenge. When you wire this circuit, I would recommend starting with the two switch inputs followed by the "!A0 · !A1" and "A0 · A1" terms and then the driver transistors for the two LEDs. When you are doing this, make sure that you build each transistor circuit in as small an area as possible.

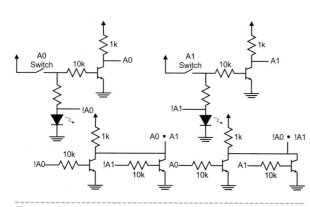

Figure 8-13   *Logic for decoding two bits to display as digits on a seven-segment LED display*

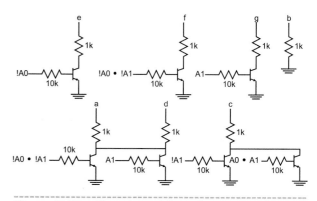

Figure 8-14   *Transistor drivers for seven-segment LED display*

Experiment 46
Optoisolator Lock and Key

Parts Bin

Assembled PCB with breadboard

Four opto-interrupters

Six Zetex ZTX649 NPN transistors

Five red LEDs

Green LED

Seven 1k, 1/4-watt resistors

Six 10k, 1/4-watt resistors

Sheet of cardboard

Tool Box

Wiring kit

Scissors

When you have moved up the robotic "food chain" and are starting to work with much heavier robots than the ones that are presented here, you will start to get into the realm of high-voltage/high-current electronics. For the most part, this is not difficult and you will find that available parts will let you work with high-power motors and batteries exactly as if they were the small devices that I use in this book. The problem will come when you want to integrate the robot's controller electronics to the motor systems; the difference in voltages and currents could result in the controller electronics being damaged (although "fried" would probably be a more accurate term). An obvious solution to this potential problem is to use relays that are driven by the control electronics to switch the motors on and off. As I have indicated earlier in the book, I do not like to use relays in robots because they are mechanical devices that require substantial current to operate and do not operate at electronic speeds that allow for effective PWM motor control.

The all-electronic solution to the problem of isolating high-voltage/current motor circuits from low-voltage/current control circuits is the optoisolator. This component (in the dotted line of Figure 8-15) consists of an LED that can shine light on a phototransistor. When the LED is on, the light causes the production of electrons in the phototransistor, which acts as a base current and allows current to flow from the collector to the emitter. The optoisolator's LED

and phototransistor are normally built together in an opaque black plastic chip package so that external light does not affect the operation of the transistor. Optoisolator's are not as fast as other transistor-based switching circuits (taking 0.5 to 5 ms to change state) and can only switch a few milli-amps at a time.

The signal passed is typically digital (which means it is either on or off), and when a high signal is passed to the optoisolator, it turns on the LED, which allows current to pass from the phototransistor's collector to the emitter, connecting the output signal to ground. In the reverse situation, if the signal passed to the LED is a low voltage, the LED is off and the phototransistor does not conduct.

A modification of the optoisolator is the opto-interrupter (Figure 8-16), in which the light path between the LED and phototransistor is opened to

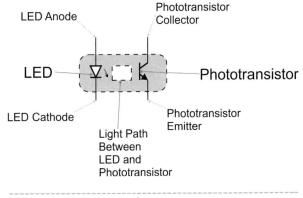

Figure 8-15 *Optoisolator*

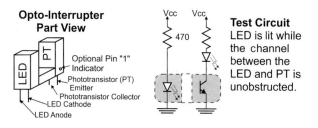

Opto-Interrupter Part View

- Optional Pin "1" Indicator
- Phototransistor (PT)
- Emitter
- Phototransistor Collector
- LED Cathode
- LED Anode

LED PT

470

Vcc   Vcc

Test Circuit
LED is lit while the channel between the LED and PT is unobstructed.

Figure 8-16 *Opto-interrupter circuitry and operation*

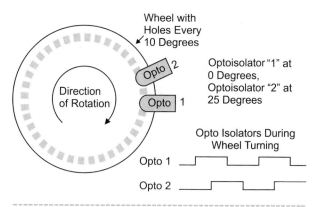

Wheel with Holes Every 10 Degrees

Direction of Rotation

Opto 2
Opto 1

Optoisolator "1" at 0 Degrees, Optoisolator "2" at 25 Degrees

Opto Isolators During Wheel Turning

Opto 1
Opto 2

Figure 8-18 *An optical interrupter on a holed wheel to detect rotary movement and direction*

allow for a physical block of the light, as a method to "switch" on and off the phototransistor based on some external, physical event. The opto-interrupter does not have the same "bounce" as a physical switch, nor does it require any force to operate, which makes it the optimal solution in some applications. A very typical application for the opto-interrupter is in a simple PC mouse; the ball turns a shaft that has a wheel with holes in it that allows light to pass through it and be sensed by an opto-interrupter as I have shown in Figure 8-17.

The drawings in Figure 8-17 show the opto-interrupters used for each wheel. In Figure 8-18, I show how two opto-interrupters are used to not only detect the movement of the wheel, but by offsetting them so that they are turned on and off at different angles; the direction of the wheel turning can also be sensed.

The reason why I am going into such detail about the operation of the PC mouse is because of the operation of the opto-interrupters used in robots to monitor its motion and position. Using opto-interrupters in

robots is part of the science of odometry in which the movement of the robot is recorded in an effort to navigate it to a specific location in space.

For this experiment, I would like to demonstrate the operation of the optoisolator/opto-interrupter by creating a lock circuit that requires the paper "key" shown in Figure 8-19. This key is cut from a piece of cardboard and should be designed when you have built the circuit shown in Figure 8-20 and know the distance between the opto-interrupters.

The opto-interrupter is used as a current switch. When light is passed to the phototransistor, current flows from the collector to the emitter; this current is used to turn on the "Tumbler Blocked" transistor. When you look at the schematic diagram for the lock circuit (Figure 8-20), the individual circuits for each opto-interrupter tumbler seem to be quite simple, but the combination of the circuits is a bit hard to understand. The outputs for each tumbler consist of an NPN transistor with its emitter tied to ground and its base controlled in a traditional manner. The collector

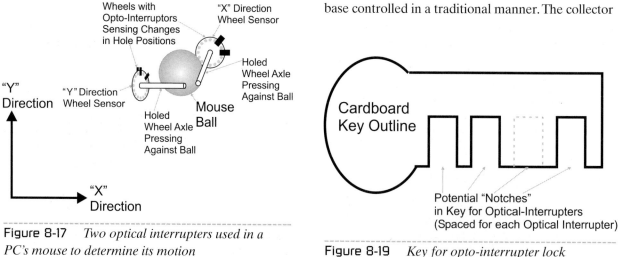

Holed Wheels with Opto-Interruptors Sensing Changes in Hole Positions

"X" Direction Wheel Sensor

Holed Wheel Axle Pressing Against Ball

"Y" Direction Wheel Sensor

"Y" Direction

Holed Wheel Axle Pressing Against Ball

Mouse Ball

"X" Direction

Figure 8-17 *Two optical interrupters used in a PC's mouse to determine its motion*

Cardboard Key Outline

Potential "Notches" in Key for Optical-Interrupters (Spaced for each Optical Interrupter)

Figure 8-19 *Key for opto-interrupter lock*

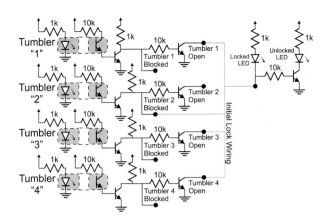

Figure 8-20 *Opto-interrupter lock circuit. Note that lock tumblers are adjustable.*

is passed to a circuit with a central resistor tied to the output voltage. This output transistor configuration is known as an *open collector* output and allows for the output of multiple transistor circuits to be tied together in a common configuration known as a *dot-*

ted AND gate. The digital AND gate only passes a high output when all of its inputs are high. The dotted AND gate shown in Figure 8-20 fulfills this requirement as any of the multiple open collector transistors wired in parallel can pull the circuit to ground (or low). In the later sections when I discuss digital logic, the AND function will become clearer, but for now, you should remember the dotted AND used here in which the output is high if and only if all the open collector transistor drivers are turned off.

In Figure 8-20, I have indicated that the opto-interrupter open and blocked transistor collector outputs for each tumbler can be wired to the locked LED (which is connected to the unlocked LED via an NPN transistor). In the figure, I wired all of the open collector connections except for one (which used the blocked collector connection) to the locked LED. You can vary this wiring or add additional opto-interrupters and transistor circuits to make the lock more sophisticated and difficult to "pick."

Experiment 47
White/Black Surface Sensor

Parts Bin

Assembled PCB with breadboard
Opto-interrupter
LED, any color
10k resistor
Two 1k resistors
Paper

Tool Box

Wiring kit
Wire clippers
Black Magic Marker
White paint pen

The most useful type of light used for mechanical sensors occurs in wavelengths that are invisible to the human eye. *Infrared* (IR) light (Figure 8-21) is used for a variety of purposes, including being a part of pass-through interrupters, as shown in the previous experiment, and a number of different sensor applications that I will show in this experiment. In Figure 8-21, I show some of the different wavelengths of IR light that are produced by different objects.

The simplest infrared sensor (and most traditionally used in robotics) is the line follower, which consists of an infrared LED and phototransistor reflecting light from a surface (Figure 8-22). When

the surface is white, you will find that the IR light from the LED reflects very efficiently, turning on the

Figure 8-21 *IR light and temperatures for different wavelengths*

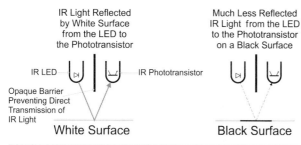

Figure 8-22   *Infrared LED and phototransistor line detector operation*

Figure 8-24   *Photograph of stock opto-interrupter and one that has been separated on prototype breadboard circuit. When cutting apart the opto-interrupter, remember to keep track of the LED and phototransistor, as well as their polarities.*

phototransistor. A black surface tends to absorb infrared light and very little is reflected, keeping the phototransistor turned off.

The traditional way of building IR line detectors is to drill multiple holes in a wood (or metal) block as shown in Figure 8-23. These holes are drilled in such a way that light from the LED can only pass to the phototransistor if it reflects off of some object. There cannot be a direct path from the LED to the photo-transistor, which can make part placement and their geometry challenging. Rather than going through the hassle of cutting up a piece of wood or metal (and potentially having to paint it black to minimize the chance for reflections), I want to use an electronic device that is basically designed for this application and with which you are already familiar.

The solution is to clip apart the LEDs and photo-transistor portions of the opto-interrupters that you used in the previous experiment (Figure 8-24). Opto-interrupters are usually designed with the LED in one tower and the phototransistor in the other. After clipping, you should be able to place the two towers side by side, and the transmitted IR light will only reach the phototransistor if it reflects off of some object. Remember to mark which tower is the LED and which is the phototransistor (as well as the

LED's cathode and phototransistor's collector) using the white paint pen before you clip apart the two parts. Along with this, make sure you keep an unmod-ified opto-interrupter on hand to help you figure out how the part is to be wired.

Once you have cut apart the two halves of the opto-interrupter as shown in Figure 8-24, you can now build the circuit shown in Figure 8-25. When you put the battery in the holder, you should find that the LED will light slightly. With a piece of white paper with black lines drawn on it slid through the slot in the middle of the breadboard, you will see the LED definitely blinking on and off.

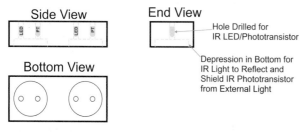

Figure 8-23   *Simple holder block for IR LED and phototransistor used for line following*

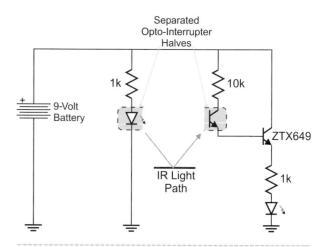

Figure 8-25   *Opto-interrupter white/black sensor circuit*

Experiment 48
Line-Following Robot

Parts Bin

Assembled PCB with breadboard

DC robot motor vase with four AA battery clip and switch

Two opto-interrupters (see text)

LM339 quad comparator

Two ZTX649 NPN bipolar transistors

Two XTX749 PNP bipolar transistors

Two LEDs, any color

Two 100k resistors

Ten 10k resistors

Two 1k resistors

Two 470 Ω resistors

Two 100 Ω resistors

Two 10k breadboard-mountable potentiometers

1/8-inch heat shrink tubing

Aluminum rain gutter end cap (see text)

22-inch by 28-inch sheet of white bristol board (see text)

Tool Box

Wiring kit

Black Magic Marker

Five-minute epoxy

Tin snips

Rotary cutting tool with carbide bit

Now that you have seen a simple IR opto-interrupter being used as a white/black detector, you can add this capability to your robot projects and give yourself a chance to compete against other robot developers. One of the most popular robot contests that you may want to enter is the "line follower," in which a robot is expected to follow a black line on a sheet of paper. With the material listed in the Parts Bin and the knowledge and skills you have developed, you can create a simple line-following robot.

To start off, you will need a line to follow. I am going to suggest that you make this first because it is useful to test the robot and its sensors as you build the circuit that I will present to you. Using a standard piece of Bristol Board and a Magic Marker, you can put down a track like the one shown in Figure 8-26.

When I lay out my tracks, I keep the line 4 inches (10 centimeters) from the edge of the board and round each corner with a radius of 3 inches

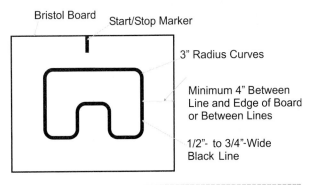

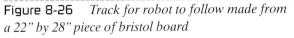

Figure 8-26   *Track for robot to follow made from a 22" by 28" piece of bristol board*

(7.5 centimeters). I try to make my lines ¹/₂ inch (1 centimeter) to ³/₄ inch (1.5 centimeter) wide. If you put on a "Start/Stop" mark, you might want to keep it away from any possible robot sensors, as they may be incorrectly identified as turns for the robot.

Once you have built your track, you will have to make a mounting plate for the cut-up opto-interrupters. In Figure 8-27, you can see that I started with a piece of aluminum gutter end cap (which I bought at a hardware store for $0.25) and cut it down so that it could be mounted on the DC Motor robot using the 1-inch (2.54 centimeter) standoffs.

I chose the aluminum gutter caps because of their low cost and the ease in which they can be formed. Looking around any "big box" hardware store will yield a number of different products that can be adapted for this purpose. I did not find a plastic product that worked as well, and beware of steel as it can be difficult to work with. Figure 8-28 shows the dimensions for the final piece of sheet metal, and I cut it out using a combination of a rotary (Dremel) tool and tin snips. Note that I just made a couple of tabs for the mounting plate to be held between the standoff and the finished plywood. I didn't put in drilled mounting holes because this would be extra work, and the tabs worked quite well.

With the mounting plate designed, I then took two opto-interrupters, cut them in half and 5-minute epoxied them 1 inch (2.54 centimeters) apart on the mounting plate. Once the glue had cured, I then added a 1k and 10k resistor to each and wired them together as shown in Figure 8-29. Once this was done,

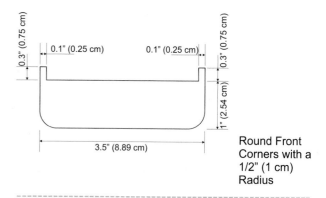

Figure 8-28 *IR sensor mounting plate made from mild aluminum*

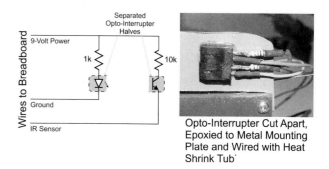

Figure 8-29 *Combined opto-interrupter white/black sensor circuit on mounting plate*

I put heat shrink tubing over the solder joints to make sure that nothing could short out. By doing this, I simplified the wiring that is passed to the breadboard to just three wires.

With the mounting plates built and the cut-up opto-interrupters glued to them and wired, I came up with the circuit shown in Figure 8-30 for the line-following robot. The two LEDs in the circuit are used to indicate when a sensor is over a black surface.

The theory behind the circuit is actually very simple; the motors are on and turning forward unless the sensor on that side detects the black line, in which case the motor turns off and the robot turns toward that side. This results in a slight wobbling motion as the robot goes forward, but allows the robot to negotiate the turns quite effectively.

The circuit itself probably seems quite complex, but it was built with two requirements in mind. The first was that LEDs had to light when a sensor detected a black line. This was actually very easy to implement by having the current pass by the

Figure 8-27 *Original aluminum rain gutter, a cut piece with opto-interrupters mounted to it, and the assembly mounted on a robot*

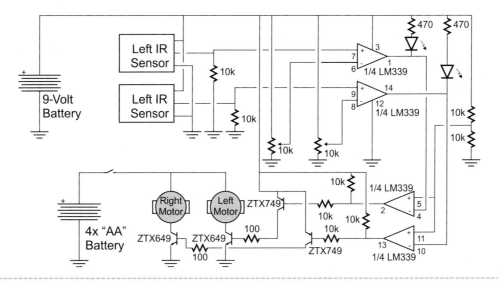

Figure 8-30 *Schematic diagram for DC motor-driven line-following robot*

phototransistors in the opto-interrupters to a 10k resistor; as the current increased, the voltage across the 10k resistor increased and could be compared using the LM339 comparator chip. The second requirement was that this circuit would have to work both on its own as well as under Parallax BASIC Stamp 2 control. This requirement resulted in this circuit, which can be powered two different ways with up to three different power supplies (9 volts from the battery, regulated 5 volts, and 6 volts for the motors), and it necessitated the use of the PNP transistors to pass current to the motor driver NPN transistors.

When you build this experiment, I recommend that you first get the LEDs to light when the opto-interrupter is over a black line. Because the phototransistor output voltage level can vary, I have

included a potentiometer that can be used to come up with the voltage transition between white and black. Once the sensors are working reliably, you can then wire the circuits to the other two comparators (which invert the signal from the first comparators) and then add the motor drivers. Remember to make sure that you have a switch on the four AA battery pack—this is an application where the robot will start running unpredictably.

The 100 Ω resistors used as the current-limiting resistors for the motor driver transistors were optimum for the motors that I chose; the robot should move at about 2 inches (5 centimeters) per second. You will have to experiment with different resistors to get the best performance with your motors and robot.

Audio Electronics

We take few things for granted more than "simple" audio electronics such as radios, stereo amplifiers, and CD and MP3 players. This lack of appreciation is unfortunate because not only do these devices have much of the same technology as high-performance computer systems to provide memory functions, but they also have advanced analog electronics that allow them to inexpensively provide high levels of analog power with very limited distortion. Even discounting digital technology provided for many of these devices, the science of audio electronics is quite advanced.

Audio electronics also go back more than 150 years to Samuel Morse's first telegraph message. Although the first experimental telegraphs wrote on paper tape (with dashes and dots printed as varying line lengths as the tape was drawn through the tele-graph), it became a practical instrument when a series of audible clicks were produced by a simple electromagnetic speaker that deforms a steel plate (or *diaphragm*) when current passes through the electromagnet, causing it to click.

The telegraph clicker was improved upon to pro-duce a speaker in which a specific voltage would cause a partial movement in the diaphragm. This *dynamic speaker* consists of a permanent magnet and a diaphragm that has a coil of wire built into it (Fig-ure 9-1). By rapidly changing the applied voltage, the position of the diaphragm changes and produces an audible sound. The first practical application for this device was the telephone.

Along with the dynamic speaker, several other devices convert voltages and currents to sound. If you have a set of headphones, then chances are you are listening to sound by use of a piezo-electric crystal-based speaker (Figure 9-2). A piezo-electric crystal produces a small but measurable change in its size when an electric current passes through it. This prop-erty is used to drive a speaker diaphragm.

The advantages of the crystal speaker over the dynamic speaker are its low cost, its responsiveness to small currents, and its robustness. The dynamic speaker is better suited for situations where high-powered signals are used to drive the speaker. If you are going to listen to AC/DC on your personal MP3 player, you are going to use headphones with crystal speakers built into them, but if you are going to the

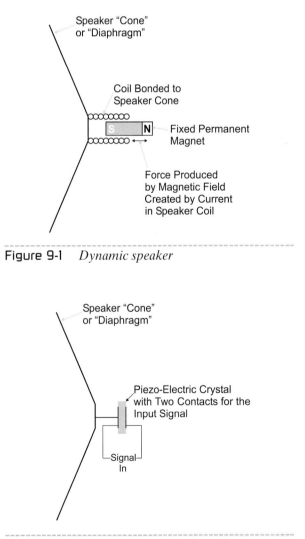

Figure 9-1 *Dynamic speaker*

Figure 9-2 *Piezo-electric crystal speaker*

concert, then the music from Angus Young's guitar will be driven out of (large) dynamic speakers.

The first practical microphone was invented in 1876 by Emile Berliner (the telegraph speaker made its first appearance in 1837) as a result of a demonstration from a telegrapher showing that the electrical current flowing from a telegraph key was based on the amount of pressure placed on the key. Berliner's microphone used carbon granules that have the property of changing resistance when the amount of pressure placed on them changes. His microphone consisted of a diaphragm connected to a plate pressing on carbon in a metal cup. Berliner had found that the resistance of this apparatus changed according to the noise applied to the diaphragm.

The change in electrical resistance of the carbon microphone was converted to a change in voltage using the circuit shown in Figure 9-3. The change in resistance was altered to a change in electrical current by applying a voltage to the microphone. This current change is amplified and converted to a voltage change by the use of a transformer.

The carbon microphone works quite well, but it is expensive (as well as dirty) to manufacture and will provide a low-level noise (or hiss) due to the carbon granules continually rubbing against each other. To get better performance, a number of different microphones have been invented. The first was the *dynamic microphone*, which works on the reversed principle of the dynamic speaker; rather than an applied voltage producing a sound, a sound produces an induced voltage by moving a coil in a magnetic field. The crystal microphone also works in reverse to the crystal speaker. An interesting experiment that you can try is to connect headphones into the microphone input of your stereo and see what happens when you talk into them. This is a useful "quick and dirty" *public address* (PA) system when you are at a party and you need to get people's attention.

Most modern microphones work with a completely different theory; rather than changing resistance or voltage based on the input sound, the value of a capacitor built into the microphone changes. This type of microphone is known as a *condenser micro-*

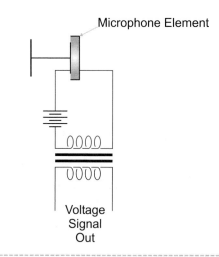

Figure 9-3   *Circuit to convert the changing resistance of the carbon microphone to voltage*

phone (condenser being the original name for capacitors), is very simple to manufacture, and offers good frequency response. Like the carbon microphone, the change in capacitance has to be converted into a change in voltage.

This circuit works well, but it is somewhat cumbersome to design as part of a complete system. The ideal would be the elimination of the power source and resistor. This is done by placing a permanently charged material (Teflon is good for this application) on one of the capacitor plates. Now, when the capacitor plates move, a small voltage is induced that can be amplified and used like the voltages produced by crystal or dynamic microphones. This type of microphone is called the electret (Figure 9-4) and is the most popular microphone used today.

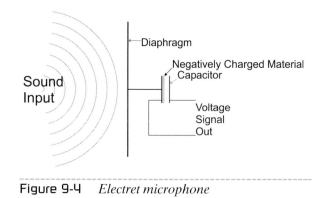

Figure 9-4   *Electret microphone*

Experiment 49
Buzzers

Parts Bin

Assembled PCB with
breadboard

3-20 volt DC buzzer

Breadboard-mountable
SPDT switch

Tool Box

Wiring kit

One of the few things that has never changed from when I was a kid is the basic electronic kit in which you can build a clanging bell that is driven by a single battery. At the time, the bells were used for almost all alarm applications, even though today they have been largely replaced by electronic alarms that are usually part of a public address system. The basic design for the bell is shown in Figure 9-5, and it is built using an electromagnet and a spring, both of which push against each other. When the spring is holding the clapper away from the bell, there is a closed circuit and the electromagnet is active. The electromagnet pulls the iron bar (along with the clapper) toward it. When the iron bar starts moving, the circuit is broken.

The process of the electromagnet pulling the clapper toward the bell, followed by the spring pulling it away, will repeat indefinitely while electrical power is applied to the bell mechanism. Practically speaking, the contact points and the copper contact of the bell mechanism (identified in Figure 9-5) will have to be changed and adjusted periodically as each time the bell rings, the contacts will become oxidized and worn by repeated action. To be fair, electrical bells have been around for more than 150 years, so they are well understood, and the maintenance required for them is minimal.

Although electric bells work well, they are not practical for many electronic projects. The reasons why they are not practical are due to the amount of space they take up, the relatively low frequencies of sound output the large amounts of current draw, and the effects they can have on a circuit. The electromagnet switching on and off can cause large voltage spikes in the circuit's power supply. For the past 50 years or so, most smaller circuits that require sound output either have an oscillator built into them that

Figure 9-5   *Electromechanical bell*

drives a speaker or have a buzzer (Figure 9-6). Buzzers generally require 50 mA or more current to operate and should be driven by a circuit such as the one shown in Figure 9-7.

When I looked at reference information on buzzers, I found the curious line that buzzers work exactly the same way as bells. This is true to the point that a part in a buzzer, just as in a bell, moves when current passes through it, and then the part returns to the original state when the current stops; however, everything else about the two devices is completely different. The buzzer does not include any magnet components as the bell does; instead, it takes advantage of the characteristics of a piece of piezo-electric material.

When used in a buzzer, the piezo-electric material is held at one end with an electrical connection while at the other end, a piece of copper is placed in contact with the material. When current flows through the piezo-electric material, it deforms and the connection to the contact is lost; this causes the material to return to its original form, which closes the connection again and the process repeats.

Internal Assembly

External Appearance

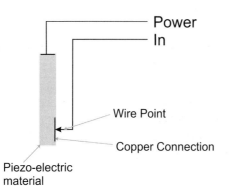

Figure 9-6   *Buzzer assembly and appearance*

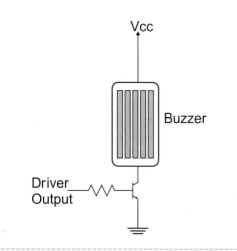

Figure 9-7   *Circuit to drive buzzer from low-current driver*

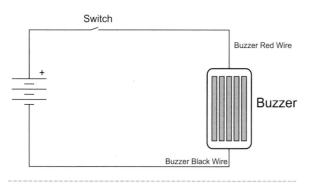

Figure 9-8   *Buzzer schematic*

You can test the operation of a buzzer using the simple circuit shown in Figure 9-8. The switch in the circuit will turn the buzzer on and off.

I would like to emphasize that you should *never* place your ear directly against the buzzer when it is operating. Along with this warning, you should make sure that you do *not* buy a buzzer that is rated at over 75 decibels (dB). The dB range is a measure of sound pressure, or the loudness of sound coming out of the buzzer.

When something is producing sound, the sound waves move away from it in a spherical pattern (which should not be a surprise). What may be a surprise for you is that as the sound waves move away from the source, their power decreases as the square of distance rather than linearly with distance. This statement may be hard to understand and conceptualize.

When you move away from an object, you will find that its apparent size changes with how far away you are from the object. In Figure 9-9, I show that as a robot gets farther away from your eye, its size is proportional to the distance it is from your eye. To prove this, place an object a certain distance from your eye and an identical object twice the distance from your eye. The further object will appear to be half the size of the closer one. If you put an object three times the distance of the closest, it would appear to be one-third the size. Put mathematically, you could say that the apparent size of the object could be defined using the formula:

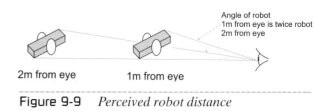

Figure 9-9   *Perceived robot distance*

$$RobotSize = RobotSize@1meter \times distance$$

"Apparent size" is a subjective term, and when it is used, you are only thinking in terms of one dimension (the object looks like it is half the size because it is half as tall). You must remember that there are two dimensions and *both* are halved, which means the actual area of the object being viewed is one-quarter when it is twice as far away.

When you are discussing sound or light, the opposite phenomenon occurs. After traveling twice the distance, the sound, is spread out over four times the area because the same amount of sound energy is now spread out over twice the distance in the x and the y directions. Put mathematically, you can state that sound pressure is defined as

$$SoundPressure = SoundPressure@1meter \times (distance^2)$$

When you get closer to the sound source, the sound pressure increases according to the same square law, and this is why I emphatically state that you should not place your ear any closer to the buzzer than a few feet. When the buzzer is close to your ear, the sound pressure applied to your ear will very quickly get to the level where your ear (and its hearing) will be damaged.

Experiment 50
Basic Transistor Oscillator Code Practice Tool

Parts Bin

Assembled PCB with breadboard

Three ZTX649 NPN bipolar transistors

8/16 Ω speaker

Momentary on switch/Morse code key (see text)

Two 1.5k resistors

4.7k resistor

Two 470k resistors

Two 0.01 μF capacitors, any type

Tool Box

Wiring kit

When I was growing up, all educational and hobbyist circuits were built from individual transistors—it wasn't until the mid- to late 1970s that chips such as the 555, LM339, LM386, and LM741 started to be commonly used as building blocks for these circuits. These chips are all very configurable, but none offer the range of operation and low cost of discrete transistors. In this experiment, I could have introduced you to a simple oscillator driving a speaker designed to be used for learning Morse code with a 555 timer, but I want to go back in time and show you how this could be done with a few transistors, resistors, and capacitors.

The oscillator circuit that I am going to work with is the basic relaxation oscillator circuit shown in Figure 9-10. Included in Figure 9-10 are the formulas for when the output is high and low, as well as an important formula indicating that the value of R1 and R2 (the time-defining resistors) must be the transistor h_{FE} multiplied by the value of the pull-up resistors (Rpu). If R1 or R2 is less than this product, then you will find that the oscillator will not start reliably and not run at a constant frequency.

The transistor relaxation circuit that I came up with is shown in Figure 9-11. If you apply the formulas given in Figure 9-10, you will discover that the

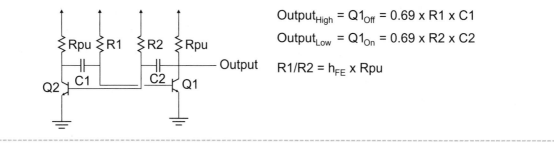

$$\text{Output}_{High} = Q1_{Off} = 0.69 \times R1 \times C1$$

$$\text{Output}_{Low} = Q1_{On} = 0.69 \times R2 \times C2$$

$$R1/R2 = h_{FE} \times Rpu$$

Figure 9-10   *NPN transistor-based astable oscillator*

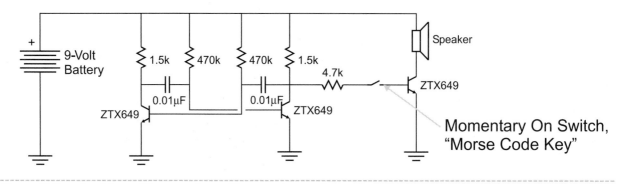

Figure 9-11   *Astable oscillator used as a Morse code practice tool*

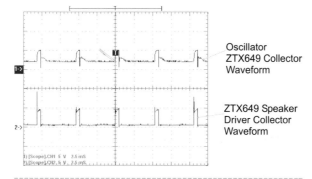

Oscillator ZTX649 Collector Waveform

ZTX649 Speaker Driver Collector Waveform

Figure 9-12   *Astable oscillator circuit waveforms*

frequency of the oscillator is 154 Hz. Looking at an oscilloscope picture of the actual oscillator signal (Figure 9-12), you will see that the signal's period is fairly close to this (182 Hz), although its appearance is probably quite a bit different from what you would expect (it is not a square or smoothly rounded wave), and it only activates the speaker transistor for a very short time, not 50 percent of the waveform as you would expect. This is normal, and if a square wave or a sine wave is required, then another type of oscillator would be used, or this signal would be filtered and modified to produced the desired signal.

Morse code has not remained static since Samuel Morse's first message ("What Hath God Wrought").

Table 9-1   International code standard

Letter	Morse Code	Letter/ Character	Morse Code
A	.-	X	-..-
B	-...	Y	-.--
C	-.-.	Z	--..
D	-..	0 (Zero)	-----
E	.	1	.----
F	..-.	2	..---
G	--.	3	...--
H	....	4	....-
I	..	5	.....
J	.---	6	-....
K	-.-	7	--...
L	.-..	8	---..
M	--	9	----.
N	-.		
O	---	. (Period)	.-.-.-
P	.--.	, (Comma)	--..--
Q	--.-	? (Question Mark)	..--..
R	.-.	; (Semicolon)	-.-.-.
S	...	: (Colon)	---...
T	-	/ (Slash)	-..-.
U	..-	- (Dash)	-....-
V	...-	' (Apostrophe)	.----.
W	.--	_ (Underline)	..--.-

Today, the international standard for code is listed in Table 9-1; a short tone (a "di") is indicated by a period (.) and a long tone ("dah") is indicated by a dash (-). Using the momentary on button (or, if you want to practice Morse code seriously, you can buy a key from either a surplus store or a "ham" supply retailer. Along with a key, many books and tapes can help you learn and apply Morse code.)

When you are practicing, you might want to start off with messages such as the following:

-- -.-- -.- . . .- - .... .- --- .-. -- .... .-.-.-

Experiment 51
Electronic Stethoscope

Parts Bin

Assembled PCB with breadboard

LM386 8-pin audio amplifier

Electret microphone (see text)

Plastic straw (see text)

10k breadboard-mountable potentiometer

10k resistor

10 Ω resistor

Two 0.01 µF capacitors, any type

220 µF electrolytic capacitor

8/16 Ω speaker

Tool Box

Wiring kit

Rubber cement

Scissors

The genesis of this experiment was trying to figure out where a rattle was located in the "wall-following robot" presented later in the book. When the robot's left wheel was running, I heard a funny noise that I couldn't find by running the robot and trying to find it by ear. The solution was to go back to an old mechanic's trick for finding the source of a funny noise in a car's engine. In case you've never seen a mechanic find the source of a noise under the hood of a car, they do it by using a length of garden hose. The hose directs the sound coming in to it along its length and to the ear of the mechanic. It's probably the simplest diagnostic tool used with modern cars, but often the most effective for situations such as isolating a knocking valve.

To find the problem (a motor with a piece of metal rattling around inside it), I used the same principle as the garden hose to the car motor trick, but scaled down for the small robot. You shouldn't be surprised that a garden hose would be too large for finding a problem with a small part, but a soda straw cut down to 3 inches (7.62 cm) in length works quite well. To avoid the potential embarrassment of having somebody walk in on me while having a short piece of soda straw stuck in my ear and holding up a running robot, I decided to come up with an amplifier circuit that would pass the sound coming through the straw to a microphone and then to a speaker. The finished "electronic stethoscope" block diagram is shown in Figure 9-13.

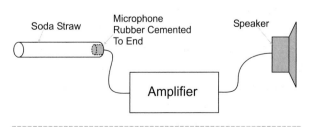
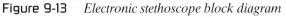

Figure 9-13 *Electronic stethoscope block diagram*

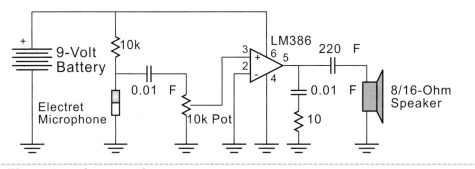

Figure 9-14 *Electronic stethoscope schematic*

The most difficult aspect of this project was finding an electret microphone that fit inside a soda straw. I did find a fairly narrow microphone at an electronics store and used a large shake straw from a fast-food restaurant. After cutting down the soda straw, I glued the microphone into it, taking care to not get any glue on the face of the microphone. Once that was done, I built the circuit shown in Figure 9-14.

The basis for the stethoscope is the LM386 (Figure 9-15). This chip is a general purpose differential high-current output driver with a 20 to 200 times gain. It is very often used for applications like this one, taking a "small signal" source and boosting it. It can also be used for a variety of other applications, including driving very small motors—instead of driving it with PWM, an analog voltage could be used with the amplifier wired as a voltage follower, as shown in Figure 9-16.

Like the 555 and some other chips that became commonly used in the 1970s, the LM386's parameters can be modified quite a bit. One of the most popular modifications is changing the gain (how much the amplifier multiplies the signal) of the chip. This is done by adding a 10µF capacitor and a resistor to the two gain pins as shown in Figure 9-17. Using the formula included in Figure 9-17, you can choose a resistor value to change the gain upwards to 200 (from the nominal 20), which means the signal output will have an amplitude 200 times greater than the input.

This level of amplification should only be used for signals that are very small; an input signal of 1 volt, even with a gain of 20 times, will be "clipped" by the two transistor driver. This driver (known as a "totem pole" driver) can only drive the output from ground to the power input to the LM386, and if the signal would go outside these limits, it is just clipped. Some-

times clipping is used in recording to get a unique sound, but for the most part, it is considered to be distortion and should be avoided at all costs.

When you build this experiment's circuit, you may discover that it wails when the speaker is brought

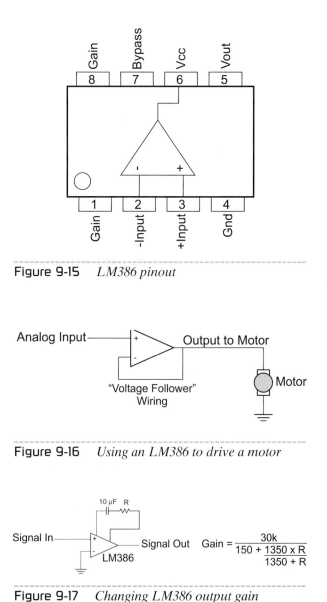

Figure 9-15 *LM386 pinout*

Figure 9-16 *Using an LM386 to drive a motor*

Figure 9-17 *Changing LM386 output gain*

close to the microphone or if you disconnect the 10k potentiometer's connection to ground. This is caused by positive feedback (Figure 9-18) in which the output from the speaker is picked up by the microphone, passed to the amplifier, and output from the speaker in a large loop. Positive feedback, such as clipping, is to be avoided whenever possible, and this is a secondary reason for using the soda straw with the microphone; it helps keep sound that is not directly in front of the microphone from being picked up and amplified.

You may have some confusion about the term "feedback" because you will hear it used in many different places. Negative feedback is often used in mechanical/electronic systems (such as an R/C servo) because it is subtracted (added negatively) from the actual position of the servo arm relative to the specified position, and if there is a difference, this arm is moved to better match the specified position. Positive feedback, such as what you get with this circuit, does not negatively add the output to the input, but adds it positively, so a small signal is amplified, picked up, and amplified repeatedly until the amplifier is in saturation and all other inputs are lost in the wail.

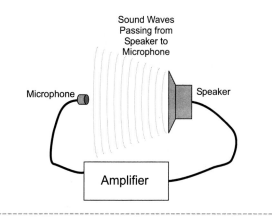

Figure 9-18 *Positive feedback in an amplifier circuit causing screeching sound*

The last feature that you should be aware of with the LM386 is that the output frequency response can be modified if you would like to boost the bass (lower) frequencies. This is shown in the chip's datasheet (which you can download from the Internet) along with a number of other possible applications, including using the LM386 as an oscillator or as part of an AM radio.

Parts Bin

Assembled PCB and breadboard

LM386 8-pin audio amplifier

LM339 quad comparator

Four bright LEDs (see text)

Electret microphone

100k breadboard-mountable potentiometer

Three 10k resistors

Nine 1k resistors

10 Ω resistor

10 µF electrolytic capacitor

Three 0.01 µF capacitors, any type

Tool Box

Wiring kit

The image of a robot that usually comes to mind is a hulking man-like machine that wreaks destruction based on the (usually verbal) commands of its creator. These commands are spoken normally, although they may be a bit louder and stiffer than normal speech and prefaced by the robot's name, to make sure it understands that the following command is for it. These commands are rarely misunderstood, and although some may be executed too literally, they are always performed with mindless efficiency. In this experiment, I would like to look at what basic speech becomes after it has been converted to electrical signals by a microphone, as well as the tools that are required to display the electrical signals.

In the previous experiments, I introduced you to the LM386 audio amplifier. This chip converts the small electrical signals from the microphone into a power signal that could be used to drive a speaker. These signals are generally displayed using a tool called an oscilloscope, which displays changing voltage levels over time as I show in Figure 9-19.

If you were to look inside a very primitive oscilloscope, you would see something like Figure 9-20. The electron gun used by the CRT would be continuously active and moving across the screen, left to right and then snapping back to the left and starting over. This movement of the electron beam is performed by a set of electrostatic deflection plates that are driven by a sweep generator. The sweep generator produces a sawtooth wave that is passed to the deflection plates and causes the beam to move from left to right.

This is a very cursory introduction to oscilloscopes, but it should help you to understand what is being shown when I put in an oscilloscope screen shot in this book. Oscilloscopes tend to be quite expensive ($500 and up). A very basic and inexpensive oscilloscope will give you the ability to observe one or two changing signals at speeds up to a few MHz. More expensive oscilloscopes (costing well over $100,000) will have more than two inputs and can observe signals in the GHz (microwave) frequencies.

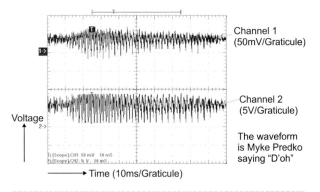

Channel 1 (50mV/Graticule)

Channel 2 (5V/Graticule)

The waveform is Myke Predko saying "D'oh"

Voltage

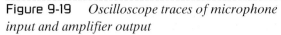

Time (10ms/Graticule)

Figure 9-19 *Oscilloscope traces of microphone input and amplifier output*

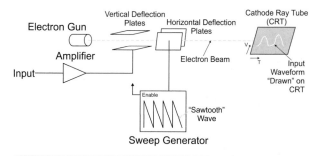

Figure 9-20   *Basic oscilloscope internal block diagram*

Rather than requiring you to purchase an oscilloscope, in this experiment I would like to give you a simple circuit that can be used to observe simple electrical signals, such as audio input. This circuit (Figure 9-21) converts small audio signals to signals that can be compared. LEDs can then be lit depending on the output level of the sound. I only provide output for four LEDs to fit on the breadboard that is built on the PCB that comes with the book.

You should be aware of a few issues when you build this circuit. First, when the circuit is operating you will find that the LEDs will turn on for very short intervals for loud noises and spoken words. To make their operation more noticeable, you should use the brightest LEDs you can find. If you look around the Internet, you will find some circuits that hold the voltage for a few hundred milliseconds in order for the LEDs to be active for a few seconds. Second, looking at the circuit, you'll see that I had to increase the gain of the LM386 by wiring the 100k potentiometer to the LM386 in such a manner that the gain would be increased over the 200 times normally produced by the capacitor across the gain pins. The actual gain of the circuit is about 5,000 times to increase the output to the point where the LEDs will indicate different levels of sound inputs. You may find that you will have to change the resistor and potentiometer values of the components connected to the LM386 for the circuit to work with the microphone that you use. Finally, a number of chips can be purchased that will perform the same function as this circuit and are very easy to wire into a circuit. I decided to build the circuit out of discrete parts to give you a better understanding of how audio signals are converted to electrical signals and how comparators work.

The LM339 (Figure 9-22) contains four open collector output voltage comparators. In this

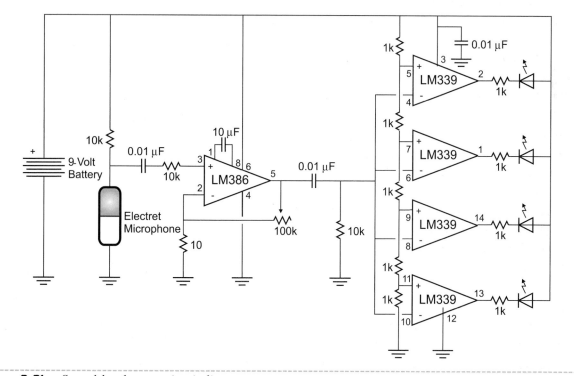

Figure 9-21   *Sound-level-meter circuit diagram*

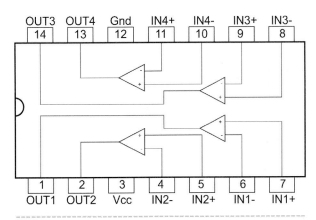

Figure 9-22 *LM339 quad comparator with open collector outputs*

experiment's circuit, I have created a resistor voltage divider "ladder" that provides a stepped set of voltages that can be used to compare against the sound-level input to the LM386. This circuit is very similar to the circuit that is built into your stereo to display the current audio output using a number of LEDs. The number of LEDs that are used to display the sound level can be increased by simply adding another LM339 chip along with four more 10k resis-

tors to provide additional voltage level "steps" for the comparator.

When sound hits the microphone, it produces a small voltage when the microphone is wired in a voltage divider as I have shown in Figure 9-23. The signal is passed through a capacitor and resistor and then amplified using the LM386. Before the final signal is passed to the LM339 comparator, it is zero based using the 0.01 µF capacitor and 10k resistor. The final voltage divider acts as a volume control and makes sure that the output signal from the LM386 is not clipped. Figure 9-23 shows both the different parts of the input circuit as well as the waveforms (displayed on my oscilloscope) that are passed along.

To get an idea of the different magnitudes of the signals (and understand what "small signal" means), note that at the bottom of Figure 9-23, the voltage levels for each signal are displayed. This voltage level is defined as the voltage range between *graticules* on the oscilloscope screen, which are the dotted lines on the oscilloscope picture (as well as any other oscilloscope pictures shown in this book).

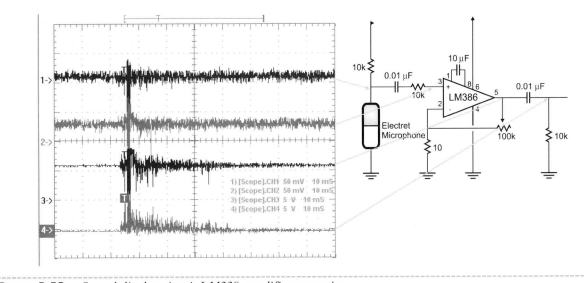
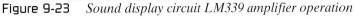

Figure 9-23 *Sound display circuit LM339 amplifier operation*

Digital Logic

The mathematics of logic came from the need to understand philosophical statements such as the following:

> Now some of these perceptions are so transparently clear and at the same time so simple that we cannot ever think them without *believing them to be true*. The fact that I exist so long as I am thinking, or that what is done cannot be undone, are examples of truth in respect of which we manifestly possess this kind of certainty. For we cannot doubt them unless we think them; but we cannot think them without at the same time *believing they are true*, as was supposed. Hence we cannot doubt them without at the same time *believing they are true; that is*, we can never doubt them.
>
> —*The Philosophical Writings of Descartes*, Volumes I and II

It may be possible to understand what Descartes is trying to say in the propositions presented in this quote by reading through it, but I am at close to a complete loss at what exactly is being said despite having spent many minutes of studying the text and trying to break it down into some simple statements. Part of the problem is that this quote is quite confusing with what seems to be extraneous statements as well as seemingly poorly constructed phrases. If you, like me, have trouble understanding what Descartes is trying to put forward here, don't worry; we are in good company because others for many hundreds of years have been trying to easily and clearly understand what a philosopher is trying to say.

One of the people who was instrumental in helping philosophers figure out exactly what other philosophers were saying was a school teacher by the name of George Boole. Boole worked to define a method of converting a written proposition (such as "All dogs have two ears and fur") into a series of mathematical statements in order to better understand them and test their validity. By restating a text proposition as a mathematical equation and making the assumption that all propositions were either true or false, he came up with a simple set of rules that became the basis for digital electronics and computer systems, as well as helped people understand the ramblings of philosophers like Descartes.

If I had an old mongrel that had gotten into a fight and lost an ear, he could not be considered to be a dog using the logic statements used to define a dog presented previously:

1. My pet has fur. (true)
2. My pet has one ear. (true)
3. My pet is a dog. (false)

The last statement is false because the original proposition was that all dogs have two ears and fur. Any animal that doesn't meet these criteria could not truthfully be called a dog. You might be thinking that by using the proposition "all dogs have two ears and fur" as the test, you could claim that a cat or a rabbit was a dog, and you would be right.

What we are interested in is the idea of applying mathematical principles to these statements. If we consider that something can either be true or false, then we can extend this knowledge, along with other pieces of knowledge, to test to see if something is true. In the example of the "dog" above, I have made the proposition that an animal is a dog if it has two ears *and* fur. If you had an animal with no fur, but two ears, it could not be considered a dog, just as if you had an animal that had three ears and no fur. Putting this into table format (known as a *truth table*), we could write out this proposition in a format that is easy to understand, as shown in Table 10-1.

In digital electronics, instead of using "true" and "false," the terms "1" and "0" or "high" and "low" are used. The "AND" truth table written as Table 10-1 could be restated in these terms using the labels "A" for "Two Ears" and "B" for "Fur," as in Tables 10-2 and 10-3. Tables 10-1 through 10-3 are examples of the "AND" function; the result of which is true if and only if both inputs are true.

When I have written out the different values for "A" and "B," notice that I have done it in such a way that only one of the two values changes at any one time. I do this because as the logical operations become more complex, the need for simplifying them becomes more important, and only changing one value at a time allows relationships between the values to become much more obvious.

Along with the "AND" operation, Boole found that two values in an "OR" statement would be true (such as "A cat is an animal with claws or a bad personality") if either or both of the conditions were true. Using the truth table format I used for the "AND," the "OR" is written out as shown in Table 10-4.

In modern colloquial English, we tend to think of "OR" as either one value or another but not both for a situation to be true. This is not an accurate way of thinking of the "OR;" if both values are true (or 1 or high) then the result will be true.

The last basic logical operation that Boole discovered was the "NOT" or negation. This operation is

Table 10-1 "Truth table" used to define a "dog"

Two Ears	Fur	Dog?
False	False	False
False	True	False
True	True	True
True	False	False

Table 10-2 Logic "AND" truth table

A	B	A and B
0	0	0
0	1	0
1	1	1
1	0	0

Table 10-3 Voltage-level "AND" gate input/output truth table

A	B	A and B
Low	Low	Low
Low	High	Low
High	High	High
High	Low	Low

Table 10-4 Logic "OR" truth table

A	B	A OR B
0	0	0
0	1	1
1	1	1
1	0	1

true if its value is false (as in "A car does not have six wheels"). "NOT" is different from "AND" or "OR" because it only has one input value (called a parameter) to test. The "NOT" truth table is shown in Table 10-5.

Boole was able to combine these three basic operations to perform very complex operations (like testing the propositions made in the paragraph quoted at the start of this section). Many years after the publication of these rules in *An Investigation of the Laws of Thought* (1854), they were discovered by computer architects and designers when developing the first electronic computers, which could sense high- and low-voltage levels cheaply and efficiently as well as perform simple operations on them.

Boole's rules became the basis for operations within the computers and became known as Boolean arithmetic. In the following experiments, I will introduce you to many of the intricacies of Boolean arithmetic, along with the basic circuits that implement them.

Table 10-5 The NOT truth table

A	Not A
0	1
1	0

Experiment 53
Basic Gate Operation

Parts Bin

Assembled PCB with
battery

Two SPDT switches

Two 1k resistors

470 Ω resistor

Three LEDs, any color

74C08, CMOS AND gate

74C32, CMOS OR gate

74C04, CMOS NOT gate

0.01 μF Capacitor (any type)

Tool Box

Wiring kit

I think the best way to understand the operation of digital logic is to actually build some circuits to test them out. The circuits presented in this experiment can be built in just a few seconds, and by using LEDS, you will see how the outputs change based on the value of the inputs. Although modern *Transistor-to-Transistor Logic* (TTL) logic chips are very easy to work with, they require a 5-volt regulated power supply. To avoid the need for a separate power supply, the circuits presented in this section will use 74Cxx logic, which can be run from 3 volts to 15 volts and is ideally suited for a 9 Volt alkaline radio battery.

TTL is a bipolar transistor-based logic technology and first invented in the mid-1960s, remaining popular ever since. The chips themselves offer fast performance (an 8-nanosecond transition time or less) and reasonable output current capabilities (approximately 20 mA). The logic functions built into the 74xx chips are called "gates" because they gate output signals based on their inputs.

The 74Cxx logic chips have the same pinout as standard TTL chips (which are presented at the end of this section), can run over a wide power input, and can source enough current to light an LED. They do not operate at the same speed as TTL and cannot source or sink as much current as TTL. For the most part 74Cxx can be put in any application that uses standard TTL, as long as you remember that *Complementary Metal Oxide Semiconductor* (CMOS) logic is voltage based, whereas TTL logic is current based. I

will discuss this important difference in more detail later in this section.

If you are familiar with electronics, you might be surprised that I did not use the 40xx series of CMOS parts. The 40xx series is the most popular CMOS logic family, but it does not have enough current drive capabilities to light an LED. If you wanted to drive an LED from it, you would have to use a bipolar transistor to amplify the output current, as I have shown in Figure 10-1, which allows current to flow through the LED when current is output from the CMOS gate to the transistor's base.

To change the voltage at the input of the logic chips, I used the circuit on the left-hand side of Figure

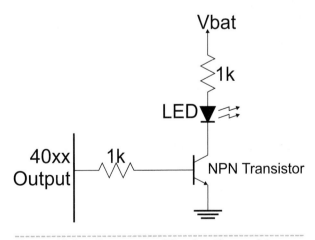

Figure 10-1   *Current amplifier for 40xx to drive an LED*

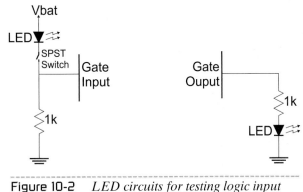

Figure 10-2   *LED circuits for testing logic input and output*

Figure 10-3   *AND gate*

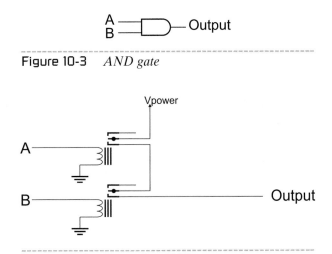

Figure 10-4   *Two relay AND gate analog*

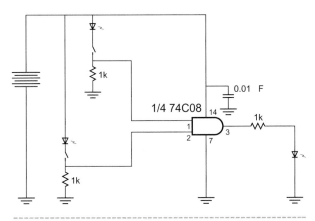

Figure 10-5   *Circuit to test operation of AND gate*

10-2. In this circuit, when the switch is closed, power passes through the LED down through the 1k resistor. The voltage at the switch/1k resistor connection will be equal to the battery voltage minus the voltage across the LED (which will always be high enough for the gate's input to recognize it as a "high" voltage). If the switch is open, the gate will be connected to ground through the 1k resistor and the LED will not be lighted. By using this circuit, a "high" voltage (or a "1") will light the LED and a "low" voltage (or a "0") will not have the LED lit. The output circuit (the right side of Figure 10-2) will light when the voltage output is "high" and not light the LED when the output is "low." These circuits will allow you to easily track the operation of the logic circuit without the need of a *digital multimeter* (DMM) or other test instruments.

The most obvious logic function or gate is "AND" (the schematic symbol is shown in Figure 10-3), which only passes an electrical signal when both inputs are "high." Figure 10-4 shows a possible circuit for an AND gate using two relays; when both inputs cause the relays to close, Vpower is output. To test the operation of the AND gate, use the circuit shown in Figure 10-5. In this drawing, I have used the convention of indicating the chip's power inputs on one of its gates.

When you have the circuit built, move the switches back and forth to test out the function of the AND gate. Using the knowledge that when the LED is on, the corresponding input is high, you can create a table like Table 10-6 to investigate the operation of the AND gate; you should discover it is the same as the introduction to this section.

When you are comfortable with the operation of the AND gate, you can look at the operation of the

Table 10-6 Truth table for testing AND and OR gates

Pin 1 Voltage	Pin 2 Voltage	Pin 3 (Output) Voltage
Low (LED off)	Low (LED off)	
Low (LED off)	High (LED on)	
High (LED on)	High (LED on)	
High (LED on)	Low (LED off)	

Figure 10-6   *OR gate*

Experiment 53 — Basic Gate Operation

OR gate (the schematic symbol shown in Figure 10-6), which outputs a high voltage if either input is high. The circuit for testing the OR gate is shown in Figure 10-7, and the breadboard wiring is identical to the one used for the AND gate.

The last basic gate is the NOT (the schematic symbol shown in Figure 10-8) and inverts the logical signal from a high to a low. The NOT gate is different from the AND and OR gates as it only has one input.

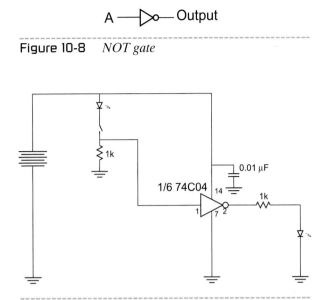

A ———▷○— Output

Figure 10-8   *NOT gate*

Figure 10-9   *Circuit to test operation of NOT gate*

Build the NOT test circuit shown in Figure 10-9 and once you have done this, create a three-row by two-column table, recording the output values for the inputs.

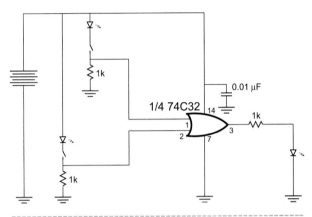

Figure 10-7   *Circuit to test operation of OR gate*

Experiment 54
CMOS Touch Switch

Parts Bin

Assembled PCB and
   breadboard
24-gauge solid core wire
74C04
10M resistor
1.5M resistor
1k resistor
LED, any color
0.01 µF Capacitor, any
   type

Tool Box

Wiring kit
Wire strippers

It is a very common misconception that digital logic always uses voltage as its input. Often it is assumed that digital inputs are keyed by voltage, but this isn't always the case. Voltage is used to check the logic levels, but, as I will show in this experiment and the next one, digital logic can be keyed by both voltage and current with the most popular logic families being controlled by current and not voltage.

When I was a teenager, it wasn't unusual for products to be built with touch switches where the user

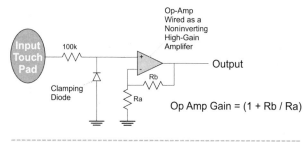

Figure 10-10   *Simple op-amp-based touch switch*

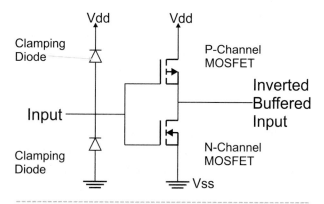

Figure 10-11   *Basic CMOS inverter circuit*

simply touched a metal pad on an electronic device that performed some kind of action. Figure 10-10 shows the original touch switch circuit; when a user would touch the metal pad, the 115-volt AC present in the room that is induced within a person is amplified by the operational amplifier and passed to other circuitry.

The diode in the circuit is used to clamp the input voltage and make sure that only the positive component of the signal is passed to the amplifier. The amplifier is wired in what is known as a *noninverting amplifying* configuration in which the amplification factor (known as *gain*) is defined by the formula on Figure 10-10. The output of the amplifier is fed back to the negative input of the op-amp through the voltage divider and amplifies any signal passed to the positive input. The gain of the circuit should be at 200 or more to ensure that the output will either be nothing when the switch isn't being touched, or will be at the positive voltage level when the switch is being touched and the voltage of the person's body is being passed to the op-amp.

CMOS logic *is* controlled by voltage and can be used to simulate the behavior of a high-gain ampli-

fier, such as the operational amplifier in Figure 10-10. The CMOS NOT (inverter) logic is shown in Figure 10-11.

The CMOS NOT gate consists of two clamping diodes that ensure the voltage input level does not exceed the power supply ranges (and potentially damage the transistors within the chip). N-channel and P-channel MOSFET transistors used in CMOS logic are very sensitive to static shocks (also known as electrostatic discharge or ESD), and the two diodes help prevent damaging voltages from being passed to the transistors.

The primary difference between MOSFETs and bipolar transistors is that MOSFETS are voltage (not current) controlled. MOSFET-based logic is cheaper to manufacture due to the elimination of resistors in its logic and requires fewer manufacturing steps.

The operation of the two transistors is shown in Figure 10-12. When a high voltage is input to the logic gate, the n-channel MOSFET connected to ground

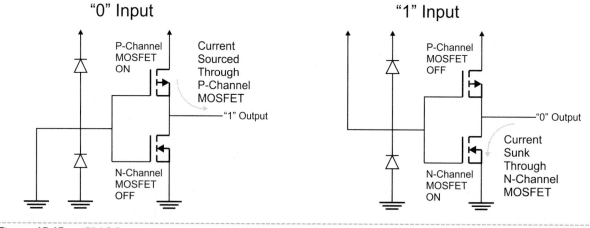

Figure 10-12   *CMOS inverter operation for different inputs*

turns on and pulls the output to ground (Vss or CMOS logic). When the input is a low voltage, the n-channel MOSFET is turned off and the p-Channel MOSFET is turned on, tying the output to the chip's power input (Vdd).

To demonstrate the operation of the CMOS inverter and how it can be controlled by just voltage, I created the circuit shown in Figure 10-13. The input to the circuit's inverter is held low by the 10M resistor tied to ground, but will change state when the input is connected to a voltage source.

When you wire the circuit, make sure that you place two pieces of stripped wire side by side as shown in the wiring diagram (Figure 10-14). These two bare wires are your touch switch; when you touch both of them, you will complete a very low current circuit and the input of the CMOS gate will be pulled to a high voltage, which will turn off the LED. Removing your finger will allow the gate to be pulled down to ground again (via the 10M resistor) and the LED will turn on.

You might want to try removing the bared wire that is connected to Vdd to see if just the induced voltage in your body would cause the inverter to change state. If you are in a room with fluorescent lighting,

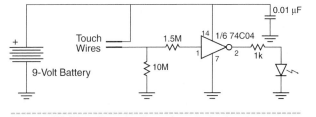

Figure 10-13   *Practical touch switch circuit*

there is a good chance that it will work without any problems. If you are in a room with incandescent wiring, which does not radiate as much energy, you may find that touching the switch will not turn off the LED; this is why I included the connection to Vdd. Your skin normally has a resistance of around 1.5k, so when you touch both the wire connected to Vdd and the 1.5M/10M resistors, you are actually creating a voltage divider circuit with the voltage at the inverter input at almost Vdd. If you do the math, you'll discover that the voltage being passed to the CMOS gate will be around 90 percent of the Vdd voltage.

You may find that the LED lights unpredictably when you touch the bared input wire or simply wave your hand over the chip. In this circuit, there may be some induced voltages from your body that are occasionally high enough that the CMOS inputs recognize

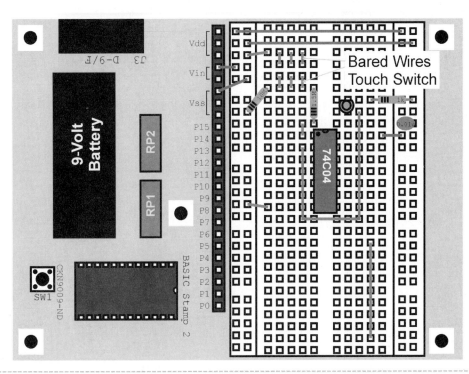

Figure 10-14   *Wiring for CMOS touch switch. Note the bared wires for the +5-volt connection and the touch sensor.*

them as high inputs. CMOS gates work very similarly to very high gain amplifiers (like the operational amplifier circuit shown in Figure 10-10) and for this reason, every unused CMOS input must be tied either to Vdd or Vss in your circuit, or you could find unexpected responses in different situations.

Please do not attempt to "charge up" your body before touching the switch by shuffling on a synthetic carpet or by touching live wires in your home. In the first case, you will be passing a massive surge of energy to the gates, and there is a very high likelihood that you will damage them, despite the clamping diodes on the input.

Experiment 55
Bipolar Transistor-Based TTL "NOT" Gate

Parts Bin

Assembled PCB with breadboard

Four ZTX649 transistors

Two 1N914/1N4148 silicon diodes

1k resistor

150 Ω resistor

2.2k resistor

1.5k resistor

4.7k resistor

Two 100k resistors

10k PCB-mountable potentiometer

SPDT switch

Tool Box

Wiring kit

Wire strippers

DMM

The most popular form of digital logic that most people start working with is called "Transistor to Transistor Logic" (TTL). TTL is created from NPN bipolar transistors and resistors that are built on a silicon chip. Unlike CMOS logic, TTL is current controlled, which gives it some different operating characteristics that I will demonstrate in this experiment.

In the section dealing with optoelectronics, I created a number of different logic functions using the resistor/NPN transistor inverter. You may have confused this with a TTL logic gate, but it actually isn't one; this circuit does work as an inverter and multiples can be combined to perform specific logic functions, but it is not a TTL gate. This gate is actually a *Resistor-to-Transistor Logic* (RTL) gate, which is a technology that is a precursor to TTL.

RTL can be used for logic functions (as I demonstrated in the earlier section), but it is not as robust as

TTL. TTL sought to rectify two problems with the RTL gate. The first problem is the limited amount of current that can pass through the resistor connected to the transistor's collector. A small value resistor can be used, but if the transistor is on, a large amount of current is passed through the resistor to ground, wasting a significant amount of power.

The second problem is the use of the current-limiting resistor between power (called Vcc for TTL) and the transistor's collector. Going back to the 555 timer that I introduced earlier in the book, you should recognize that the resistor will be part of the RC network used to time the operation of the 555, and signals will take some time to rise and fall. This delayed rise and fall slows down the time in which signals can be sent, lessening the usefulness of RTL logic.

A TTL inverter (Figure 10-15), although it looks more complex, is actually a very elegant circuit as I will explain in this experiment. In fact, it has some features that make it easier to use than the CMOS logic chips I primarily use in this section.

The first feature that I would like to bring to your attention is that TTL is *not* voltage controlled the same way that CMOS logic is. A TTL is active when current is drawn from it. The input diode, resistor, and transistor can be modeled as the three diodes and resistor. You should see that current can be drawn from the input, and it comes from a current-limiting resistor connected to the chip's power.

Current cannot be driven into the input pin of a TTL gate because of the reverse-biased diode built into the emitter of the input transistor. This means that while also being current controlled, the current used for control is provided within the chip, minimizing any worrying that you might have about properly sourcing the correct amount of current to the gate.

When current is not being drawn from the TTL inverter gate input, current follows the path shown in Figure 10-16. When you follow the current path, you will see that the current will ultimately turn on the bottom-right transistor, connecting the gate's output pin to ground (low voltage output). When current is drawn from the TTL input pin (Figure 10-17), the current that ultimately turned on the bottom right transistor is taken away, due to the in a different path for currents taken within the gate. This change in current flow ultimately turns on the top-right transistor,

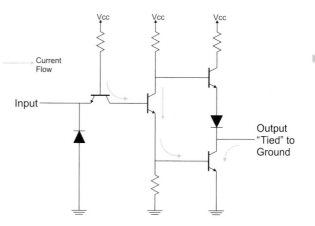

Figure 10-16   *TTL inverter with a "1" or floating input*

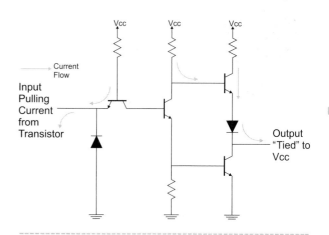

Figure 10-17   *TTL inverter with a 0 input*

effectively tying the output to power and driving out a high voltage.

To see the operation of the TTL inverter, you can build it as a circuit as shown in Figure 10-18. By connecting the SPDT switch to ground, you will be allowing current to flow from the TTL inverter to ground, causing the upper-right transistor to turn on, passing current to the LED to turn it on. If you checked the voltage of the input pin while the switch was open or connected to 9V, you would find that there is no voltage; this is due to the effectively reversed diodes of the gate that block current flow. When there is no current flow, there is no voltage.

With this done, you can now look at the analog aspects of the circuit by putting a potentiometer in the circuit, as I have shown in Figure 10-18, and adjust it until the LED flashes on and off. Measure the voltage across the potentiometer and then

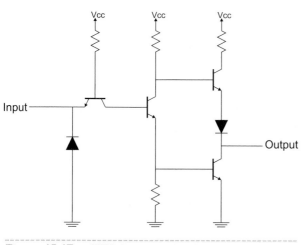

Figure 10-15   *TTL inverter circuit*

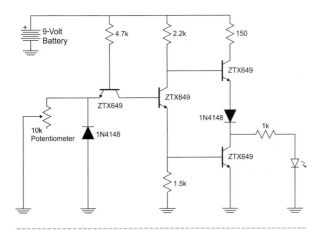

Figure 10-18   *Circuit to test current draw for TTL operation*

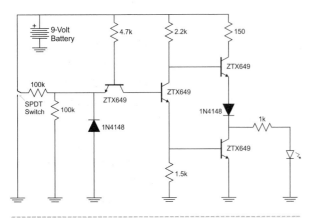

Figure 10-19   *Circuit to test voltage control of TTL operation*

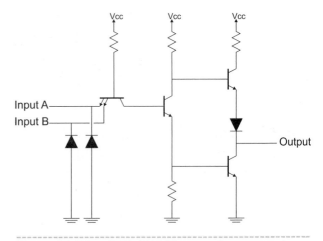

Figure 10-20   *TTL NAND gate using two-emitter transistor*

remove the potentiometer from the circuit and measure the resistance (see Figure 10-19). Using Ohm's law, you will find that the inverter circuit that was built has a *threshold* current of around 1mA.

The TTL inverter is useful for the NOT operation, but you are probably wondering how it is relevant to the multiple input gates built from TTL logic. At the risk of stealing the thunder from a later experiment, I just want to say that all TTL logic is based on this circuit. Instead of using an input transistor with a single emitter, the basic TTL logic gate uses a transistor with multiple emitters. This modification changes the inverter into a NAND gate, which is the TTL gate that all others are built from (see Figure 10-20). CMOS logic is built from NOR gates.

Experiment 56
Sum of Product Circuits

Parts Bin

Assembled PCB with
  battery
Three SPDT switches
Four 1k resistors
Four LEDs, any color
74C08, CMOS AND gate
74C32, CMOS OR gate
74C04, CMOS NOT gate
Three 0.01 µF capaci-
  tors, any type

Tool Box

Wiring kit

When I presented the basic logic circuits, one thing that may not have been clear is that they are building blocks and not necessarily complete functions needed for an application. When these basic functions are combined, they allow multiple inputs to create specific outputs to meet specific requirements. For example, when you are starting the "program" function of your VCR, you can either press the Record and Play buttons on the front panel of the unit or press the Record button on the remote control. You could write this out as the statement:

> Start recording if the Record and Play buttons on the front panel of the VCR are being pressed or if the Record button of the remote control is pressed.

This is not quite as profound as the logic statements put forward by the great philosophers, but it follows the same rules of the logic as discussed at the start of this section. If we were to assume that the buttons were true when they were pressed, and replaced AND with a dot (·) and OR with an addition sign (+), the function could be written as an equation such as:

Start Recording = ("Panel Record" ·
"Panel Play") + "Remote Record"

These two symbols for ANDing and ORing are based on the two functions' operations. AND behaves like a multiplication operation. Using binary values, the result is 1 only if both inputs are 1 (if one input is zero, then zero multiplied by anything is zero; the product of 1 and 1 is 1). OR is represented as addition because the result is not equal to zero if either input is not equal to zero. The exclamation point (!) is used for the not function.

Looking at the equation above, you should see that the first thing done is to bring together the inputs that all have to be true for the output to be true. These ANDed outputs are combined together in an OR gate to produce the final output. Using the nomenclature introduced above, the AND output could be represented as a product and the OR output as a sum.

This is where the term *sum of products* comes from. This is a very intuitive way to represent complex logic functions and will be used throughout this book (and most others). In this experiment I will show how the logic functions are combined as a sum of products to create a complex function.

In Figure 10-21, I have shown a memory map for a fictional computer that only has eight memory locations. Four of them are located in a single chip. They are the shaded regions and are accessed when a pin has a high (1) voltage applied to it.

The table in Figure 10-21 could be considered the truth table for the decoder function. The first is that I treat the three address line inputs like a three-bit binary value (I will explain this in more detail later in the book) and change each value in an orderly progression, making sure that each possible value is represented in the table.

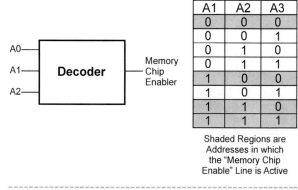

A1	A2	A3
0	0	0
0	0	1
0	1	0
0	1	1
1	0	0
1	0	1
1	1	0
1	1	1

Shaded Regions are Addresses in which the "Memory Chip Enable" Line is Active

Figure 10-21 *Simple three-address-line memory map*

Table 10-7 Output results

A2	A1	A0	Output	Comments
0	0	0	1	Output is high if !A1 · !A0 is true
1	0	0	1	
0	1	0	0	
0	1	1	0	
0	0	1	0	
1	0	1	0	
1	1	0	1	Output is high if A2 · A1 is true
1	1	1	1	

Combining A2 AND A1 to get the output high is one of the "products" I was discussing earlier. The truth table can be rearranged with the two input sets that result in "Output High" to see if there is anything common about them as well. In Table 10-7, I have moved the fifth entry to be beside the first entry and found that the output is high if !A1 AND !A0 is true.

For this case to be true, the values for A1 and A0 have to be inverted, so zeros are changed to ones before they are ANDed together. This is accomplished by the NOT gate, and the logic circuit to produce this function is shown in Figure 10-22.

You might be wondering if there is such a thing as a "product of sums" method of presenting logic. Yes, there is, but it is quite a bit more difficult to come up with, because thinking in terms of ORing inputs together and ANDing the result is not intuitive. Figure 10-23 shows what a product of the sum logic circuit for the decoder designed for this experiment would look like. The circuit is quite straightforward, although to really see how it works, you will have to create a truth table and work through the gates to confirm that it really does work.

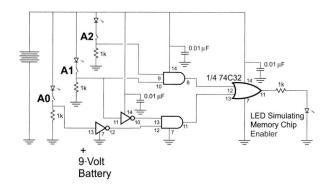

Figure 10-22 *Three-address-line decoder circuit*

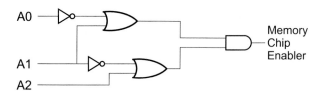

Figure 10-23 *Product of sums providing the same functions as the sum of products used in this experiment*

Experiment 57
Common Logic Built from the NOR Gate

Parts Bin

Assembled PCB with
  battery

Two SPDT switches

Two 1k resistors

470 Ω resistor

Three LEDs, any color

74C02, CMOS NOR gate

Two 0.01 µF capacitors,
  any type

Tool Box

Wiring kit

When you purchased the chips for the previous two experiments, you probably assumed that they were built from something called AND, OR, and NOT gates. Just as I showed how a complex circuit could be built from these basic gates in the previous experiment, these gates are built from even simpler gates; CMOS logic technology uses the NOR gate (see Figure 10-24).

The NOR gate outputs high (1) when its two inputs are low (0), and it can be built very easily in CMOS technology. Looking at it from the perspective you have been given so far, the NOR gate consists of an OR gate in which the output has been inverted. The small circle at the end of the OR symbol indicates that the output is inverted. !(A + B) is the written format for the gate.

It is probably surprising to hear that the NOR gate can be built very easily in CMOS technology. Figure 10-25 shows how a two-input NOR gate is implemented using four MOSFET transistors. When laid out on a silicon chip, this function can be implemented in a very small amount of area without requiring any aluminum traces that could interfere with gate interconnections on the chip.

The operation of the NOR gate is shown in Figure 10-26. In the left drawing, I have shown what happens when both inputs are low. In this case, the two p-

channel MOSFET transistors are turned on and provide a direct path for current to flow from the chip's power source to the output. When the inputs are low, the n-channel MOSFET transistors are turned off and there is no ground connection. In Figure 10-26's drawing on the right, one of the inputs is high and the p-channel MOSFET transistor that it is connected to is turned off while the N-channel MOSFET is turned on. This blocks current from the power source and connects the output directly to ground.

To implement the three basic functions that I introduced in this section using the NOR gate is quite simple, and this will be the point of this experiment. Before going on, I suggest that you wire the breadboard as I have shown in Figure 10-27; from here, you will create custom wiring to create each different gate. Note that in Figure 10-27 I have wired an LED to act as the circuit output; this output will be used for each circuit presented in this experiment. For all

Figure 10-24   *NOR gate*

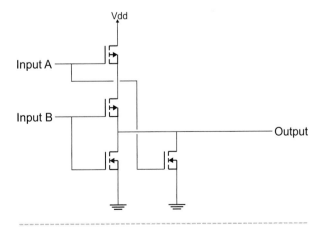

Figure 10-25   *CMOS NOR gate*

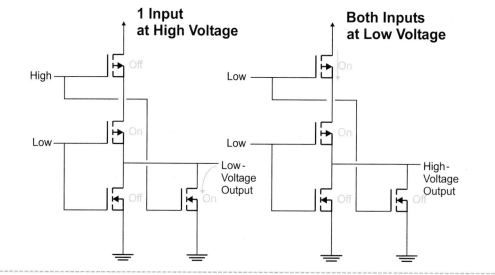

1 Input at High Voltage

High — Off

Low — On

Low — Off — On — Low-Voltage Output

Both Inputs at Low Voltage

Low — On

Low — On

Off — Off — High-Voltage Output

Figure 10-26 *CMOS NOR gate operation with different inputs*

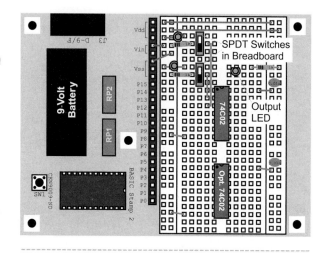

Figure 10-27 *Basic wiring for testing out different functions built from the two-input NOR gate*

these circuits, I recommend that you create a truth table, as in the section's first experiment, to validate the results and make sure the combined gates function as the basic gates.

The most basic logic gate is the NOT, and it is implemented as shown in Figure 10-28. With one input tied to ground, the NOR gate's output will be dependent on the value of the single input. You may see some circuits that have both inputs tied together instead of only using one input and the other tied to ground. I have tied one input of the NOR gate to ground to minimize the load the circuit driving the input will have. For this circuit, where the load is pro-

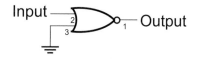

Figure 10-28 *NOR gate used to manufacture a NOT gate*

vided by a switch connected to the PCB's power supply, this is not an issue at all. When you are driving inputs from a CMOS output, the number of inputs connected to the output must be considered and ideally minimized to no more than two inputs per output as a rule of thumb.

Using the ability to negate a signal, two NOR gates can be used to create a single OR gate as shown in Figure 10-29. The operation of the circuit should be quite easy to see; the output of the NOR gate is inverted in order to get a positive OR gate function.

The final of the basic three gates is the AND gate and is implemented as shown in Figure 10-30. The realization that it performs the same function as the AND gate is probably more surprising. To create an

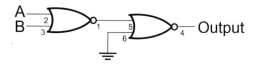

Figure 10-29 *Two NOR gates used to manufacture an OR gate*

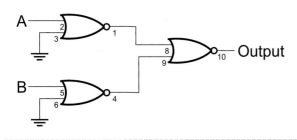

Figure 10-30   *Three NOR gates used to manufacture an AND gate*

AND gate using NOR gates, I used the laws listed at the end of this experiment. Knowing that I wanted an AND function and all I had was a negated OR, I started with a double-negated AND statement along with De Morgan's Theorem.

The following basic rules for Boolean arithmetic will allow you to create virtually any function using a technology's basic gates or whatever you have on hand. I admit that I apparently came up with the AND gate equivalent from the NOR gates very quickly, but as you become more comfortable with digital logic and the rules below, you will be able to bend the laws to come up very complex circuits using basic NOR (for CMOS logic) and NAND (for TTL logic) gates.

Identity Functions
$A \cdot 1 = A$
$A + 0 = 0$

Output Set/Reset
$A \cdot 0 = 0$
$A + 1 = 1$

Double Negation Law
$!(!A) = A$

Complementary Law
$A \cdot !A = 0$
$A + !A = 1$

Idempotent Law
$A \cdot A = A$
$A + A = A$

Commutative Law
$A \cdot B = B \cdot A$
$A + B = B + A$

Associative Law
$(A \cdot B) \cdot C = A \cdot (B \cdot C)$
$(A + B) + C = A + (B + C)$

Distributive Law
$A \cdot (B + C) = (A \cdot B) + (A \cdot C)$
$A + (B \times C) = (A + B) \times (A + C)$

De Morgan's Theorem
$!(A + B) = !A \cdot !B$
$!(A \cdot B) = !A + !B$

Parts Bin

Assembled PCB with battery

Two SPDT switches

Four 1k resistors

Four LEDs, any color

74C08, CMOS AND gate

74C86, CMOS XOR gate

Two 0.01 µF Capacitors, any type

Tool Box

Wiring kit

So far in this section, I have introduced you to five of the six basic logic gates (AND, OR, NOT, NAND, and NOR). These gates can provide you with almost all of the capabilities that you will require for any digital circuit. The last type of gate is required for providing basic math functions in digital logic, as I will demonstrate in this experiment.

The XOR gate outputs a 1 any time one but not the other input is equal to 1. In many texts and references, the symbol you will see is a plus sign (+) with a circle around it. In this book, I will use the symbol ^ to indicate the XOR function (which is the same symbol as used in PBASIC). The XOR operation is

```
Output = A ^ B
```

This operation has the truth table (see Table 10-8) and symbol shown in Figure 10-31.

Table 10-8 XOR gate truth table

A	B	A ^ B
0	0	0
0	1	1
1	1	0
1	0	1

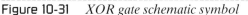

A
B —)D— Output

Figure 10-31   *XOR gate schematic symbol*

XOR is not a basic operation in any type of logic. It is normally written out as a compound operation in the sum of products form:

```
XOR(A, B) = (!A · B) + (A · !B)
```

It can also be written out in the product of sums form as

```
XOR(A, B) = !((A + !B) · (!A + B))
```

To prove this is true, you can use the Boolean arithmetic laws presented in the previous experiment, or you can construct a truth table, such as Table 10-9, in which each intermediate value for the input values is calculated and used to find the final value. This type of tool is useful for seeing how a logic equation works (or why it doesn't) and only takes a few moments to create.

The XOR gate is useful when you are confronted with situations such as adding two binary numbers together. When I start discussing programming, I will present the concept of binary numbers in more detail, but for this experiment, when I add together two binary values, I would like to display the results using two LEDs; if just one binary value is high (or 1), I would like to turn on just the 1 LED. If both binary values are high, then I would to turn on just the 2 LED (the 1 LED should not be on because two input values are high).

Turning on the 2 LED is a pretty simple proposition—you just AND the two inputs together to turn on the LED if both are high. Turning on just the

Table 10-9 Truth table explaining operation of XOR gate

A	B	A + !B	!A + B	(A + !B) · (!A + B)	!((A + !B) · (!A = B))
0	0	1	1	1	0
0	1	0	1	0	1
1	1	1	1	1	0
1	0	1	0	0	1

1 LED when only one input is high is a bit more difficult. Before going on with the experiment, take a pencil and paper, and using the Boolean arithmetic rules that I have presented, try to figure out how to just turn on the LED with AND, OR, NOT, NAND, and NOR gates.

If you came up with something that did the job, it would be an implementation of the XOR gate that I have presented here. Although I have shown two equations for the function, you may also have discovered the following:

$$XOR(A, B) = !(!(!A + !B) + !(A + B))$$

This is the formula used to implement the XOR gate in CMOS logic (using NOR gates) and it is the basis for the 74C86, which is used in the circuit shown in Figure 10-32 to turn on the 1 LED when only one input is active. The circuit in Figure 10-32 is known as a *half adder* and will turn on the 1 LED only when one input is high and the 2 LED only when both inputs are high.

When the circuit shown in Figure 10-32 is used as part of an adder circuit, the 1 LED output is labeled "S" for "Sum" and the 2 LED output is labeled "C" for "Carry." The sum outputs the bit result for adding the two input lines, and the carry outputs whether or not the result is greater than 1. This circuit is the basis for the adder circuits built into computers, but for it to be useful in a computer some more circuitry has to be added to it.

The reason for having to add circuitry to this circuit is in the case where there is a lesser bit providing its carry value to it. In this case, the bits actually have three inputs and must somehow add them all together and pass out an "S" and "C" bit, just as the

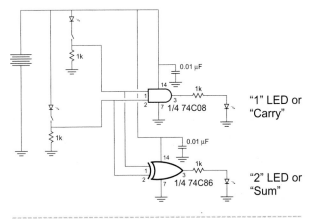

Figure 10-32 *Half adder circuit*

half adder did for two bits. To do this, two half adders are wired together as shown in Figure 10-33.

When several full adders are combined to form an adder for two multibit numbers, the time required for the correct answer is the longest amount of time a signal takes to pass through all the different half adders to get to the most significant bit. Because the result changes as the signal passes through the circuit, it is known as a *ripple adder*. For many applications, this type of adder and its delay is acceptable, but for high-performance computing applications (such as

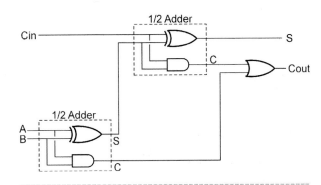

Figure 10-33 *Full adder circuit*

the microprocessor in your PC), the time required for the signals to pass through the adder is just too long, and another type of adder is used (known as a *carry look ahead*) that produces its carry output by testing

the input values while the output values is being generated and does not require a signal to ripple through.

Experiment 59
Pull-Ups/Pull-Downs

Parts Bin

Assembled PCB with breadboard

Three LEDs, any color

Three 1k resistors

10k resistor

74C00 CMOS quad NAND gate

0.01 μF capacitor, any type

Three breadboard-mountable SPDT switches

Tool Box

Wiring kit

When you design your first "practical" digital circuit, you will probably discover that you will have to set some of the digital inputs to high logic levels. The obvious solution to this dilemma is to tie the inputs directly to Vcc (to get a high logic level) or ground (to get a low logic level) as I have shown in Figure 10-34. This will work, but there are reasons why these direct connections should never be used, as I will explain in this experiment.

The problem with tying the input directly to power is that it can never be changed. Instead of tying a

logic input to the power input, I recommend passing it through a 10k resistor as shown in Figure 10-35. This will hold both CMOS and TTL inputs high, and a switch (or even a simple wire) can be used to pull the input down if it has to be changed from a high input to a low input. By using the switch or a wire, if a pull-down is actually required, then just 100 μA will flow through the resistor to ground.

Being worried about how much current is flowing through a pulled-up gate probably seems to be strange, but it is very important when you are designing products to be manufactured. The *In-circuit Tester* (ICT), which provides a multitude of pins (known as

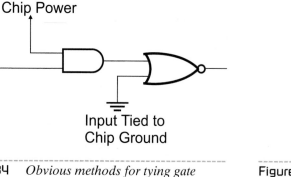

Figure 10-34 *Obvious methods for tying gate inputs to logic highs and lows*

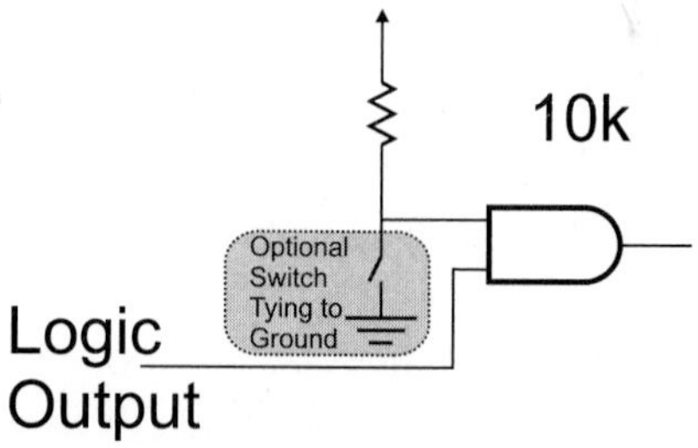

Figure 10-35 *The best method for implementing a pull-up*

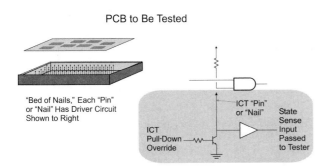

PCB to Be Tested

"Bed of Nails," Each "Pin"
or "Nail" Has Driver Circuit
Shown to Right

ICT "Pin"
or "Nail"

ICT
Pull-Down
Override

State
Sense
Input
Passed
to Tester

Figure 10-36 *InCircuit Tester (ICT) bed of nails connection to PCB*

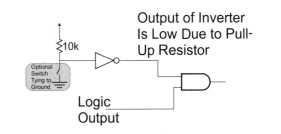

Output of Inverter
Is Low Due to Pull-
Up Resistor

10k

Optional
Switch
Tying to
Ground

Logic
Output

Figure 10-37 *The method I recommend to pull down a logic input*

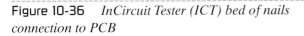

a "bed of nails" and shown in Figure 10-36) that contact the product's PCB, is a very common manufacturing test. In its simplest version, each pin probe (or "nail") has a logic input, and an open collector driver with each pin is connected to one of the connections (or "nets") in the circuit. To test the operation of the gate, the ICT can either sense the current logic value or drive the pin low to see what the result is at a "downstream" test pin.

Even if you are thinking that it is not very likely that your applications will be manufactured, it is still a good idea to use resistor pull-ups as shown in Figure 10-35. There will be cases where you may want to disable different sections of your circuit, and having a pull-up resistor in place will allow the addition of a direct or switched connection to ground. Many times over the years I've been thankful to have had the forethought of putting in a pull-up instead of directly connecting a pin to power or ground, which would have meant I would have had to unsolder the pin and add a resistor.

Following a logical progression, you are probably thinking that putting in a resistor to ground should be used when you want to "pull down" a logic input pin. Actually, this is something that you will want to avoid; instead use a pulled-up inverter output as I have shown in Figure 10-37.

Although a single resistor pull-down (no more than 470 Ω) could be used for tying both CMOS and

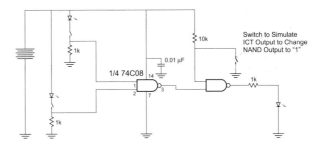

Figure 10-38 *Circuit to demonstrate pull-up control of an AND gate*

TTL inputs low, the problem is its impedance is lower than many ICTs can overcome and drive to a high voltage level. This may seem like actually more trouble than its worth (and for virtually all hobbyist projects I would agree with you), but if you ever hit the big time, you should know what changes should be made to your circuitry to make it easy to manufacture.

For this experiment, I would like to look at the operation of the pull-up as I have discussed so far. The circuit that I would like use is the AND gate built from two NAND gates shown in Figure 10-38. When the switch on the 10k pull-up is open, the output will behave like a standard AND gate. When the pull-up switch is closed, the output will be high (and the LED lit), regardless of the state of the other two switches.

Experiment 60
Mickey Mouse Logic

Parts Bin

Assembled PCB with breadboard

Three LEDs, any color

470 Ω resistor

Three 1k resistors

Three 10k resistors

74C08 CMOS quad AND gate

Two 1N914 or 1N4148 silicon diodes

Two ZTX649 NPN transistors

Three PCB-mount SPDT switches

0.01 µF capacitor, any type

Tool Box

Wiring kit

One of the most frustrating aspects of designing digital electronic circuits is that when you finish them, you will often discover that you are a gate or two short, and you are left with the question of whether or not you should add another chip to the circuit. In most cases, a gate can be "cobbled" together with a few resistors, diodes, and maybe a transistor.

These simple gates are often referred to as *Mickey Mouse Logic* (MML) because of how unorthodox and simple they seem to be. To be used successfully, MML must be matched to different logic families and should not result in long switching times, which will affect the operation of the application.

The most basic MML gate is the "inverter" and should not be a surprise. The circuit for the MML inverter can be built out of two 10k resistors and an NPN transistor. This inverter outputs a high voltage when it is not being driven by any current (and very low current). When current is passed to the gate, the transistor turns on and the output is pulled to ground (with good current-sinking capability).

To test the operation of the RTL inverter, use the circuit shown in Figure 10-39. The shaded area is used to indicate the location of the MML gate, and when I present different gate circuits this area with its connections should be used.

The circuit that I have presented here probably seems more complex than is required because of the use of the CMOS AND gate to provide the inputs as well as drive the output LED. This circuit (as well as the other MML gates I will present in this experiment) cannot handle high voltage or current inputs and outputs, as well as commercially available logic gates and the need to be buffered. The need for buffering the MML gate's inputs and output is an important point to note when considering using an MML gate in an application. As a rule, MML gates must be placed in the middle of a logic string rather than at the input or output ends to ensure that if you are expecting certain characteristics (such as the ability to drive an LED), the circuit will be able to do so.

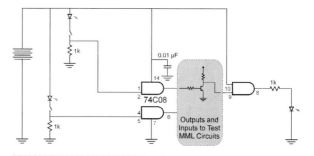

Figure 10-39   *Mickey Mouse logic test circuit. The inputs tied to Vdd are for convenience.*

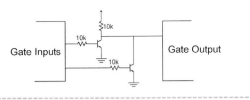

Figure 10-40   *RTL NOR gate*

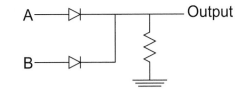

Figure 10-42   *OR gate analog circuit*

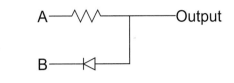

Figure 10-43   *AND gate analog circuit*

When you have assembled the circuit, test it out by changing the switch positions and seeing how the output changes. You should also create a simple truth table for this circuit as well as the following ones to make sure that they behave as you would expect.

Once you have tested out the operation of the inverter, you can simply modify the circuit by adding another transistor and resistor as shown in Figure 10-40 to create an RTL NOR gate. The RTL NAND gate is shown in Figure 10-41.

Implementing an AND or OR gate in MML is a bit more complex and requires a good understanding of the parameters of the logic families. In Figure 10-42, I have shown a sample design for an OR using two diodes and a resistor. The use of a pull-down resistor is probably surprising, but it was chosen to allow the gate to be used with both CMOS and TTL logic. In this case, if neither input has a high voltage, then the resistor will pull the input to ground. If the input is a CMOS gate, then the input will behave as if it were tied to ground. A 470 Ω resistor will allow the TTL input current to pass through ground, and it will behave as if the input was at a low logic level. In either case, when one of the inputs is driven high, the input pin will be held high, and the gate connected to the output of the OR gate will behave as if a high logic level was applied to it.

An MML AND gate (Figure 10-43) is the simplest in terms of components. The diode and resistor work together to provide a high voltage when both inputs are high, but when one of them is pulled low, then the

voltage level will be pulled down and current will be drawn from the input gate to which it is connected.

Although the MML AND and OR gates presented here will work in virtually any application, you may find that you will want to use a 470 Ω resistor with TTL and a 10k one in CMOS logic applications. The reason for doing this is to minimize the current drawn by the application; with a 470 Ω resistor, roughly 10 mA will be drawn when the output of the gate is low. This current draw decreases to 100 μA when a 10k resistor is used instead of the resistors in these two gates.

For Consideration

You should be aware of two myths that are commonly believed about TTL chips and make sure you understand why they are inaccurate. TTL gates are often said to be internally pulled to power (Vcc). This is not accurate, although the gate with no connection to an input pin does behave identically to an input tied directly to Vcc. The second myth is related and states that you *have* to tie an unused TTL input to Vcc because electrical noise can affect its operation. TTL is controlled by current that is sourced within the chip, and external voltage variations will not lead to inadvertently incorrect inputs. CMOS logic (even 74Cxx chips), on the other hand, can be affected by noise and is not internally pulled up. Any unconnected CMOS inputs will behave as if they are tied to a low voltage input and they can be affected by

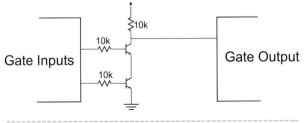

Figure 10-41   *RTL NAND gate*

Table 10-10 Logic chips

Chip Type	Power Supply	Transition	I/P Threshold	"0" O/P	"1" Output	Output Sink
TTL	Vcc = 4.5 to 5.5 V	8	N/A	0.3 V	3.3 V	12 mA
L TTL	Vcc = 4.5 to 5.5 V	15	N/A	0.3 V	3.4 V	5 mA
LS TTL	Vcc = 4.5 to 5.5 V	10	N/A	0.3 V	3.4 V	8 mA
S TTL	Vcc = 4.75 to 5.25 V	5	N/A	0.5 V	3.4 V	40 mA
AS TTL	Vcc = 4.5 to 5.5 V	2	N/A	0.3 V	Vcc - 2 V	20 mA
ALS TTL	Vcc = 4.5 to 5.5 V	4	N/A	0.3 V	Vcc - 2 V	8 mA
F TTL	Vcc = 4.5 to 5.5 V	2 ns	N/A	0.3 V	3.5 V	20 mA
C CMOS	Vcc = 3 to 15 V	50 ns	0.7 Vcc	0.1 Vcc	0.9 Vcc	3.3 mA*
AC CMOS	Vcc = 2 to 6 V	8 ns	0.7 Vcc	0.1 V	Vcc - 0.1 V	50 mA
HC/HCT	Vcc = 2 to 6 V	9 ns	0.7 Vcc	0.1 V	Vcc - 0.1 V	25 mA
4000	Vdd = 3 to 15 V	30 ns	0.5 Vdd	0.1 V	Vdd - 0.1 V	0.8 mA*

electrical noise. You should always tie unused CMOS inputs to voltage or ground as a general rule. (Preferably through a pull-up as described later in this section, and it's not a bad idea to always tie unused TTL inputs to voltage or ground, again through a pull-up.) If you always tie up your unused inputs, then you don't have to worry about when you have to and when you don't.

To aid you in looking at different logic chips, I have compiled a list of basic chips that you will want to consider for your own use. For the different varieties of TTL, C, AC, and HC/HCT logic families, the part number starts with 74. For the 4000 series of CMOS chips, the chip has a four-digit part number,

starting with 4. Table 10-10 lists the different aspects of the different types of logic chips that you will want to use.

The output sink currents are specified for a power voltage of 5 volts. If you increase the power supply voltage of the indicated CMOS parts (marked with a "*"), you will also increase their output current source and sink capabilities considerably.

In the table, I marked TTL input threshold voltage as "not applicable" because, as I demonstrated in this section, TTL is current driven rather voltage driven. CMOS logic is voltage driven, so the input voltage threshold specification is an appropriate parameter.

Power Supplies

For most of this book, I do a pretty good job of specifying electronics that are very tolerant of differing power supplies and work quite well with simple battery packs. This was accomplished by only specifying parts that can run from a variety of different supplies (battery packs) that produce power at voltages from 2.4 volts to over 9 volts. This makes the book experiments easy to build and increases the possibility that they will work for you the first time, but in the real world, you will have to work with components that must have power supplied at a set voltage and that does not vary by more than a few tens of millivolts or it could stop working. In these cases, you will have to include a circuit, called a power supply, which regulates the voltages and supplies enough current for the circuits to work reliably.

I have presented you with the DC power equation:

$$P = V \times I$$

In this equation, "P" is the power consumed (in watts), "V" is the voltage applied to the circuit, and "I" is the current (in amperes or amps) drawn from the power supply. For example, in a +5 volt *Transistor-to-Transistor Logic* (TTL) circuit, if 0.15 amps are drawn from the power supply, then the circuit is dissipating 0.75 watts or 750 milliwatts.

Earlier in the book, I presented the concept of the voltage divider (see Figure 11-1), which is used to convert an input voltage to a smaller output voltage. You might be thinking that you could use this circuit to convert a higher voltage into something that is useable by your electronics, but this is quite problematic. To illustrate the problems with using a simple voltage divider in an electronic circuit, consider the situation where a robot's 18-volt battery source is used to provide 5 volts at 100 mA to TTL electronics.

The resistor values can be specified using the formula for the voltage divider:

$$Vout = Vin \times (Rs/(Rn + Rs))$$

And:

$$Vout = 5\ Volts$$
$$Vin = 18\ Volts$$

To find Rn and Rs, I could go back to Ohm's law (assuming I wanted 100 mA at +5 Volts):

$$R = V/I$$

Where:

$$V = 5\ volts$$
$$I = 100\ mA$$
$$R = 50\ ohms$$

This R is actually Rs in the voltage divider formula, and it is actually the resistor equivalent load to

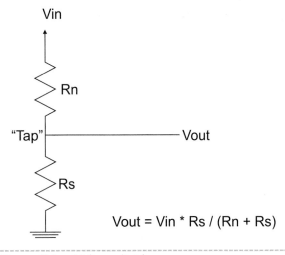

Vin

Rn

"Tap" ——————— Vout

Rs

$$Vout = Vin * Rs / (Rn + Rs)$$

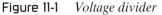

Figure 11-1 *Voltage divider*

the electronics running at 5 volts. This value (even though it is never used) must be calculated, so that the current-limiting resistor (Rn) can be calculated:

$$Vout = Vin \times (Rs/(Rn + Rs))$$

$$5 \text{ volts} = 18 \text{ volts} \times (50 \text{ ohms}/(Rn + 50 \text{ ohms}))$$

$$Rn = (18/5) \times 50 \text{ ohms} - 50 \text{ ohms}$$

$$= 130 \text{ ohms}$$

Finding or producing a 130-ohm resistor isn't difficult; at least it isn't for a standard power dissipation resistor like the 1/4-watt resistors that have been used for the projects in this book. The problem comes when you start looking for a resistor that will dissipate the amount of power that this one does. If you use this formula (recognizing that there is a 13-volt drop across the 130-ohm resistor), you will be dissipating 1.3 watts of power. Although a 1/4-watt resistor can be purchased for a penny or less, you're going to find that a 2-watt resistor will cost you several dollars and it will get surprisingly hot in operation (lost power that could have been used by the robot).

Another problem with this circuit is what happens if the load changes or the battery output changes. As I will discuss in this section, battery output changes over time and in this example of a TTL, the operating voltage range is 4.5 volts to 5.5 volts. This means that if the robot's battery outputs less than 16.2 volts or more than 19.8 volts, the electronics will no longer work (and could be damaged). This kind of variation isn't unusual in a robot (especially when motors are being turned on and off).

The issue of the changes in current drawn by the electronics is of even more concern. If you had a number of *light-emitting diodes* (LEDs) in your robot, you will find that turning them on or off will drastically affect the effective load of the electronics and, in turn, the voltage being applied to it. Assuming that a resistor requires 5 mA to light, when an LED is added to the circuit, the voltage drop through the 130Ω resistor increases by 5 percent, or from 13 volts to 13.65 volts, resulting in 4.35 volts for the electronics. Since 4.5 volts is the minimum voltage the elec-

tronics will work at, the robot's logic will fail by turning on one LED.

Rather than trying to come up with work-arounds for these problems, let me just suggest that you look at using a voltage regulator, which will convert one voltage to another (one that can be used by the electronics in your circuit) and, more important, will be tolerant of changes in the supply voltage and in the current load. In this section, I will introduce you to some simple power supply circuits that have the following characteristics:

- They are safe for their users and designers.

- They are relatively efficient in terms of the amount of power that is lost converting voltage levels.

- They provide very accurate voltage levels independent of the voltage input or the current required by the application.

- They are inexpensive.

- Their design can be optimized for the application that they are providing power for.

A few important points about what is presented in this book: I will be focusing on creating power supplies for applications that require 1 amp or less of current. The 250-watt power supply used for your PC requires methodologies and circuits that are quite a bit different from what is required for the simple power supplies presented here.

Although I would consider several different methods for powering the controller circuitry in the robot, I would not be so open-minded with regard to power-filtering circuitry built into the application. At a minimum, I would suggest that you place a high-value (10 μF or greater) and a medium-value (0.001 to 0.1 μF) capacitor on the controller power input. The large capacitor will be responsible for filtering out low-frequency power upsets, whereas the smaller capacitor will help take care of high-frequency transients (such as caused by motors turning on and off). Along with the capacitors, you might want to consider a series inductor to filter out any current transients.

Experiment 61
Zener Diodes

Parts Bin

Assembled PCB with breadboard

220 Ω, 1/4-watt resistor

200 Ω, 1/4-watt resistor

330 Ω, 1/4-watt resistor

5.1 V Zener Diode, any type

LED, any color

Tool Box

Wiring kit

DMM

At the start of this book, I demonstrated electrical concepts using water analogies, but quickly stopped doing this as I started looking at semiconductors. These devices can rarely be modeled effectively using water analogs. Some semiconductor-based circuits, like Zener diode power supplies (Figure 11-2), do lend themselves to being modeled using water analogs (Figure 11-3). The Zener diode power supply works as a shunt regulator, applying a specified amount of current to a circuit at a rated voltage and shunting the rest as wasted power.

When the term "shunt" is used, it is simply saying that excess is passed through the Zener, away from the circuit, and this concept is illustrated in the water pressure regulator shown in Figure 11-3. This water regulator simply consists of a catch basin with a hole at the bottom; water coming out of the hole is at a pressure that is determined by the depth of water in the basin. To maintain this depth (and bottom pressure), even with water being drawn from the hole at the bottom, water is continually poured into the basin. More water is pouring in than is expected to exit through the hole in the bottom, with the excess leaking out over the side.

This is exactly how the Zener diode works, except that extra current does not "leak out over the side" but is passed (or shunted) through the diode. The diode itself is expected to be reverse biased when it is wired into the circuit, and it will pass current through it to maintain a set voltage level at its anode (positive terminal). This property is known as *breakdown* and it is not unique to the Zener diode. All diodes will break down when a high-enough reverse-bias voltage is applied to them. A diode will resist passing current as voltage is applied to it until it reaches a point where the reversed PN junction is overwhelmed and the diode starts conducting in the reverse-bias direction. The breakdown voltage for a Zener diode is usually specified to be in the range of 1.5 to 25 volts, whereas the breakdown voltage for a typical diode (say, the 1N4148/1N914 that I usually use) is 75 to 100 volts.

Specifying a Zener diode for use as a power supply in an application isn't very difficult, but it will require you to understand what your incoming power specifications are, as well as what the required

Figure 11-2   *Zener diode voltage regulator*

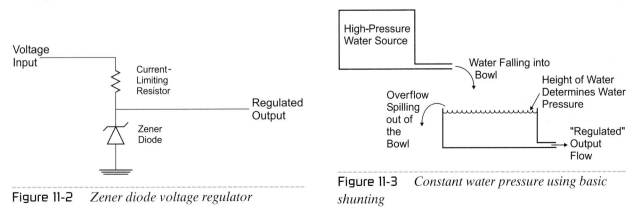

Figure 11-3   *Constant water pressure using basic shunting*

current is for the circuit being powered. The powered circuit's voltage should be the same as the rating of the Zener diode. For 5-volt circuits, I use a Zener diode rated at 5.1 volts. Specifying the resistor that is to be used with the Zener diode, as well as the Zener diode's power rating, can be somewhat complex. Care must be taken to ensure that the circuit has enough current to be powered in all circumstances, including if the input power *sags* (if it is powered by a battery that is discharging). To do this, some kind of margins must be designed into the circuit.

For this experiment, I would like to use a 5.1-volt Zener diode to act as a power supply for an LED circuit. The circuit is shown in Figure 11-4, and before it can be assembled, the value for the Zener diode's current-limiting resistor R must be determined. For a Zener diode power supply to be 100 percent efficient in terms of current (no current is shunted through the Zener diode), R must be chosen so that the voltage drop through it will allow the same amount of current as the powered circuit uses to pass through it. In this application, I am going to assume that the LED has a 2-volt drop, so using the basic electrical formulas, I can determine the current through the LED:

$$i = V/R$$

$$= (5.1 V - 2 V)/330 \Omega$$

$$= 9.39 mA$$

Assuming that the battery produces an even 9 v, the value of R can be calculated:

$$R = V/i$$

$$= (9 V - 5.1 V)/9.39 mA$$

$$= 415 \Omega$$

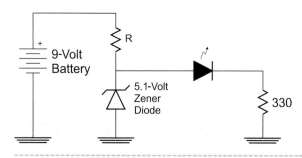

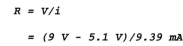

Figure 11-4   *Zener diode voltage regulator experiment circuit*

No standard 415 Ω resistors are available, but I can make a 420 Ω resistor using a 200 Ω and a 220 Ω in series. This will result in a current of 9.29 mA (a difference of about 1 percent from the targeted value.

When I built this circuit, I used a cheap 9-volt carbon battery (with 9.12 V output) and measured 9.4 mA passing through the LED.

To check on the stability of this circuit, I left it running overnight and during the morning, and then I found the LED was no longer lit. I measured 6.3 volts being output from the battery and 2.5 mA being passed through the LED (which is less than the 5 mA that is normally required to light it). The voltage across the Zener diode and 330-ohm resistor was still 5.1 volts, but not enough current was passing through them to light the LED. This is the major problem with the Zener diode power supply; it is not very tolerant of sags in the input voltage. When I design Zener diode power supplies, I normally design them so that twice the expected required current passes through the resistor (and so that the Zener diode will shunt 50 percent of the current being passed to the circuit). For this experiment, I replaced the series 420 Ω resistance with a single 220 Ω resistance, and I found that the LED lit again, even though the battery was driving out 6.3 volts. I measured the current through the LED to be 5.1 mA.

By allowing 50 percent of the current to be shunted through the Zener diode, you will have to make sure you calculate the amount of power being dissipated by its resistor and Zener diode. A maximum of 70 mW will be dissipated through the resistor, and 90 mW will be dissipated by the Zener diode. Although these power levels are quite low and easily dissipated by 1/4-watt parts, you can find yourself in the situation where the power dissipation requires parts rated at a 1/2 watt or more.

Looking at the power dissipated in the different parts of this circuit, 30 mW has dissipated in the LED/330-ohm resistor diode circuit, and 160 mW has dissipated in the Zener diode and its resistor. This makes this Zener diode regulator application only 15.8 percent efficient in terms of power, or put another way, five-sixths of the power supplied to the Zener diode, resistor, and LED application is lost. For this reason, Zener diode regulators are rarely used in applications that are battery powered; they

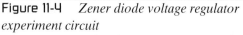

are suited for applications that use household wall power.

One of the aspects that makes Zener diode regulars attractive for use in power supplies is the Zener diode's current-limiting resistor. If a short or unexpectedly high current drain is in the circuit being powered, this resistor will prevent a damaging amount of current being passed to the circuit.

Experiment 62
Linear Power Supply

Parts Bin

Assembled PCB and breadboard

78L05 +5 volt regulator, in TO-92 package

220 Ω resistor

220 Ω 9-resistor SIP

10 μF electrolytic capacitor

0.01 μF capacitor, any type

LED, any color

Eight-position SPST switch

Tool Box

Wiring kit

DMM

When I described the Zener diode regulator acting like a basin of water in which the unused current was simply lost, I'm sure that many people grimaced because they knew of devices that are much better at regulating fluid pressure. If this book was written in the 1980s (or earlier), just about everybody would know about the commonly used fluid regulator that is used in older cars called a carburetor (Figure 11-5).

The carburetor is a very clever device that provides fuel on demand. In Figure 11-6, I have drawn the situation where no fuel is being drawn from the carburetor; a float is connected to a simple valve that closes when the bowl that the float is full of fuel.

When fuel is drawn from the bowl, the fuel level within the bowl drops (along with the float) and the valve opens, allowing more fuel into the bowl.

The carburetor acts as a regulator, just providing the volume of fuel (current) as required, and the shallow bowl will result in lower pressure (pressure regulation) than what was available from the high-pressure source (the fuel pump). An electrical version of the carburetor would look like Figure 11-7; current from the high-voltage source is switched through a PNP bipolar transistor with the control of the transistor being the output of the comparator. The comparator's inputs are the current voltage level of the regulator's output, and the specific "output" voltage

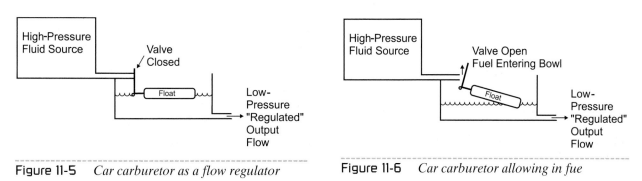

Figure 11-5   *Car carburetor as a flow regulator*

Figure 11-6   *Car carburetor allowing in fue*

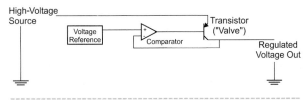

Figure 11-7   *Simple regulator controlling voltage*

that comes from some kind of voltage reference. The voltage reference is usually a Zener diode that has a miniscule amount of current passing through it.

You can buy a number of different linear regulator chips that were designed from this block diagram; I would recommend that you stay away from these parts when you are first starting working with electronics because they lack two features that I consider critical. These two features will shut down the regulator if its operation goes outside of normal operating limits. The term usually used for the regulator shut off is a *crowbar*, and it should be enabled if the current drawn from the regulator exceeds its rated value or if the temperature of the regulator exceeds the normal operating range.

The most popular positive linear voltage regulators that provide the crowbar features are the 78xx and 78Lxx series. The 78xx shown in Figure 11-8 ("xx" stands for the voltage, so a 5-volt regulator is a 7805) can normally source up to 500 mA and up to 1 A with heat sinking. The heat sink is used to dissipate the power and keep the temperature within the regulator less than 125 C, which is the crowbar temperature. For lower-current applications (up to 100 mA), the 78Xxx (Figure 11-9) can be used. For either

device, the input voltage should be at least 2 volts above the regulated output voltage. When wiring the regulator in circuit, you should include at least 10 µF of capacitance on the input and a 0.1 µF capacitor on the output (Figure 11-10).

To demonstrate the operation of a linear regulator, along with its capability to shut itself down if the current or temperature parameters are exceeded, build the circuit on the PCB's breadboard as shown in Figure 11-11. This circuit consists of a 9-volt battery's output being regulated to 5 volts using a 78L05. The 78L05 powers an LED with a current-limiting resistor along with eight switch-controlled 220 Ω resistors in a "SIP" Package.

After building the circuit, open all of the switches and connect the 9-volt battery. The LED should turn on, and by measuring the voltage across it and the current-limiting resistor, you will see that it is outputting 5 volts (with some slight deviation). The regulated output will remain quite stable, even if you close one the switches, increasing the current being drawn through the regulator. With the top of a finger, check the temperature of the 78L05 voltage regulator. With a 9-volt input and one switch closed, the 7905 should be slightly warm. Next, start closing the switches, one by one, while waiting two or three minutes for the

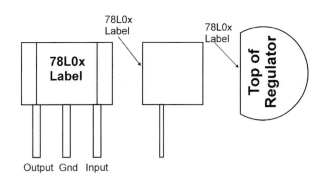

Figure 11-9   *78L0x voltage regulator in TO-92 package*

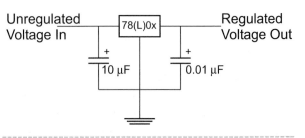

Figure 11-10   *Using a 780x as a voltage regulator*

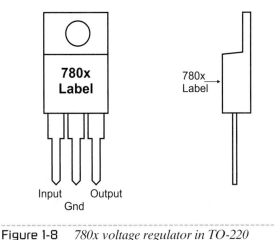

Figure 1-8   *780x voltage regulator in TO-220 package*

Experiment 62 — Linear Power Supply

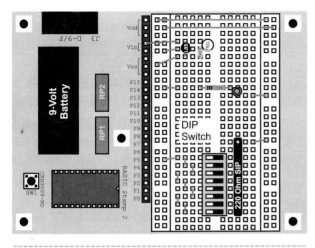

Figure 11-11   *Test circuit wiring to find the point where the 78L05 stops operating*

78L05's temperature to stabilize. If you have a DMM with a temperature sensor, you will see the case temperature of the 78L05 rise each time a switch is closed. At some point, the LED will go out (it was with seven switches closed for me); this is the point at which the crowbar became active. By opening two or three of the switches and waiting a few minutes, the LED should come back on as the 78L05 starts working again. The crowbar is usually not latched. To test to see if it is, when the LED is back on, short the output of the 78L05 directly to ground. When you remove the short and the voltage regulator does not come back on until the 9-volt battery is removed and plugged back in, the crowbar circuit is latched.

Experiment 63
Switch Mode Power Supply

Parts Bin

Assembled PCB with breadboard

LT1173CN8-5 switch mode power supply

Two 74LS123 dual multi-vibrator chips

1N5818 Shottkey diode

LED, any color

Dual C switched battery clip

100 uH coil

Two 100k resistors

10k resistor

470 Ω resistor

100 µF electrolytic capacitor

Three 10 µF Electrolytic capacitors

Two 0.01 µF capacitors, any type

Tool Box

Wiring kit

Although the Zener diode and linear power supplies presented so far in this chapter are useful and easy to work with, they do have two characteristics that can make them problematic when they are being used in a robot. First, they require a higher voltage than the regulated output. This can be an issue when you want to use very simple power such as two AA cells for a robot—remember that minimizing weight should always be an important goal. Second, they are not

terribly efficient. It isn't unusual if 80 percent or more of the power input to the Zener diode power supply is lost, and 40 percent or more is lost in the linear power supply. What is required is a power supply circuit that is very efficient and will "step up" voltages.

Although these two requirements seem impossible, they can actually be achieved very easily through the use of the *switch mode power supply* (SMPS). The basic SMPS circuit (Figure 11-12) is quite simple and

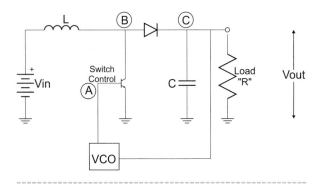

Figure 11-12   *Switch mode power supply circuit*

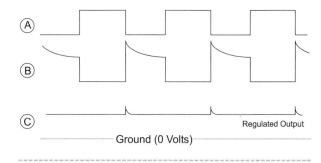

Figure 11-13   *Switch mode power supply operation*

relies on the energy-storing characteristic of the coil, which I have not yet explained in detail other than to say it is a problem with motors turning on and off. Whereas the capacitor stores energy in the form of charge, the coil stores energy in the form of a magnetic field that is maintained by current running through the coil. When this current is shut off, the magnetic field produces a voltage spike (which I called kickback when discussing magnetic devices) that can be used as the basis for an output voltage.

Using the circled letters in Figure 11-12, I have drawn the waveforms (Figure 11-13) that you can expect to see in the SMPS. The control signal is a PWM produced by a *voltage-controlled oscillator* (VCO). A VCO oscillates at a different frequency based on the voltage at an input. The input to the VCO used in the SMPS is the output voltage of the power supply; the VCO frequency will change according to the power supply output to ensure the output stays as stable as possible at the required voltage. The output of the VCO is the base of a transistor that periodically pulls one side of the coil to ground, allowing current to flow through it. When the transistor connected to the coil is turned off, current flow through the coil stops and the magnetic field "kicks back," producing a higher voltage.

The operation of the VCO PWM output along with the coil's response and the output voltage is shown in Figure 11-13. When the VCO is turning on the transistor, the coil (symbol "L") is tied to ground and current flows through it. When the transistor is off, the coil kickback can be seen, and any voltage greater than the current voltage output from the supply passes through the diode and is stored in the output capacitor. As I said, if the output voltage is more or less than the target voltage, the VCO frequency

changes along with the transistor control PWM, bringing the output voltage into line.

To determine the correct coil value as well as the PWM parameters, the following three formulas are used once the output voltage (V_{out}) is known, along with the expected output current draw (I_{out}) and the input voltage (V_{in}). These formulas are used repeatedly until the values for "L" (the coil value), T_{on} (time the transistor is on), and T_{off} (time the transistor is off) are values that can be produced by reasonable hardware.

$$I_{peak} = 2 \times I_{out} \times (V_{out}/V_{in})$$

$$T_{off} = L \times I_{peak}/(V_{out} - V_{in})$$

$$T_{on} = (V_{out}/V_{in}) - 1$$

Designing an SMPS is not a trivial exercise. Although you may think you can do it using something like a 555 timer, I'm going to recommend that you use a commercially available chip that provides the function for you such as the LT1173-5. This chip can be used to create five volts (necessary for TTL and many CMOS logic chips) from three volts as shown in the basic circuit in Figure 11-14.

To demonstrate the operation of the SMPS bringing up 3 volts from two AA batteries, I wanted to show something useful working, not just an LED or a 555 LED flashing an LED (a 555 could cause an LED to flash while being powered by two C batteries). Looking over the information I provided for the combinatorial logic and sequential logic chips, I noticed one type of logic function that I haven't mentioned, and it can be extremely useful in some situations.

This function is known as a multivibrator or programmed delay and is extremely useful when you have to sequence functions within a chip. I like to

3-Volt to 5-Volt Step-Up Circuit

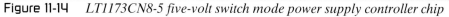

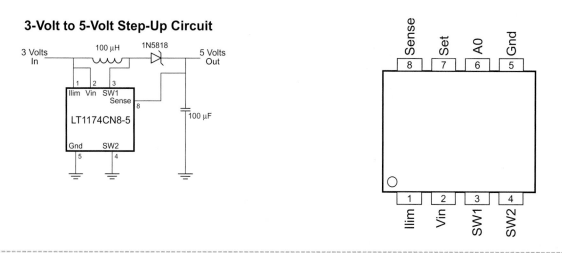

Figure 11-14   *LT1173CN8-5 five-volt switch mode power supply controller chip*

keep a number of 74LS123 dual multivibrator chips on hand in case I have to provide a delay within my circuit. Each 74LS123 consists of two multivibrators, with three inputs (the delay will be triggered when the A input is falling or when B or _CLR is rising) and two outputs. The Q output is normally low with a positive pulse, the duration of which is specified by the external resistor and capacitor, whereas _Q is normally high and provides a negative pulse of the same duration. The two multivibrators in the chip cannot be wired together to form an astable oscillator, but the delays in two saprate chips can be used to produce an oscillator as shown in Figure 11-15 (the

rising edge of each multivibrator delay triggers the other multivibrator, resulting in a 50 percent duty cycle or square wave signal).

The flashing LED may seem like a trivial application (especially because it could be demonstrated using a 555 timer, a few resistors, and capacitors) instead of the SMPS and multivibrator circuit presented here. The purpose of this circuit is to demonstrate that one or two alkaline or rechargeable batteries can be used to power the digital electronics of your robot, rather than relying on a larger battery voltage that may make your robot larger and heavier than required.

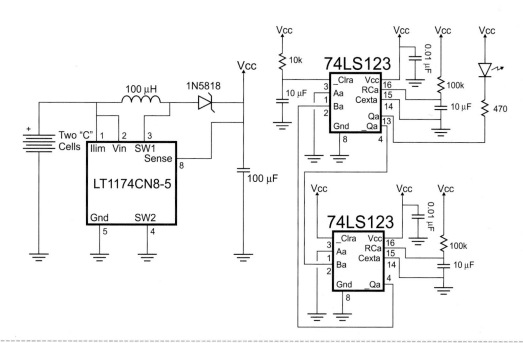

Figure 11-15   *Flashing circuit*

Section Twelve

Sequential Logic Circuits

When you first start working with digital electronics, playing around with simple logic and building different functions is fun. I've always enjoyed using the different tools (truth tables, Karnaugh maps, and Boolean arithmetic equations) to look at the different ways of providing a logic function and to see how much I could simplify it. Unfortunately, simple logic functions (which are known as *combinatorial logic circuits*) are not that useful in the real world because they do not change on their own over time or provide a sequence of operations.

Virtually all logic circuits that you will work with are known as *sequential logic circuits* because they are designed to provide a sequence of operations. A good example of a sequential logic circuit is a digital clock; the current time is stored in a memory function and is updated at different points in time. Along with using the current time as input data, the combinatorial logic circuits responsible for formatting the data output use button inputs for setting the time. Figure 12-1 shows a possible block diagram for a digital clock.

The block diagram format is one that I use quite a bit when working with sequential logic circuits, and it is somewhat related to the computer programming flowchart that will be discussed in later sections. Its

purpose is twofold: first, to show the path electrical signals take through the circuit, and second, to break down the operation of the circuit into small parts that can be easily designed.

The "Time Memory" block of Figure 12-1 is a memory unit that is updated on command from a 1 Hz oscillator input. The triangular feature on the "Time Memory" block connected to the 1 Hz oscillator is known as a clock input and is a common symbol used in logic diagrams to indicate the input pin used by the memory device to store the logic values at the I/P (input) pins. For the digital clock, this happens once every second, so the incremented seconds (and minutes and hours) are saved while the "Time Update Circuitry" and "Output-Formatting Circuitry" perform logic functions on the value stored in the "Time Memory."

The 1 Hz oscillator input in Figure 12-1 is often called a *clock* when it is a regular signal, as in this case. You may also see it referred to as a timebase or trigger, depending on its function within the circuit. In most sequential circuits, the clock input is used to store the updated values for the memory circuitry as is done in this application.

The_RST input to the Time Memory block is an input that will reset memory values to an initial value so the circuit will start out at a known, valid state. For this circuit (and most sequential circuits that I work with), I use an RC network to provide a delayed rising signal; using a 10k resistor and 10 μF capacitor, I will get a logic low for 10 to 20 milliseconds. Although this signal is low, most digital logic memory devices will hold the memory units reset regardless of the other input signals.

The "Time Update Circuitry" is a conventional combinatorial logic circuit that increments (adds one to) the second counter if the button is released or increments the minute counter if the second button is

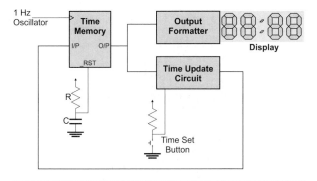

Figure 12-1   *Digital clock diagram*

183

pressed. Assuming that the "Time Memory" contains values for seconds, minutes, and hours, the operation of the time update circuitry could be modeled with the statements:

- If the Time Set button is pushed, then Seconds is set to 59; otherwise, Seconds is set to Seconds plus 1.

- If Seconds is equal to 59, then Seconds is set to 0 and Minutes is set to Minutes plus 1.

- If Minutes is equal to 59, then Minutes is set to 0 and Hours is set to Hours plus 1.

- If Hours is equal to 11, then Hours is set to 0.

These statements execute very quickly (over a few billionths of a second), so when the updated time is to be loaded into the Time Memory block, no danger of incorrect values being saved exists. In the state diagram shown in Figure 12-2, I show that the Time Memory blocks are updated when the signal from the 1 Hz oscillator transitions from a high to low voltage level. After the transition, the output from the Time Memory block changes, and this value flows through the Time Update Circuit. The new value is driven to the Time Memory block's inputs and will not be saved until the next high to low transition of the 1 Hz oscillator.

The "Time Update Circuit" statements are similar to the programming statements that I will introduce later in the book. I have used them here because the actual Boolean logic functions are actually quite complex. For example, to clear or increment the least significant hours bit you would have to implement a function such as

$$Hours.0.Input = ((Hours.0.Output \wedge 1)$$
$$\cdot \; !((Hours.0.Output \wedge 0)$$
$$\cdot \; (Hours.1.Output \wedge 0)$$
$$\cdot \; (Hours.2.Output \wedge 1)$$
$$\cdot \; (Hours.3.Output \wedge 0))$$

In this function, the least significant bit of hours is XORed with 1. If it and the other three bits that make up the current number of hours do not equal 11 decimal (1011 binary) and are not equal to zero, then 1 is stored as its value in the bit.

The Output Formatting Circuit of Figure 12-3 would be defined in exactly the same way as the Time Update Circuit, and its output, instead of being passed back to the Time Memory block, is used to drive four seven-segment LED displays.

The general form for a sequential digital circuit has the block diagram shown in Figure 12-3. You should see quite a few similarities to the clock diagram of Figure 12-1. The difference is that I have assumed that the memory unit that I am going to use can be cleared, and input is used as part of the next state (or time) update circuitry.

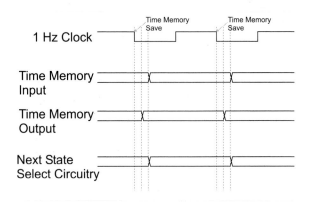

Figure 12-2   *Digital clock timing diagram*

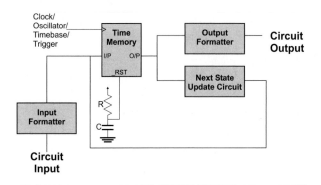

Figure 12-3   *Basic sequential circuit block diagram*

Experiment 64
RS Flip Flops

Parts Bin

Assembled PCB with breadboard

Four 1k resistors

Four LEDs, any color

Two SPDT breadboard-mountable switches

74C02

Tool Box

Wiring kit

Earlier in the book, I presented the idea of building a memory device using a two-coil relay. This device could be set to one of two states, depending on which relay coil was last energized, and pull the contact towards it. Once electricity to the coil is stopped, then the state will stay until the other coil is stabilized. This device works very similarly to the most basic electronic memory device that you will work with, the *Reset-Set* (RS) flip flop.

Whereas the relay device relies on friction to keep the saved value constant, the electronic memory unit takes advantage of feedback to store the value. When I discussed the radio control servo, I presented the concept of analog feedback (in the servo, the current position of the control arm is compared against the specified position of the servo and if they do not match, the arm is moved toward the specified position). The process of sensing the actual position of the control arm and passing it back to be compared against the specified position is known as *feedback*. The current output value is used to determine whether or not the servo arm should move. This is an example of analog feedback. The position of the arm that is returned can be a range of values, not a specific on or off (true or false, 1 or 0). Analog means there is a range of values, usually expressed as a fraction from 0 to 1.

Digital feedback can only be one of two values, so its use in circuits probably seems like it is much more limited than that of analog feedback. This is true, except when it is used as a method to store a result in a circuit like the NOR flip flop shown in Figure 12-4. Normally, the two inputs are at low voltage levels, except to change the circuit's state, in which case one of the inputs is raised to a high logic level.

If you are looking at this circuit for the first time, then it probably seems like an improbable device. The device may seem like one that will potentially oscillate, because if the output value of one gate is passed to the other and that gate's output is passed to the original, it seems logical that a changing value will loop between the two gates. Fortunately, this is not the case; instead, once a value is placed in this circuit, it will stay there until it is changed or power to the circuit is taken away. Figure 12-5 shows how by raising one pin at a time, the output values of the two NOR gates are changed.

When the R and S inputs are low, only one signal left will affect the output of the NOR gates, and that is output of the other NOR gate. When Q is low, then a low voltage will be passed to the other NOR gate. The other NOR gate outputs a high voltage because its other input is low. This high signal is passed to the original NOR gate and causes it to output a low voltage level, which is passed to the other NOR gate and so on.

The outputs of the flip flop are labeled as Q and _Q. Q is the positive output, whereas _Q is the negative value of Q—exactly the same as if it were passed through an inverter. The underscore character (_) in front of the output label (Q) indicates that the signal

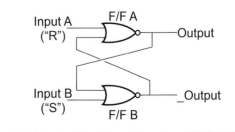

Figure 12-4   *NOR gate-based flip flop*

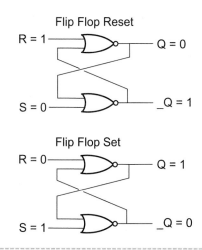

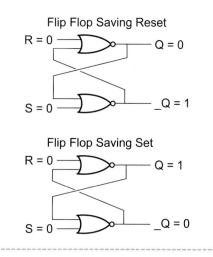

Figure 12-5   *Different states of NOR flip flop*

is inverted. When inverted outputs are presented in circuit diagrams, they are either identified with the underscore prefix or by a horizontal line on top of the label. You will not see a ! character in front of the label because this implies that the signal has passed through an inverter. When you look at some chip diagrams, you will see some inputs that have the underscore before or the line above the pin label. This indicates that the pins are active when the incoming signal is low. I will explain negative active inputs later in this section.

The R and S input pins of the flip flop are known as the Reset and Set pins, respectively. When the R input is driven high, the Q output will be low, and when S is high, the Q output will be driven high. These values for Q will be saved when R and S are returned to the normal low-voltage levels. Q_0 and $\_Q_0$ are the conventional shorthand to indicate the previous values for the two bits, and the notation indicates that the current values of Q and _Q are the same as the previous values. Truth tables are often used to describe the operation of flip flops, and the truth table for the NOR flip flop is shown in Table 12-1. You can build your own NOR RS flip flop that has its state set by two switches as I show in Figure 12-6

In the truth table, I have marked that if both R and S were high, while the outputs are both low, the inputs would be invalid. The reason why they are considered invalid is because of what happens when R and S are driven low. If one line is driven slower than the other, then the flip flow will store its state. If both R and S are driven low at exactly the same time

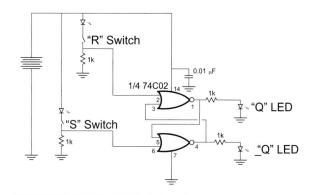
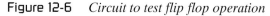

Figure 12-6   *Circuit to test flip flop operation*

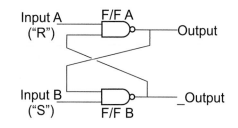

Figure 12-7   *NAND flip flop*

Table 12-1   NOR RS flip flop state table

R	S	Q	_Q	Comments
0	0	Q_0	$\_Q_0$	Store current value
1	0	0	1	Reset flip flop
0	1	1	0	Set flip flop
1	1	0	0	Invalid input condition

(not a trivial feat), then the flip flop will be in a *metastable* state, Q being neither high nor low, but anything that disturbs this balance will cause the flip flop to change to that state. The metastable state, although seemingly useless and undesirable, is actually very effective as a charge amplifier—it can be used to detect very small charges in capacitors.

Along with building a flip flop out of NOR gates, you can also build one out of NAND gates (Figure 12-7). This circuit works similarly to the NOR gate except that its metastable state occurs when both

Table 12-2 NAND RS flip flop state table

R	S	Q	_Q	Comments
0	0	1	1	Metastable input state
0	1	0	1	Reset flip flop
1	0	1	0	Set flip flop
1	1	Q_0	$\_Q_0$	Save current value

inputs are low and the inputs are active at low voltage levels, as shown in Table 12-2.

Experiment 65
Edge-Triggered Flip Flops

Parts Bin

Assembled PCB with breadboard

Two 74C00 quad dual input NAND chips

Three LEDs, any color

10k resistor

Three 1k resistors

47 µF electrolytic capacitor

Two 0.01 µF capacitors, any type

Two SPDT breadboard-mountable switches

Tool Box

Wiring kit

The RS flip flop is useful for many ad hoc types of sequential circuits in which the flip flop state is changed asynchronously (or whenever the appropriate inputs are active). For most advanced sequential circuits (like a microprocessor), the RS flip flop is a challenge to work with and is very rarely used. Instead, most circuits use an *edge-triggered flip flop* that only stores a bit when it is required. You will probably discover the edge-triggered flip flop (which may also be known as a *clocked latch*) to be very useful in your own applications and easier to design with than a simple RS flip flop.

The operation of the clocked latch is quite simple (Figure 12-8). A data line is passed to the flip flop along with a clock line. While the data line stays constant, the contents of the flip flop don't change. When the clock line goes from high to low, the data is stored in the flip flop; this is known as a *falling edge clocked*

flip flop and is the most common type of flip flop that you will have to work with.

The edge-triggered flip flop is based on the RS flip flop. For this experiment, I will show you how to build an edge-triggered flip flop using NAND gates. Instead of always calling this circuit a rising edge-triggered flip flop or clocked latch, this circuit is normally known as a *D flip flop*. The organization of the

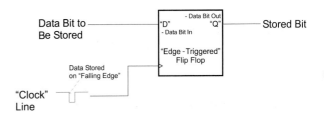

Figure 12-8 *Clocked latch memory device, a single bit of data stored using a clock line*

flip flops used in this circuit may seem complex, but their operation is actually quite simple; the two input flip flops condition the clock and data lines and only pass a changing signal when the clock is falling, as I show in Figure 12-9.

Note that in Figure 12-9, I have marked the flip flop states *before* their bit values are established as being unknown. This is actually a very important point and one that you will have to keep in mind when you are designing your own circuits. You cannot expect a flip flop to be at a specific state unless it is set there by some kind of reset circuit (which is discussed in the next experiment). The output of the edge-triggered flip flop stays unknown until some value is written in it. If you look at the signals being passed to the right flip flop, you will see that the inputs are unknown until the data line becomes low, at which point the two inputs to the right flip flop become high, and the unknown bit value is stored properly in the flip flop.

The first value written into it is a zero; the data line is pulled down before the clock line and it does *not* change the state of either flip flops' output. When the clock line goes low, it forces out a 1 to be passed to the right flip flop, keeping it in its current state.

When the clock line goes high, the right flip flop is then loaded with the current data value. After the clock line has gone high, the state of the right flip flop cannot be changed by the data line going high or low.

In this experiment, I would like you to build the edge-triggered flip flop shown in Figure 12-10. The 47 μF capacitor across the clock line resistor is meant to debounce the switch's signal and make the rising edge operation of the D flip flop more easily seen.

The operation of the edge-triggered flip flop should become very obvious as you test the operation of this circuit, although there might be one wrinkle. You may find that the data is latched on both the falling and rising edges as the clock switch is turned up or down. This is due to the switch on the clock line bouncing. Although the 47 μF capacitor will minimize this phenomena, you might want to add a very long delay 555 monostable (with an active period of one or two seconds) to actively debounce this signal.

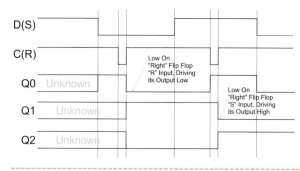

Figure 12-9   *D flip flop operation/waveforms*

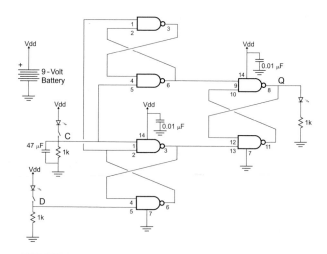

Figure 12-10   *D flip flop test circuit*

Experiment 66
Full D Flip Flop

Parts Bin

Assembled PCB with breadboard

78L05 +5-volt regulator

74LS74 dual D flip flop

Five LEDs, any color

Five 470 Ω resistors

10k resistor

47 µF electrolytic capacitor

10 µF electrolytic capacitor

Two 0.01 µF capacitors, any type

Four breadboard-mountable SPDT switches

Tool Box

Wiring kit

I find the D flip flop to be the flip flop that I build into my circuits most often. It is simple to work with and can interface to microcontrollers and microprocessors very easily. It is, however, quite awkward to wire, especially when you want to work with the full circuit, which is shown in Figure 12-11. This circuit not only has data stored on the rising edge of the clock line, but two other lines, _CLR and _PRE, will force the flip flop's output to a 0 (low voltage) or a 1 (high voltage), respectively, when they are pulled low. This allows for a number of different options when using the D flip flop in your circuit that can allow you to pull off some amazing feats of digital logic.

Looking at Figure 12-11, you might be thinking that the circuit could be easily upgraded to perform the full D flip flop function, but before you do, I want to point out that things are not as simple as you may have thought. If you wanted to convert a two-input AND or OR gate to a three-input gate, you could simply pass two input signals and its output to the input of a second gate in which the third input is passed to its second input. I neglected to point out that this trick does not work for NAND or NOR gates. For example, if you wanted to use two input NAND gates to create a three-input NAND gate, you would have to create the logic function shown in Figure 12-12; after two inputs are NANDed together, you would then have to invert the output so that they are ANDed together and then passed to the NAND gate with the other input. A three-input NOR gate would also be built the same way.

Using three gates to produce one two-input NAND gate, 18 NAND gates would be required to implement the full D flip flop function, which would require four and a half 7400 chips. To demonstrate the operation of the circuit, you could build it out of two 7410 (three three-input NAND gates), or be lazy

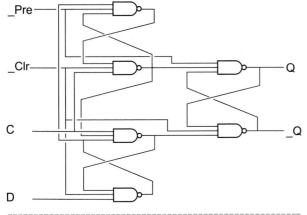

Figure 12-11 *Full D flip flop with set/reset controls*

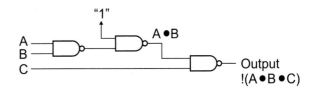

Figure 12-12 *Three input NAND gate built from two input NAND gates*

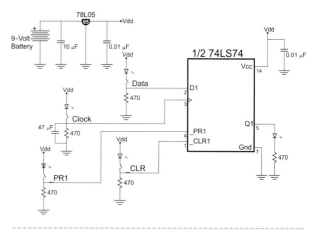

like I am and just use one 74LS74 to experiment with the different functions of the full D flip flop.

The 7474 chip consists of two D flip flops with both the Q and _Q outputs passed to the chips pins. All four inputs shown in Figure 12-11 (data and clock as well as two pins that provide you with the ability to set or reset the state of the flip flop without the use of the data or clock pins) are provided for each of the two flip flops built into the chip. The 7474 is a very versatile chip and can be used for a wide range of applications.

To experiment with the 7474, I decided to go with the 74LS74, which necessitates the use of a 5-volt regulator along with the four switches for the different inputs shown in Figure 12-13.

With the circuit built, you can now experiment with the operation of the 7474's D flip flop. Before powering up, move all four switches to the Up (closed with the LEDs turned on) position to keep the _PR or _CLR lines from affecting the operation of the flip flop. Once you have done this, by toggling the data and clock switches, you should see data being saved in the flip flop in exactly the same manner as when you built the flip flop out of NAND gates.

Once you have become comfortable with the operation of the flip flops with the data and clock switches, place both of these up (closed with the LEDs turned on), open (move to down) the _PR1 and _CLR switches, and observe the results with the LED connected to the Q output of the flip flop. You

Figure 12-13   *D flip flop test circuit*

should see that when the _PR1 LED is off, the flip flop's output is high. If you were to try and save data using data and clock while _PR1 is low, you will find that you cannot change the state of the flip flop. When you test the _CLR switch, you will see a similar operation as to the _PR1 switch; when the _CLR switch is low, the output is low and any incoming data being clocked in will be ignored. As a final check, switch off both the _PR1 and _CLR switches (LEDs off) and observe what happens. If you look back at Figure 12-11, you should be able to figure out why the flip flop behaves in this way (hint: follow the _Pre line and look at where it is connected).

Once you are finished with the experiment, don't disassemble the circuit; you will need it for the next one.

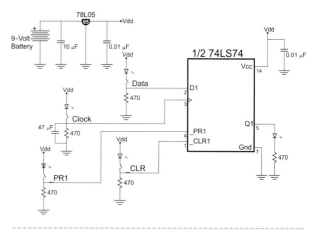

Experiment 67
Flip Flop Reset

Parts Bin

Assembled PCB with breadboard

78L05 +5-volt regulator

74LS74 dual D flip flop

Five LEDs, any color

10k resistor

Five 470 Ω resistors

47 μF electrolytic capacitor

Two 10 μF electrolytic capacitors

Two 0.01 μF capacitors, any type

Four breadboard-mountable SPDT switches

Tool Box

Wiring kit

If you turned power on and off to the D flip flop built in the previous experiment several times, you will have noticed that the initial state (or value) can be either 0 (LED off) or 1 (LED on), with no way of predicting which value it will be. This is normal because when power is applied to the flip flop, if there is any kind of imbalance in the circuit (for example, residual charge or induced voltage) on the inputs of either NAND gate, the flip flop will respond to it and this will be its initial state. Often, this random initial state is not desired; instead the circuitry should power up into a specific known state for it to work properly.

Specifying the state when the circuit is powered up is known as *initialization* and is required for more than just sequential logic circuits. When I discuss programming, I will be discussing how the internal parameters to a program (known as *variables*) must be initialized to specific values for the program to run correctly. Initialization normally takes place when the application is reset, or waiting to start executing. In this experiment, I will show how the D flip flop presented in the previous experiment can be modified so that every time it powers up it, outputs a 0.

To avoid confusion later, I should clarify the two types of resets described in this book when I talk about digital circuits. Earlier, when I was talking about simple combinatorial circuits, I also called a low or 0 voltage level as being reset (and high or 1 as being set). In this experiment, when I use the term

reset, I am describing the state when the circuit is first powered up or stopped to restart it from the beginning. When you read the term reset later in the book (as well as in other books), remember that if a single bit or pin is being described, the term reset means that it is 0 or at a low level. If a sequential circuit (like a microcontroller) is "held reset" or "powering up from reset," I mean that it is being allowed to execute from a known state.

The 74LS74 used in the previous experiment can always be powered up with its output value being a 0 by replacing the switch, 1k resistor, and LED on the _CLR pin with the resistor/capacitor network shown in Figure 12-14.

The _CLR pin is known as a negative active control and is active when the input is at a low voltage

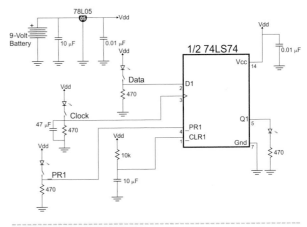

Figure 12-14   *D flip flop with RC reset circuit*

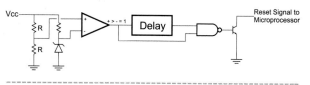

Figure 12-16   *Possible microprocessor reset circuit*

level (as demonstrated in the previous experiment). To make this pin active during power up yet allow the chip to function normally, the resistor/capacitor network on the *Transitor-to-Transistor* (TTL) input pin delays the rise of the pin (as shown in Figure 12-15) so that the pin is active low while power is good. When the signal on _CLR goes high, and the clear function is no longer active, the chip can operate normally with it being in an initial known state.

The time for the RC network to reach the threshold voltage can be approximated using the equation:

$$Delay\ Time = 2.2\ x\ R\ x\ C$$

When you work with microprocessors and microcontrollers in robots, you will want to implement a more sophisticated reset circuit. The BS2 that I will be presenting you with has a comparator-based reset circuit like the one shown in Figure 12-16. This circuit controls an open collector (or open drain) transistor output pin that will pull down a negative active reset pin when power dips below some threshold value. This circuit is often available as a processor reset control chip and is put into the same black plastic package as a small transistor.

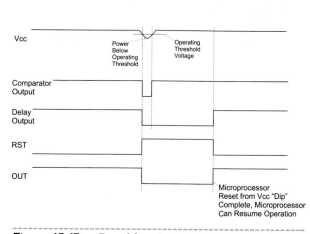

Figure 12-17   *Possible microprocessor reset circuit*

Processor reset control chips are available for a very wide variety of different cutoff voltages, ranging from 2.2 volts and upwards. Figure 12-17 shows the operation of the internal parts of the processor reset control chip when the input voltage drops below the set value; the comparator stops outputting a 1 and a delay line is activated. This delay line is used to filter out any subsequent glitches in the power line and makes sure that the power line is stable before allowing the processor to return from reset and continue executing. When the comparator outputs a low value or the delay line is continuing to output a low value, the output of the NAND gate they are connected to is high, and it turns on the open collector output transistor, pulling the circuit to ground.

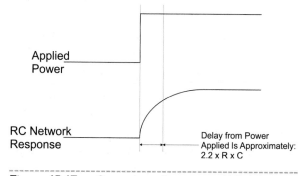

Figure 12-15   *Operation of power-up RC delay circuit*

Experiment 68
Parallel Data

Parts Bin

Assembled PCB with breadboard

78L05 +5-volt regulator

74LS174 Hex D flip flop

Six LEDs, any color

Eight 10k resistors

Six 470 Ω resistors

Two 10 µF electrolytic capacitors

Two 0.01 µF capacitors, any type

Breadboard-mountable eight-switch DIP module

Tool Box

Wiring kit

The first microprocessor (the Intel 8008) handled four data bits at a time. This grouping of four bits is known as the processor's "word" size, and if you were to chart the progression of the different personal computers over the years, you would see how the word size has grown over the past 25 or so years. The first popular personal computer, the Apple, had a processor that had a word size of eight bits. Five years later, the IBM PC used the 8088 processor that could process data 16 bits at a time (to avoid getting angry emails later, I want to point out that while the external data buses were 8 bits, the processor itself handles 16-bit data words). Another five years after the introduction of the first IBM PC, the first 32-bit word Intel 80386-based PCs were on the market. Today, if you follow the computer press, you'll know that the latest computer systems and servers are equipped with 64-bit Intel Itanium or AMD Opteron microprocessors. As the word size of the processors increases, their ability to quickly perform complex mathematical operations improves, as does their ability to handle large amounts of data. This increase in word size is attributed to what is known as Moore's law. This observation states that the number of transistors (and by inference their complexity and ability to process data) doubles every 18 months.

So far in this book, when presenting you with digital electronic circuits for the experiments, I have been focusing on processing data one bit at a time. Extremely complex circuits can be created that work with just single bits, but instead of designing and working with complex circuits, it is usually easier (and faster if speed is a criteria) to work with multiple bits in parallel. These reasons are why more powerful systems generally handle more bits at a time as part of the processor's word size.

For the remainder of this book, I will be working with multiple bits that are grouped in parallel in some way. You might be thinking that the circuits will become unmanageable because of the need for multiple chips to take care of the multiple bits of data, but I will be taking advantage of some of the many available products that can handle multiple bits simultaneously. A good example of this type of chip is the six (or "hex") D flip flop chip, the 74174.

The D flip flops used in the 74174 are similar to the D flip flops contained within the 7474 that I presented in the previous experiments, except that the clock and reset controls are common to all six D flip flops in the chip. The reason for "commonning" these pins is ostensibly to save pins on the chip (a six D flip flop chip with all 6 lines for each flip flop made available to the user would require 38 or more pins), but it actually serves the purpose of allowing you to save data in 6 independent bits or clear them at a one time.

So far in the book, I have been working with data one bit at a time. For this experiment, I would like to repeat the first D flip flop experiment, but with six bits. To test this application out, I came up with the circuit in Figure 12-18.

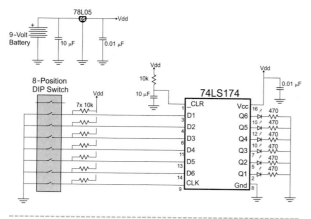

Figure 12-18 *Hex D-type flip flop test circuit*

It is quite a bit of fun to flip the switches back and forth and see the LED outputs change with the tog-gling of the clock switch. The increase in data-handling power will probably not be immediately apparent, but if you were to compare the circuit in this experiment to the first 74LS74 experiment circuit, you should see that the number of switch movements required per bit to save data has been reduced. This is because after setting each of the bit values, you only have to pulse the clock signal down and up once. A further savings in time is that each bit can be set at the same time (in parallel) with the other ones. Using individual D flip flops, in order to save six bits, it would take 12 switch cycles (six data sets and six clock pulses) where the same amount of data could be stored in just two clock cycles (one to set the data and one clock pulse).

Experiment 69
Traffic Lights

Parts Bin

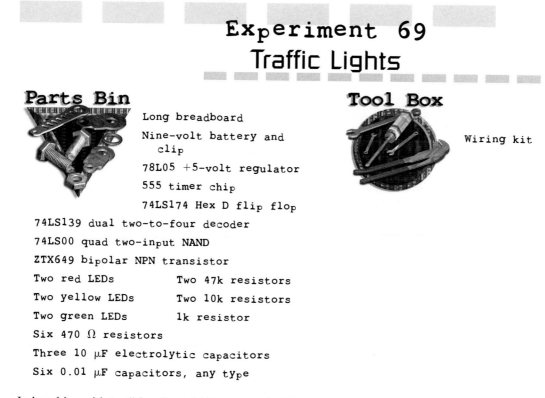

Long breadboard

Nine-volt battery and clip

78L05 +5-volt regulator

555 timer chip

74LS174 Hex D flip flop

74LS139 dual two-to-four decoder

74LS00 quad two-input NAND

ZTX649 bipolar NPN transistor

Two red LEDs

Two yellow LEDs

Two green LEDs

Six 470 Ω resistors

Three 10 µF electrolytic capacitors

Six 0.01 µF capacitors, any type

Two 47k resistors

Two 10k resistors

1k resistor

Tool Box

Wiring kit

In breaking with tradition, I would like you to build the circuit shown in Figure 12-19 wired on a long breadboard (it won't fit on the traditional small breadboard) as in Figure 12-20. Once you have built the circuit, you will have a sequential circuit that can be used to implement toy traffic lights running in two different directions. You will discover that while one set of lights is red, the other set will turn green for a second and then yellow for another second before turning red and starting the process over. This circuit could be the basis for a set of traffic lights used in a model train layout.

I'm sure that by having you build this circuit without any kind of introduction, you should recognize almost all of the pieces while looking at the schematic, but you are probably confused as to how

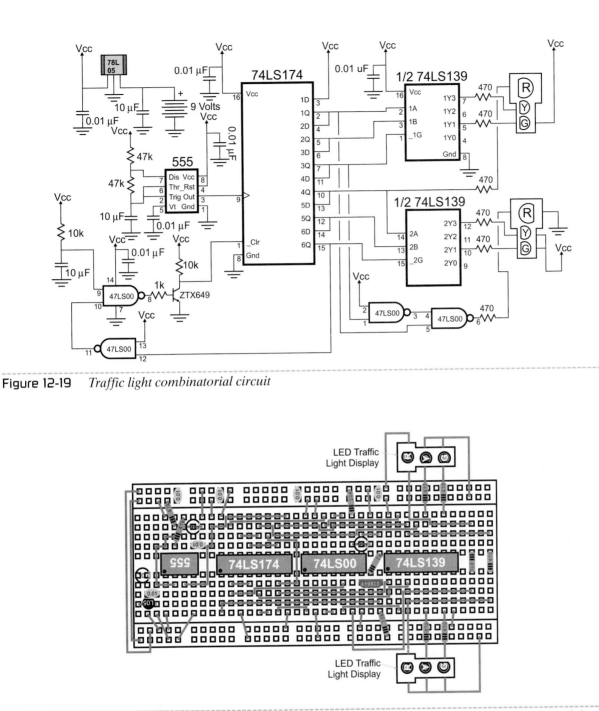

Figure 12-19 *Traffic light combinatorial circuit*

Figure 12-20 *Traffic light circuit built on a long breadboard because it is too large for the breadboard that fits on the books PCB*

they all work together. Looking at the start of this section, I gave you a very simple block diagram describing how a combinatorial circuit works. The circuit in 12-19 doesn't match it in any particular way.

The reason for giving you this circuit is to show you how easily you can end up with a combinatorial circuit that has been optimized in some way that only

makes sense to the designer. Although I would stress that you should design your combinatorial circuits using the block diagram presented earlier, I wanted to use this circuit to explain how to look at seemingly complex circuits and figure out what they are doing. This traffic light circuit should be a fairly good example of this problem because when I designed it, I was trying to optimize it to the point where it would fit on the small breadboard that is used with the PCB that comes with the book. Unfortunately, I failed and moved to the larger form factor, but I never got around to changing the circuit (which is a common occurrence when you are moving circuitry to a new wiring solution).

To understand the circuit, try to break it into functional blocks and note whether or not you understand them. If I were given this circuit, I would break out the blocks (with comments) as

1. Power supply. 9-volt battery and 78L05, standard 5-volt, 100 mA supply.

2. 555 clock. Plugging R and C into the monostable oscillator formulas, I can see that it has a period of roughly one second.

3. LED outputs. 470 Ω current-limiting resistor used with LEDs.

4. Reset circuitry. The 10k resistor and 10μF capacitor are low while the circuit powers up. The reason for the NAND gate fed by the resistor and the capacitor as well as the inverted D flip flop output is confusing. Also, the reason for the open collector transistor driver is confusing.

5. The operation and use of the 17LS174 is very confusing. It also seems to be connected to the reset circuitry.

6. What is a 74LS139? Looking at the circuitry, it looks like it is some kind of combinatorial circuit, but its purpose isn't obvious.

So, with this quick review of the major blocks of the circuit, I can see that I am confused about three parts. I would start with Number 3, but it seems to be integrated with the operation of 74LS174, so let's start with the 74LS174 and see if we can figure out how it is supposed to work.

The 74LS174 is wired with the outputs of the different flip flops wired to the inputs. The first input is tied to positive power, so it is always going to load the first flip flop with 1. Because each flip flop's output is connected to its input, you would expect the 1 to pass from flip flop to flip flop. The only problem is what happens when they are all loaded with 1s. If you look at the output of the last flip flop in the chain, you will see that it is being inverted, and through a NAND gate and the open-collector driver, it is con-

nected to the chip's flip flop reset. If you follow through the line, you will see that the _CLR pin of the 74LS174 goes low when 6Q goes high, resetting all the flip flops in the chip. If all the flip flops in the chip are reset, then the process of loading a 1 will resume, as I show in Figure 12-21. The "glitch" in Figure 12-21 is the output of 6Q going high and then being reset along with all the other bits.

You now know what part of the purpose of the reset circuit is (circuit block number 3), and you should see that it combines with the RC network to clear all the 174's flip flops on power up or if they all get set. The only question remaining is the use of the open collector driver. Since the open collector driver is only connected to one input, I would guess that it is a *Mickey Mouse logic* (MML) inverter, because the four NAND gates of the 74LS00 are already used.

The only piece left is the 74LS139 and what its function is in the circuit. When you are confronted with a chip that you have never seen before, I suggest that you look it up and see if you can figure out its function. You can either do a Google™ (www.google.com) search or look at a parts distributor (I recommend Digi-Key® at www.digikey.com because they provide links to the parts data sheets). Looking at the 74LS139's part description, you will see that it is a dual two-to-four decoder (which may or may not be a help to you).

The 74LS139 and the three-to-eight decoder (74LS138) are incredibly useful chips when you need an arbitrary logic function and don't want to wire in a bunch of ANDs, ORs, and NOTs. Decoders (also known as *demultiplexors*) convert binary values into individual output lines and are primarily used to decode memory addresses to individual chips. In Fig-

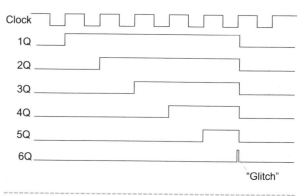

Figure 12-21   *Traffic light D flip flop operation*

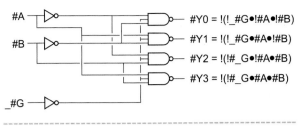

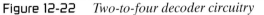

#Y0 = !(!_#G•!#A•!#B)
#Y1 = !(!_#G•#A•!#B)
#Y2 = !(!#_G•!#A•#B)
#Y3 = !(!#_G•#A•#B)

Figure 12-22 *Two-to-four decoder circuitry*

Table 12-3 Traffic light operation

State	North/South LEDs	East/West LEDs
$Q_0 = 0$	Red on	Red on
$Q_0 = 1$	Red on	Green on
$Q_1 = 1$	Red on	Yellow on
$Q_2 = 1$	Red on	Red on
$Q_3 = 1$	Green on	Red on
$Q_4 = 1$	Yellow on	Red on

ure 12-22, I show the logic built into each one of the two-to-four decoders built into the 174 along with the output equations.

The use of the decoders probably doesn't make sense, but it will once I explain how I designed the circuitry and its different operating states. By allowing the 1 to "flow" through the 74LS174's D flip flops, I have created a set of different bits that I can use to determine which of the traffic lights should be on. Using the states in Figure 12-21, I made up Table 12-3 to specify which lights should be active. From this table, I could specify equations for when the different LEDs would be on.

$$N/S \ Red = !Q_3$$

$$N/S \ Green = Q_3 \cdot !Q_4$$

$$N/S \ Yellow = Q_3 \cdot Q_4$$

$$E/W \ Red = !Q_0 + Q_2$$

$$E/W \ Green = Q_1 \cdot !Q_2 \cdot !Q_3$$

$$E/W \ Yellow = Q_1 \cdot Q_2 \cdot !Q_3$$

Looking at these equations, you will see that four LEDs are quite simple products (and can be wired using the decoder); one LED can be connected directly to a data bit. One LED (East/West Red) has a fairly complex expression that is going to have to be analyzed because I do not want to add any more chips to the circuit than I have to (in an effort to build the circuit on a short breadboard).

Fortunately, I can use DeMorgan's theorem to change the expression to use NAND gates:

$$E/W \ Red = !(Q_0 \cdot !Q_2)$$

I still have to invert one value, but this can be done easily with a leftover NAND gate or an MML inverter as I did with the reset circuit.

Experiment 70
Shift Registers

Parts Bin

Assembled PCB with
  breadboard

7805 voltage regulator

555 timer chip

74LS74 dual D flip flop

74LS174 hex D flip flop

Eight LEDs, any color

10k resistor

Two 2.2k resistors

Eight 470 Ω resistors

100 µF electrolytic capacitor

10 µF electrolytic capacitor

1 µF electrolytic capacitor

Four 0.01 µF capacitors, any type

Tool Box

Wiring kit

For the remainder of this section (and much of the book), I will be looking at circuits that pass data back and forth, from one chip or block of chips to another. This requirement to pass data is not always between computer systems; often it is between chips or subsystems within a robot. This data usually consists of multiple bits, each of which is part of the whole that is being sent. When multiple data bits are being passed back and forth, you should be asking the critical question: should the data be passed in parallel or serially? So far in the book, I have shown circuits in which multiple bits of data are transmitted in parallel; each bit is given its own wire or pin. Parallel data can be transmitted quite quickly, but there needs to be a clocking or enable bit to indicate to the receiver that the data is ready to be processed. Parallel data can become quite cumbersome when you have a large number of bits (16 bits are a *lot* more trouble than four times the trouble of 4 bits).

The alternative to sending data in parallel is to strip out each bit and send each one serially along a single line as I show in Figure 12-23. This is known as serial data transmission and is used for virtually all of the interfaces built into your PC. The only parallel data interfaces remaining in a modern PC consist of the processor's front-side bus, PCI buses, and the parallel port (if it is present). A serial data transfer, even though it requires hardware to serialize and then deserialize the data, is preferred for virtually all applications.

To send six bits in parallel, a half-dozen transmitting drivers and an equal number of receivers are required. To send six bits serially, just a single driver and receiver are required, but the sending circuit must have a shift register transmitter and the receiving circuit must have a shift register receiver. The parallel data can be sent in the time required for just one bit, whereas the serial data requires enough time to send each of the six bits individually.

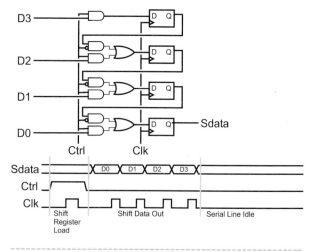

Figure 12-23 *Parallel to serial data conversion*

It probably looks like transmitting data serially requires a lot of overhead and it slows down the data transfer. You should consider a number of factors before making this assumption. The first is that most chips are not made out of individual logic gates as the simple chips presented here are; they are usually very dense circuits consisting of thousands of gates, with the impact of adding serial shift registers being very minimal. Another issue to consider is that it can be very difficult to synchronize all the parallel bits to arrive at the receiver at the same time in high-speed circuits. Finally, multiple wires can take up a lot of space and be quite expensive. If chips or subsystems could have shift registers built into them, it often makes sense (both practical and economic) that data be transferred serially.

Data can be transmitted serially three different ways (Figure 12-24). If you are familiar with RS-232 (which will be discussed with the Parallax Basic Stamp 2), then you will be familiar with asynchronous serial data transmission in which one line is used to send the data bits. When data is sent asynchronously, each bit is exactly the same length and a start bit indicates a data packet is coming. Synchronous or clocked serial data requires two lines, a data line and a clock line that indicates when the data value is correct. As I will show in this experiment, the receiving flip flops are edge-triggered flip flops with the incoming data saved on a transition of the clock (Figure 12-25). The final method of serial data transmission is known as Manchester encoding and indicates a data value based on how long the signal is

high or low. Manchester encoding is a very popular encoding format for infrared TV remote controls. For most simple applications in which data is being passed serially between two chips, you will be using a synchronous serial data stream that will be demonstrated in this experiment. Synchronous serial receiver and transmitter circuits are easy to build and do not require any of the specialized synching equipment of the asynchronous or Manchester encoding methodologies. Looking at Figure 12-25, you will see that if the data line changes to an invalid value and returns to the correct value before the falling edge of the clock (when data is latched in), the invalid value is not recorded. If an invalid value like this occurred in an asynchronous or Manchester-encoded serial line, there would be an error.

In Figure 12-26, I have shown a simple synchronous serial circuit that will shift a 1 through the shift register continuously by passing data from the output of one D flip flop to the input of another. When the circuit powers up, all the connected D flip flops of the shift register will be cleared, except for one D flip flop that will be loaded with a 1. This value will be shifted through the D flip flops at a rate of about five per second.

In this experiment, I have tied the output of the shift register back to its input, so the value within it is never lost. In a typical circuit, the transmitting shift register's input is tied to Vcc or ground (so after the data is sent, the known data continues to be sent), and the receiver shift register's output bit does not pass data to another shift register.

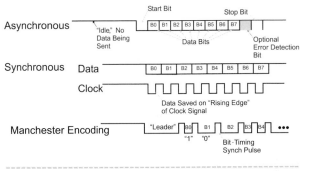

Figure 12-24 *Asynchronous, synchronous, and Manchester-encoded serial data streams*

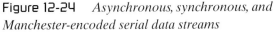

Figure 12-25 *Close up detail showing how synchronous serial works*

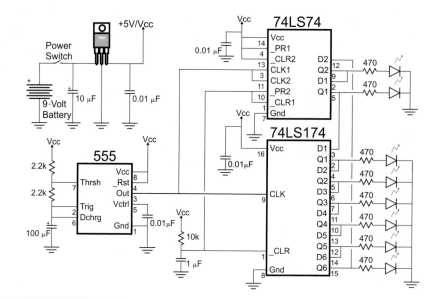

Figure 12-26 *Rotating LED using shift register*

Experiment 71
Christmas Decoration

Parts Bin

Prototype PCB

Nine-volt battery clip

Nine-volt battery holder

7805 +5-volt regulator

NE555 timer chip

74LS174 hex D flip flop

74LS74 four-bit adder

74LS86 quad two-input XOR gates

Eight LEDs, any color

Two 47k resistors

10k resistor

Eight 470 Ω resistors

Three 10 μF electrolytic capacitors

Six 0.01 capacitors, any type

Panel mount SPST switch

Tool Box

28-32 gauge wire
wrap/prototyping wire

Soldering iron

Solder

Clippers

Rotary cutting tool (see text)

Drill with bits

Programmable calculator with binary
arithmetic capability

In virtually all the books I have written, I have included a circuit that will turn on and off LEDs in a random fashion and have suggested that the reader build the circuit as a Christmas tree, with the flashing LEDs being used as decorations for the tree. The reason for including this circuit is not to spread holiday cheer so much as to illustrate a very useful binary logic circuit known as the *Linear Feedback Shift Register* (LFSR, Figure 12-27).

The simple LFSR illustrated in Figure 12-27 feeds back bits 5 and 7 of the shift register through XOR

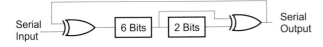

Figure 12-27 *Basic 8-bit linear feedback shift register with serial input*

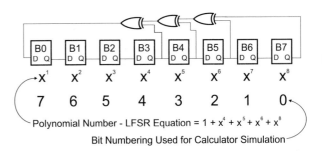

Polynomial Number - LFSR Equation = $1 + x^4 + x^5 + x^6 + x^8$

Bit Numbering Used for Calculator Simulation

Figure 12-28 *Practical 8-bit LFSR for generating pseudo-random numbers*

gates to the input. This changes the bit values in the shift register according to the formula:

$$Bit_0 = Bit_{in} \text{ XOR } (Bit_5 \text{ XOR } Bit_7)$$

The LFSR is typically used for three purposes:

- Create a checksum value known as a *Cyclical Redundancy Check* (CRC), which is a unique value or *signature* for a string of bits. Both the transmitter and receiver will pass the data through LFSRs, and at the end of the process, the CRC produced by the transmitter will be compared to the CRC produced by the receiver. If there is a difference in the CRCs, then the receiver will request that the transmitter resend the data.

- Encrypt a string of bits. LFSRs can be used as an encryption/decryption tool with part of the encryption being the initial value in the LFSR. The value output from the LFSR is dependent on the initial value loaded into the LFSR. Decrypting data is also accomplished by using an LFSR, but configured as the complementary function.

- Produce pseudorandom numbers. One of the most challenging computer tasks that you will be given is to come up with a series of random numbers. Computers are generally thought of as being deterministic, which means that what they are doing at any given time can be calculated mathematically. This property is important for most applications (nobody wants a computer to boot differently each time or have a word-processing program respond randomly to keystrokes), but it is a problem for many robot applications in which the robot has to start moving.

In all of these applications, the LFSR is an ideal choice as a solution because it can be built very simply from just a few gates (meaning low cost and fast operation). The LFSR can also be implemented in software, as I will show in this experiment.

In this experiment, I will have you build the eight-bit LFSR shown in Figure 12-28. If you were going to express this LFSR to somebody else, you could send a graphic something like Figure 12-28 or you could express it according in the polynomial format such as the following:

$$f_x = 1 + x^4 + x^5 + x^6 + x^8$$

The polynomial format is the traditional way of expressing how an LFSR works and is used by mathematicians to evaluate LFSR operation.

You should be aware of a few important facts about LFSRs:

- The LFSR can *never* have the value zero in it. If it contains zero, then it will never have any of the bits set.

- The ideal LFSR implementation will be able to produce $2^n - 1$ different values.

- A poorly specified LFSR may have the situation where it ends out with a value of zero.

For this experiment, I used the LFSR specified in Figure 12-28 to create a Christmas tree decoration. This experiment consists of a 555 timer driving an eight-bit LFSR with an LED on each bit used as a light. The circuit diagram is shown in Figure 12-29.

You should notice a few points about this circuit. I enable the clear circuitry for the 74LS174 and one D flip flop in the 74LS74. In the other D flip flop of the 74LS74, I have enabled the set circuitry to ensure that at least one bit is set to 1 on power up and not all zeros, which will result in a circuit that never changes.

Before cutting out a Christmas tree shape in a prototyping PCB and building it, I decided to test the LFSR in two ways. The first was to use my programmable calculator (which has the ability to manipulate binary values and perform Boolean arithmetic operations) to make sure that I would have 255 different values.

Another way to test the circuit designis to build it on a breadboard. This circuit is similar to the traffic Light experiment built earlier (but with the different voltage regulator). After building the circuit, I let it run for 10 minutes to make sure that it didn't stop for any reason (a bad LFSR design). This is where I discovered that I should use a 7805 rather than a 78L05.

Once I verified that the circuit worked using both the calculator as well as the prototype circuit, I then built the Christmas decoration to show off the operation of the LFSR. To cut the Christmas tree shape, I used a carbide cutting wheel on my rotary (Dremel) cutting tool. When making the cuts, make sure you are wearing eye protection and a protective mask. Once the shape was cut, I mounted the battery (and power switch), followed by the chips and finally the LEDs around the outside of the "tree" as flashing lights. Figure 12-30 shows how the battery is mounted on the PCB and used as the tree's "stand" along with a rear view of the point-to-point wiring used for the circuit.

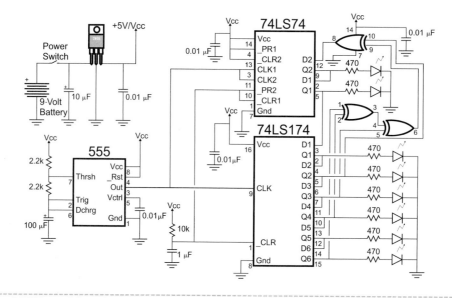

Figure 12-29   *LFSR used for Christmas tree decoration*

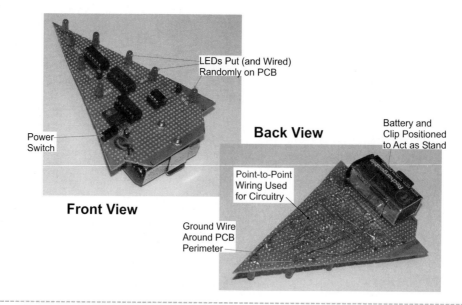

Figure 12-30   *Christmas tree prototype details*

Experiment 72
Random Movement Robot

Parts Bin

Assembled PCB with breadboard

DC motor base with four AA battery clip

555 timer chip

74LS174 hex D flip flop

74LS86 quad dual input XOR gate

Two 1N914 or 1N4148 silicon diodes

Three 10k resistors

Two 100 Ω resistors

47 μF electrolytic capacitor

1 μF capacitor, any type

Three 0.01 μF capacitors, any type

Tool Box

Wiring kit

The ability to create an LFSR is one that is actually quite important when you want to create your own robot. Often you will want a robot to move randomly about a room after completing a task so that it will be ready for the next one, or if it has boxed itself in a corner, a very successful way of getting it out is to move about randomly and restart the activity that got it stuck. Later in the book, I will show how the BS2 can provide you with a random statement that you can use to move your robot randomly about the room, but for this experiment, I would like to use the linear feedback shift register that I presented in the previous experiment and use it to drive the DC Motor control base randomly about the room.

In designing the LFSR circuit that I would use in the robot, I wanted to make sure that I came up with one that could be located comfortably on the breadboard and robot base that I have been using for the experiments presented in this book. As in the previous experiment, many logic circuits cannot be confined to the small breadboard, so as simple a circuit as possible would have to be created. The LFSR that I came up with is shown in Figure 12-31 and could be implemented just using a six-bit D flip flop and a chip with two XOR gates on them. Both these functions are available on chips that you have already worked with.

The six-bit LFSR's operation (the position of the taps) was verified using a programmable calculator program like the one presented in the previous experiment.

When I discussed shift registers earlier in the book, I said that I preferred using basic D flip flop chips and building them into a circuit as required. This circuit and the Christmas decoration are good examples why this is a good philosophy; the 74LS174 can be built into a shift register, an 8-bit LFSR, or this 6-bit LFSR directly.

In this experiment, I have created a simple robot that will move left, right, and forward randomly under the control of the LFSR shown in Figure 12-32. To allow the entire circuit to be placed on the small breadboard included with the robot, I only turn the

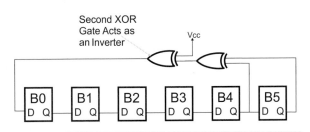

Figure 12-31 *Six-bit LFSR for controlling robot's pseudo-random motion*

motors forward. Turning is accomplished when one motor is off and the other is active. This is something less than what I consider to be a true robot, but it illustrates the random motion possible from an LFSR. The circuit that I came up is shown in Figure 12-32.

One unusual aspect of the robot (for me) is my use of a single battery pack instead of driving the robot's motors with the AA battery pack and using the 9-volt battery and a voltage regulator for the logic

(and 555 timer). I probably could have put in a 7805 (and the required resistors), but this would have made the circuit wiring denser, and in the end the robot works quite well with a single power supply.

Another unusual aspect of this circuit is that I use an XOR gate, with one input tied to Vcc as an inverter. Earlier, I discussed how NANDs and NORs can be used as inverters, but I neglected to note that an extra XOR gate can be used in the same way.

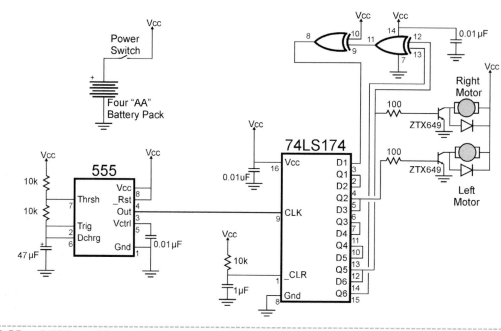

Figure 12-32   *LFSR used for random-movement robot*

Experiment 73
Counters

Parts Bin

Assembled PCB with
  breadboard

7805 +5-volt regulator

555 timer chip

74LS174 hex D flip flop

74LS283 four-bit adder

Four LEDs, any color

Three 10k resistors

Four 470 Ω resistors

Two 47 µF electrolytic capacitors

1 µF capacitor, any type

Four 0.01 µF capacitors, any type

Tool Box

Wiring kit

One of the most useful prepackaged sequential logic circuits that you can buy is the counter (Figure 12-33). The counter consists of a multibit latch that passes its output to an adder. This adder increments (adds one to) the latch output and passes it back to the input of the latch and this incremented value is saved in the latch each time the clock is cycled. Counters provide a variety of different purposes and it is important to remember that the clock can be a constant frequency clock (in order to time an event) or it can be an external event, the number of which is recorded using the counter.

In Figure 12-33, you should note that I have included the carry output of the counter's adder. The carry bit becomes active when the latch value plus one is greater than the value that can be stored in the latch. This output can be cascaded to another counter (as shown in Figure 12-34) to drive it and provide double the number of bits being counted. When you

are creating low-level computer programs, this carry function is critical for you to be able to create mathematical functions for numbers that require more bits than the processor can handle.

In this experiment, I will show how a counter is built, using separate adder and latch chips. When you add counter functions to your own applications, you will probably use prepackaged TTL and CMOS functions such as the 74161 and 74193. Both chips are four-bit counters, but the 74161 can only count up, and value changes (clear and load) are latched in with the clock. The '193 can count up or down, and value changes are processed immediately (asynchronously) and not with the input clock (synchronously).

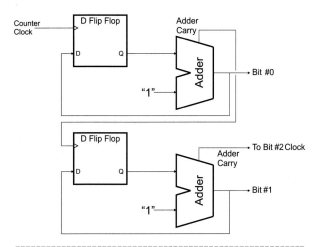

Figure 12-34   *Cascaded counter carry of one bit passed to the next one*

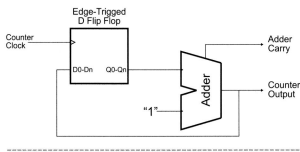

Figure 12-33   *Basic counter design*

The 74160 is identical to the 74161 but only counts to a bit value of 9, rather than 15 as in the case of the 74161 (the 74192 only counts to 9 as well). This allows for easy creation of decimal counter displays (using seven-segment LED displays and standard display drivers) for "non-evil genius" users.

Along with the 74161 and 74193, a number of different TTL and CMOS chips provide counting functions that you may want to consider for your designs. I personally use these two chips because they can be used in a wide variety of different applications and can be easily cascaded into larger number counters. In most TTL counters, the value is updated on the ris-

ing edge of the clock; this is important to note to make sure that the carry value has a rising edge at the appropriate instant to make sure the more significant bits counter is incremented at the right time.

In this experiment, the operation of the counter circuit will be demonstrated using a 555 outputting a 1 cycle per second (1 Hz) clock that is driving a clock circuit built from a 74LS174 hex D flip flop latch and a simple adder (Figure 12-35). This very simple circuit will just increment the value displayed on the LEDs until all are lit (displaying a value of 15 or %1111) and then roll over back to zero.

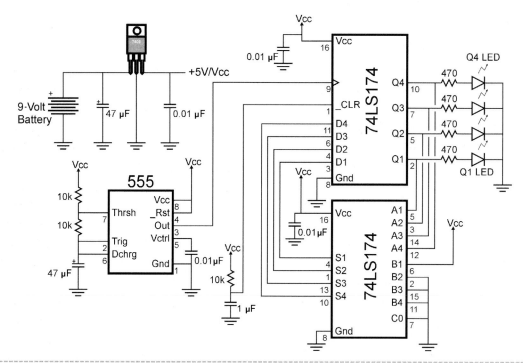

Figure 12-35 *Counter circuit using discrete parts*

Experiment 74
Schmitt Trigger Inputs and Button Debounce

Parts Bin

Assembled PCB with breadboard

78L05 +5-volt regulator

74LS191 four-bit up/down counter

74LS14 hex Schmitt trigger input inverters

Two 10k resistors

Four 470 Ω resistors

Two 10 μF electrolytic capacitors

Three 0.01 μF capacitors, any type

Momentary push buttons

Breadboard-mountable SPDT switch

Tool Box

Wiring kit

I consider the issue of debouncing switches and buttons to be one of the most important and vexing problems that you will have to deal with when you are developing robot applications. Although you might think that electrical connections happen instantaneously, you might be surprised to discover that the contacts within a switch actually bounce a few times before the switch makes a constant contact. This is shown in Figure 12-36.

When I introduce you to the Parallax Basic Stamp 2, I will show how a switch can be read and the bounces filtered out. Passing switch bounces directly to an application is a problem because the multiple bounces of the switch are usually handled as if the switch was opened and closed multiple times.

One method of debouncing a switch is to create a small flip flop that can have its state changed by a double-throw switch like the one in Figure 12-37. The switch will tie the input to the right logic inverter to whatever state it is in and this output value is passed to the left inverter, changing its state and putting the circuit in equilibrium. The inputs and outputs of the two inverters hold the value of the flip flop when the switch bounces and is not in contact with either positive voltage or ground.

The debounce circuit that I recommend you use is shown in Figure 12-38. This circuit consists of a resistor-capacitor network that charges over a given amount of time or discharges quickly through a closed switch or button. Figure 12-39 shows the filtering of the bouncing; it is not perfect, but it is much better than what we started with.

The inverters with the funny symbols in Figure 12-40 are Schmitt trigger input inverters and provide

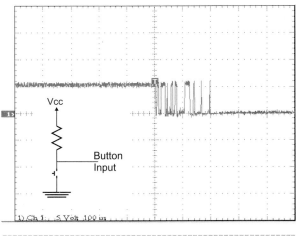

Figure 12-36 *Oscilloscope picture of a switch bounce*

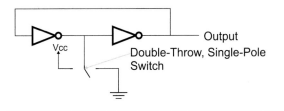

Figure 12-37 *Flip flop-based switch debounce*

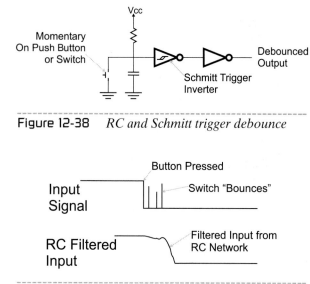

Figure 12-38   *RC and Schmitt trigger debounce*

Figure 12-39   *RC debounce filtering action*

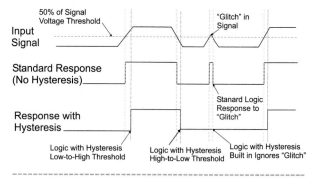

Figure 12-40   *Logic signal hysteresis*

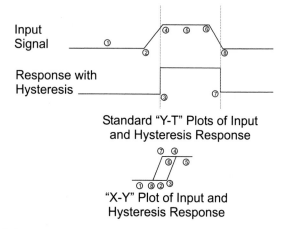

Figure 12-41   *Plotting logic signal to hysteresis*

an extra measure of filtering the button input. Schmitt trigger inputs are designed to change state on the rising or falling edge of a signal with hysteresis as shown in Figure 12-40. *Hysteresis* is the property of the Schmitt trigger inputs in which the threshold point for the rising edge of the signal is different than the falling edge. Looking at Figure 12-40, you can see that the rising edge threshold is above the normal gate voltage threshold, whereas the falling edge threshold is less.

These changing threshold values are the reason for the strange symbol on the inverters, indicating Schmitt trigger inputs. Figure 12-41 shows what the input versus the gate response is on an X-Y chart. The X axis is the input voltage with rising voltages to the right, and the Y axis represents the response of the Schmitt trigger input. By following the numbers, you should see the response of the input and that it forms the same symbol that I put on the inverter gates. For comparison, a traditional logic gate does not use this symbol; the response threshold is the same for rising and falling edge signals.

For this experiment, I am going to assume that you do not have access to an oscilloscope. To demonstrate button bouncing and the filtering of the circuit shown in Figure 12-38, I want to pass the input of a button to a counter as shown in Figure 12-42. In this circuit, by pressing the button, the counter input will be pulled down, and if the button bounces, the counter will be incremented several times. To test the circuit repeatedly, I have added a clear button. The clear button does not need to be debounced because we don't

care how many times the counter is cleared each time the button is pressed.

After you have built the circuit in Figure 12-42 and tested it (recorded the number of bounces per switch press), add in the 74LS14 inverters and 10 μF capacitor to debounce the button input.

Using a particularly noisy push button and passing the signal directly to the 74LS191 counter, I found that after 10 pushes of the button, I got an average of 10.1 bounces per press with a range of 1 to 14 for the binary values displayed on the LEDs. Adding the 74LS14 Schmitt trigger inputs, I found that this average dropped to 1.2 with a range of 1 to 2. Although this debounce circuitry wasn't perfect, it is significantly better than what I had with no circuitry, and you will find that these improvements come with a cost of just a few cents. I know quite a few people who make sure to put some extra chips with Schmitt trigger inputs (such as the 74LS14) in their designs to allow them to debounce user inputs if they are required later.

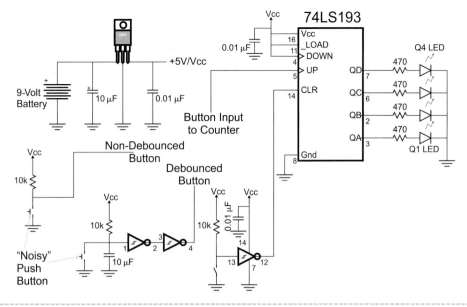

Figure 12-42   *Counter circuit to test debouncing strategies*

Experiment 75
PWM Generation

Parts Bin

Assembled PCB with breadboard

78L05 voltage regulator

555 timer

74LS191 counter

74LS85 4-bit magnitude comparator

Four-bit DIP Switch

LED, any color

Two 4.7k resistors

470 Ω resistor

10 µF electrolytic capacitor

Five 0.01 µF capacitors, any type

Tool Box

Wiring kit

In this book, I will show you a number of different ways of generating PWM signals to control motor speed or LED brightness. Earlier in the book, I showed how a 555 timer can be used to provide a simple PWM, although it has the problem that a full-on or full-off signal cannot be produced. When I discuss the Parallax Basic Stamp 2, I will show how the built-in PWM statement works, but this PWM operates at only 1 kHz, which has the potential for producing an audible whine and does not run

continuously. Another problem with the Basic Stamp 2 PWM statement is that it cannot execute in the background while other program statements are executing. What is needed is a simple PWM generator that will provide varying duty cycles from zero to 100 percent and will run continuously without any intervention from a controller.

Because it is needed, in this experiment, let's look at a design for building this type of PWM generator circuit. The PWM generator I would like to work with will be based on a binary counter driven by a 555 timer. The counter output will be continuously compared against a bit value, and when the bit value is greater than the counter value, a 1 will be output. The block diagram for the circuit that I envisioned is shown in Figure 12-43 and, amazingly enough, worked quite well when I first built it.

When you study Figure 12-43, there will probably be one point that won't make sense to you; I show that the counter ranges from 0 to 14 and not 0 to 15, as you would expect for the typical four-bit counter. I wanted the counter to reset itself at 14 rather than 15 so that when I compared the binary values, I could produce a 100 percent duty cycle as well as a 0 percent duty cycle by outputting when the set value was greater than the counter value. If the counter ran from 0 to 15, then the circuit would not be able to produce a PWM with a 100 percent duty cycle by simply outputting a 1 when the input value is greater than the counter value.

To produce the bit range from 0 to 14, I used the 74LS191 chip counting down and tying the _LOAD pin to the _RIPPLE pin and driving the inputs to 14. The _R (Ripple Output) pin becomes active when the chip is rolling over from one extreme to another, and the _LD pin moves the value at the input pins into the counter's latches when it is active. Normally, when a four-bit counter is rolling over as it counts

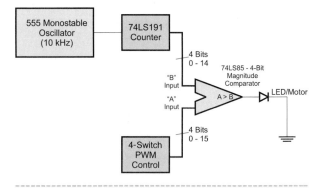

Figure 12-43   *Block diagram of PWM generator*

down from 0 to 15, but by tying the _R pin to the _LD (negative active load) pin of the 74x191, you can load in a new value when the counter reaches zero and is about to roll over. This feature is ideal for this application as it ensures the count stays between the range of 0 and 14.

Converting the block diagram to a schematic is very straightforward (see Figure 12-44) and wiring it onto the PCB's breadboard is tight, but not really a challenge. The PWM output value is specified by the four position DIP switch.

I used TTL chips (with the included 78L05 regulator) rather than CMOS chips because I found that it is difficult locating 74C85 chips. An advantage of using TTL instead of CMOS for this circuit was that I

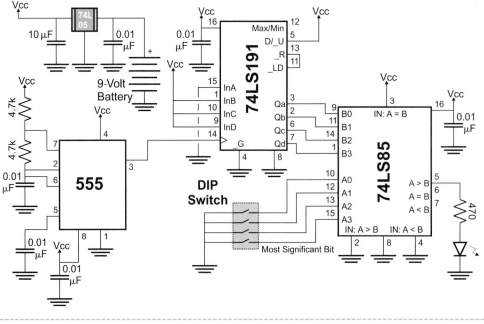

Figure 12-44   *PWM generator circuit*

could simply pull the comparator inputs to ground without a pull-up resistor. If you were to build this circuit with CMOS chips, make sure that you have pull-up resistors on the DIP switch to ensure a high voltage is passed to the comparator.

Once you have built the circuit, you will find that the LED's brightness will be dependent on the value on the DIP switch. It will be confusing as the value on the DIP switch will seem to be the opposite to the behavior of the PWM. When all the switches are on (as marked on the switch), the LED will be off and when all the switches are off, the LED will be full on. This confusion is a result of the on marking indicating when the switches are closed, not when the signal is a 1 or high (which can be mistaken for being on); when the switches are closed, the comparator input is pulled to ground and has the value 0.

The operation of the PWM generator can be observed with an oscilloscope as in Figure 12-45 and 12-46. Figure 12-45 shows the PWM output when the least significant switch is off and the three more significant switches are on. In this case, the counter value compared to is 1, and current will be passed to the LED when the counter has a value of zero (the only value which is less than 1). Figure 12-45 shows an intermediate value. I did not show the PWM values when the switches were all on (0 percent duty cycle) or all off (100 percent duty cycle) simply because the PWM output is simply a flat line.

Looking at Figures 12-45 and 12-46, you might be thinking that I'm a pretty lucky guy; it looks like I have a five-channel (or more) oscilloscope. I don't, but I was clever with the PC software that reads the trace's oscilloscope. I triggered on the most significant bit of the counter (Bit 3) and then moved another probe to the different signals to provide the five-channel display.

If I were going to use this circuit in a robot, rather than the DIP switches, I would replace it with a shift register that is driven by the controller. A simple differential robot could be implemented with two of

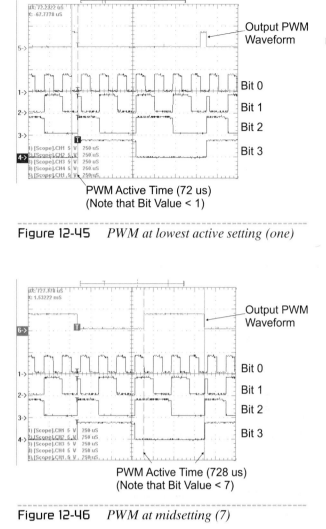

Figure 12-45 *PWM at lowest active setting (one)*

PWM Active Time (72 us)
(Note that Bit Value < 1)

Figure 12-46 *PWM at midsetting (7)*

PWM Active Time (728 us)
(Note that Bit Value < 7)

these signals and a single eight-bit shift register. The 16 levels should be adequate for most applications, but if they aren't, then you could cascade another counter and a magnitude comparator to the circuit. This would give you up to 256 PWM duty cycles to choose from, giving you much more precise control over motors or other devices. A downside to doing this, however, is that you will have to increase the clock frequency. To create a 20 kHz PWM signal for this circuit, you will have to provide a 300 kHz clock; a 20 kHz PWM for a 256-level PWM generator will require a 5.1 MHz clock.

Learning to Program Using the Parallax BASIC Stamp 2

Over the years, a number of different tools have been developed for teaching both programming and programming robots. Right now, many people are introduced to programming using LEGO® Mind-Storms and Spybotics, which are programmed via a graphical interface. This method of programming is reminiscent of developing a flowchart for planning out a program and is reasonably easy for people to understand. People who have never programmed in their lives have created their own MindStorms programs with very little effort—LEGO® has done a wonderful job of reducing the fear in developing a robot application for their product.

To give you an idea of what a flowchart looks like and how a robot program can be developed using it (or a graphical interface), I have shown a flow chart (Figure 13-1) for a differential robot that will behave like a moth and move towards the brightest point in the room. The ovals are the start and end (or stop) points to the program. The diamonds are decision points for the program. The square boxes are where something happens in this program or application. The flow of the program between the different boxes, ovals, and diamonds is illustrated by the arrowed lines in Figure 13-1.

Looking at the flow chart, it should be easy to see that the program first stops for 200 msecs. Next, the robot checks (or polls) its collision sensors, and if something is directly in front of the robot, the motors are turned off and the application stops. If there isn't anything in front of the robot, then the light sensors are checked and the motors are turned on so that the robot turns toward the side that has the brightest light (if the left side is brighter, the right motor runs forward to move the robot toward the light). After the motors are turned on, the robot returns to the start of the program and runs for 200 msecs before

polling its sensors again. Even without this explanation, chances are you will have figured out how the program works on your own—it is probably easier to understand the diagram than this written description.

I'm not trying to be cruel when I say that this will be the last time full applications will be illustrated using flowcharts in this book. Unfortunately, graphical programming using flowcharts has a number of drawbacks that make it a very inefficient method for developing the different kind of robot applications that are presented in this book. Some of the drawbacks include the following:

- Difficulty illustrating complex applications. In Figure 13-1, I have just about reached the limit for readability.

- Difficulty in updating or modifying the application. In Figure 13-1, you can see that I had to "fold" the direction of execute upward when a collision was detected. Traditionally, execution flows downward. I could have changed the flow in this section of the diagram, but this would have involved making significant changes to the entire drawing.

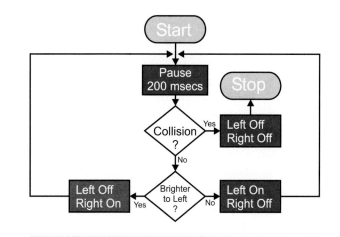

Figure 13-1   *Flowchart of a robot Moth program*

- The expense of creating software that will convert the graphical image into a program a computer can execute. A good part of the cost of a LEGO® MindStorms kit is the cost of the software that runs on your PC.

- The inability to encompass different hardware. In the LEGO MindStorms kit, remember that it will only work with LEGO-defined hardware devices. For your robots, you will be using many different devices, including what you can scrounge around for in the home (and, ironically, even using some LEGO pieces). In MindStorms, if you want to turn on and off motors, simple command blocks are used, whereas in your own robot you will have to plan out how this is done and develop different controls for the robot's speed and turning motions.

Experiment 76
Loading BASIC Stamp Windows Editor Software on Your PC

Parts Bin

Tool Box

PC

For the rest of the book, I will be working with the Parallax BS2 microcontroller for controlling both the different hardware experiments and the complete robots. The BS2 microcontroller is a self-contained unit with the power regulation, processor, reprogrammable application memory, variable memory, clocking, and I/O built into a 24-pin, DIP format *printed circuit board* (PCB)/package as I've shown in the block/pinout diagram in Figure 13-2.

If you haven't already done so, now is the time to buy yourself a BS2. A number of different models of the BS2 are available, but for the projects in this book, I recommend that you stick with the entry-level BS2, which you can buy at a reduced cost directly from Parallax using the information printed on the book's cover. The other BS2s offer faster application execution and some additional features, but the inexpensive entry-level BS2 gives the best price/performance ratio and provides all the features that are required for the projects in this book. As I work through the BS2 applications, I will be emphasizing the importance of thinking through your applications and not relying on faster microcontrollers

with more memory to get applications that work. You may later want to test the different BS2 models, but for now use the basic one.

Before you can start working with the BS2, the development software suite (known as the Stamp Windows Editor) will have to be downloaded and a

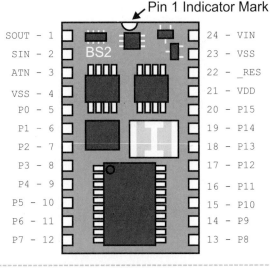

Figure 13-2   *Basic stamp two layout*

connection made to the BS2. In this experiment, I will go through the process of loading the Stamp Windows Editor onto your PC. In the next experiment, I will walk you through loading an application and downloading it into a BS2 mounted onto the PCB that is included with this book.

Along with showing you how to download the software, I am going to point you to a manual that you should download and two Yahoo! Groups that you should join to help support you with the book.

The PC that you are using must be running some version of Microsoft's Win32 operating system with a minimum of 100MB free disk space (you should also have at least 32MB of main memory). I would recommend Windows 98 Release 2, Windows/Me, Windows/NT 4.0, Windows/2000, or Windows/XP. If you are running Windows/NT, Windows/2000, or Windows/XP, you will have to have administrator rights. You will also need access to the Internet on your PC for downloading the Stamp Windows Editor software as well as for looking up information on the different electronics parts that are used.

To download the Parallax Stamp Windows Editor, open your PC's Internet browser and go to www.parallax.com.

When you have loaded the Parallax web page, move your mouse over Downloads and a pull-down menu may appear or another page will come up. When the selections appear, click on BASIC Stamp Software.

After clicking on BASIC Stamp Software from the Downloads pull-down, you should get the download web page. Left click on Download for the BASIC Stamp Windows Editor software. I recommend that you download the complete file rather than the one that "requires Internet connection during install." Other *Beginner's All-Purpose Symbolic Instruction Code* (BASIC) Stamp applications will catch your eye, but for right now, you should just download and install the Stamp Windows Editor (see Figure 13-3).

You will be prompted with a dialog box asking you to open or run the application or save it. I recommend that you click Open or Run and let it download and install itself. Follow the instructions and select the typical install as well have the program appear on your desktop.

After the software has been downloaded, you will be asked to select how the software is to be installed. Just continue with the defaults (and the typical installation), but make sure that an icon will appear on the Windows desktop; this will allow you to quickly start the BASIC Stamp Windows Editor software. When the BASIC Stamp Windows Editor is installed, you will get an icon on your desktop. To start it up, double-click it; to close it down, simply click the "X" in the upper right-hand corner (as you would with any Windows dialog box). The first time you start up the editor, you will be asked to assign files with the extension ".bs2," as well as others, to the editor. By clicking "yes" when you select a BS2 source file from an Explorer window, the BASIC Stamp Windows Editor will start up automatically, ready for the file. The BASIC Stamp Windows Editor is a Windows dialog box that looks like Figure 13-4.

When the Stamp Windows Editor first starts up, it will present you with a new hint regarding programming the microcontrollers. I recommend that you read through each one to better understand how the BS2 works and is programmed. In Figure 13-3, I downloaded the Beta version of the software. When you download the BASIC Stamp Windows Editor, it should be a release version and may look a bit different from what is shown in the different screen shots.

The Stamp Windows Editor package includes a tool that will allow you to uninstall the software if

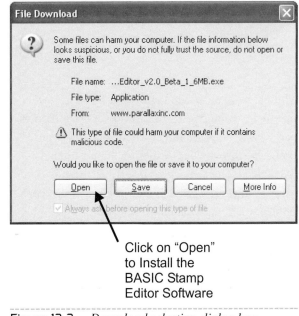

Click on "Open" to Install the BASIC Stamp Editor Software

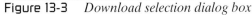

Figure 13-3   *Download selection dialog box*

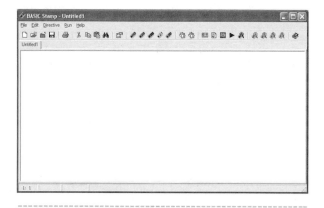

Figure 13-4 *BASIC Stamp Windows Editor operating*

you are going to download a new version of it. The uninstaller is accessed from the "Start" menu.

Once you have the Stamp Windows Editor installed, I recommend that you go back to the Parallax Web page, move your mouse to "Downloads," and

click on "Documentation." You should download and print out the BASIC Stamp User's Manual. This 300-plus-page document will explain the different electrical features of the BS2 and the PBASIC programming language in much more detail than was possible in this book. Parallax also makes this manual available in a preprinted and bound format that you can buy.

As the last step in getting your PC ready for working through the robot experiments presented in this book, you should join the Evil Genius Robots Support Yahoo! group that I have set up, as well as Parallax's BASIC Stamp Yahoo! group. The URLs for these groups are the following:

- http://groups.yahoo.com/group/evilgeniusrobotssuport/
- http://groups.yahoo.com/group/basicstamps/

Experiment 77
Connecting the PCB and BS2 to Your PC and Running Your First Application

Parts Bin

Assembled PCB with BS2

Tool Box

PC
RS-232 cable

With the software loaded onto your PC, it is now time to hook the BS2 up to it, power it up, and download your first application. I will assume that you have gone through each experiment and you have assembled (soldered parts onto) the PCB and installed the battery in it. Now is the time to go back to the earlier experiments, assemble the PCB, and install the BS2 and 9-volt battery.

Before starting, I just want to say that the BS2's software and hardware is very robust; it has been continually updated for several years now and it has been thoroughly debugged by people with even less

mechanical skills than you have. So relax, you have very little chance of damaging the BS2 or your PC.

The procedures for installing the BS2 and connecting it to your PC are reasonably straightforward, although you may have to work through some issues depending on your PC and the operating system that you are running.

As shown later, plug the BS2 into the PCB that came with the book (in the orientation shown on the PCB) and plug in a fresh 9-volt alkaline or recently charged NiMH battery. I recommend that you remove the battery and make sure that there are

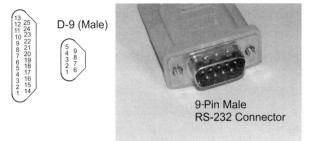

DB-25 (Male)

D-9 (Male)

9-Pin Male
RS-232 Connector

Figure 13-5   *IBM PC DB-25 and D-9 pin RS-232 connectors*

no connections on the PCB's breadboard before plugging in the BS2.

Next, connect the PCB to one of your PC's RS-232 serial ports using the straight-through 9-pin serial (often called a serial extender) cable. As I write this, most desktop and minitower PCs have built-in RS-232 serial ports. The RS-232 port can be either a 9-pin or 25-pin male connector and should be connected to a cable with a 9-pin male output pin like the one shown in Figure 13-5. For most PCs, you will have to buy a 9-pin D-Shell male-to-female straight-through cable.

Some PCs have a male 25-pin RS-232 port connector. This is the standard connector for RS-232; IBM came up with the 9-pin connector for the PC/AT in 1984, when they found there wasn't enough space for two 25-pin connectors (one for the parallel port and one for the serial port) in one adapter card slot. If you have a PC that has a 25-pin RS-232-port connector, then you can either buy an extender cable that converts from a 25-pin PC RS-232 port to a 9-pin connector or use a 25-pin to 9-pin converter. If you go to an electronic or computer store, they should have a number of different options for you to choose from to get the male 9-pin D-Shell connector for the PCB.

Most modern laptops (and some home PCs) do not have RS-232 ports built into them. If your PC does not have a 9-pin RS-232 port, I recommend that you buy a USB RS-232 interface. I have used the BS2 with both true RS-232 USB interfaces (one with four RS-232 9-pin connectors) and a simple USB interface that is designed for personal digital assistants (Palm Pilots) successfully with the BS2.

I do not recommend buying an RS-232 port ISA or PCI card for your PC (unless your PC does not have USB capability), as this will involve you opening up the PC and possibly manually configuring the serial port. USB adapters are fairly inexpensive (usually cheaper than a PCI card) and are easy to install, just requiring a CD-ROM and a few minutes of your time.

The RS-232 port you are using should only be used for programming the PCB's BS2. Windows does not share resources very well and if you share the port with another device (like a Palm Pilot), you will have problems with the operating system trying to determine whether or not the Basic Stamp Editor should be able to access the port.

Once you have established where the RS-232 port on your PC is, connect it to one end of your cable and the other to the PCB as shown in Figure 13-6. With this done, start up the Stamp Windows Editor software (Figure 13-7) and enter the following program into the tabbed white text box:

```
'   The first application
'{$STAMP BS2}
'{$PBASIC 2.5}

        debug "Hello World!"

        end
```

Either click Run or the leftward-pointing triangle and the program should be downloaded and started up. After a status dialog box that pops up and disappears, the "Debug Terminal" will appear.

Figure 13-6   *PCB with BS2 and battery ready to go!*

Figure 13-7 *Screen shot of the BASIC Stamp Windows Editor with the first program entered in*

The Debug Terminal dialog box is a small display that the PBASIC "debug" statement of the program writes to for passing status information back to the user. In this section, I will be using it to demonstrate the results of different programming operations.

There is a good chance that your "Hello World!" will work without any problems if you have followed the instructions I have laid out here. On the chance that your application doesn't work, there are a few things that you can look for. The first is to make sure that the RS-232 port is not used for *any* other applications in the PC.

The PCB is wired in such a way that the Basic Stamp Editor software should detect the BS2 and automatically start downloading to it. Another device may be wired similarly, and the Basic Stamp Editor software could incorrectly recognize it as the PCB. In this case, click "Properties," then "Debug Port," and manually select the port that you are using. Manually selecting the port has the additional advantage that the Basic Stamp Editor software does not have to search for the port with the BS2 on it (which means downloading will be faster in PCs with a number of serial ports).

Experiment 78
Saving Your Applications on Your PC

Parts Bin

Assembled PCB with BS2

Tool Box
PC
RS-232 cable

You are now able to create and download applications into a Parallax BASIC Stamp 2. In the previous experiment, you connected your PCB to your PC, started up the Basic Stamp Editor software, entered in a simple application, downloaded it into the BS2, and ran it. In this section, I want to explain a bit about how this process works and suggest a method for you to save the application for later use that will allow you to quickly and easily retrieve it.

Depending on your experience with PCs, you may not be aware that your PC has a number of different ways of storing information on its hard drive. The most basic method of saving a file is to put it on your desktop. This will keep the file somewhere visible to

you at all times, and if you have set up your PC with the proper file extension assignments, when you double-click on these files, the appropriate software will come up to read or process the information. If you download MP3s from the Internet, chances are your desktop is already full of files.

Under the latest versions of Windows (Windows/Me, Windows/2000, and Windows/XP), you can save your files as long, descriptive strings. For the "Hello World!" example application that I gave you in the previous experiment, you could give it a file name something such as Evil Genius—First BS2 Program—Hello World!

This is descriptive and accurate and should make knowing what it does simple. If you modify it, you might want to change the name to something such as Evil Genius—First BS2 Program Modified 11.19.02—Hello World!

This is not very efficient and can be very confusing. The problem of using this method is that your desktop gets full of different files very quickly, and when they are displayed on the desktop, much of the information is hidden until you single-click the file icon.

The Stamp Editor Software defaults to saving your files in a folder (it shows my age when I call them "directories" or "subdirectories") in your PC's hard file. If you click File and then Open, you will see that you can choose from a number of example BS2 programs. You could save the "Hello World!" application in this folder, but it may be difficult to find it again quickly (especially when you come back a few months later and aren't 100 percent sure of the file name).

To make it simpler for you to save all the files that are provided in this book, I would like to take advantage of the tree directory file structure built into Windows (and most other operating systems). The tree directory file structure provides a way of grouping and labeling files (and folders) into simple-to-search

categories. If you look at the "C" drive of your PC, it is arranged something like Figure 13-8.

In Figure 13-8, I have started from the place when you double-click "My Computer." From there, I have gone to the "Local Disk (C:)" and then to the Evil Genius folder and the Hello World! folder within it. As you can see in Figure 13-8, I have already saved the application in the C:\Evil Genius\Hello World! folder.

The space used in the different folders is dynamically allocated; you do not have to worry about how much space is used in each one. This means that you can store as much information as you need (up to the amount of available space on the hard file) in any folder.

To create a unique folder to store the BS2 applications presented in this book, click "My Computer" followed by "Local Disk (C:)." To create the Evil Genius folder, click New and then Folder in the Local Disk (C:). A new folder will be created and you will be prompted to enter in a name for the folder. Give this new folder the name Evil Genius. Double-click the Evil Genius folder and create a folder called "Hello World!" within it using the same procedure as you used to create the Evil Genius folder in the Local Disk (C:). You are now ready to save applications.

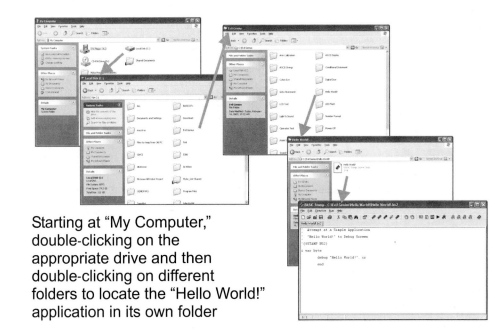

Starting at "My Computer," double-clicking on the appropriate drive and then double-clicking on different folders to locate the "Hello World!" application in its own folder

Figure 13-8   *File and folder organization in a PC*

After the "Hello World!" application has been entered into the Stamp Windows Editor, as discussed in the previous experiment, click File and then Save As to save the file. A new dialog box, like the one shown in Figure 13-9, will come up.

To save the "Hello World!" application in the "Hello World!" folder that is located in the Evil Genius folder, you may have to navigate within the Local Disk (C:) from the "Save in" toolbar at the top of the "Save As" dialog box (Figure 13-9). This toolbar provides you with a pull-down that will allow you to select folders within the current one or move from the current folder to the folder that it is located in.

The system of folders may seem complex, but once you start working with it, you will find that it is quite easy to create, save, and (most importantly) find files. The practice of saving all your files on the PC's desktop is one that I would like to discourage because it is difficult to keep track of the different applications and keep them separate from other files (like MP3s) that may be on your desktop. I like keeping each experiment's application in its own separate folder so I can modify the code and save it in the same folder,

Figure 13-9 *Saving a file in its folder*

making it easier to find and keep track of the updates. For every application in this book, I have saved the application code for each experiment in its own folder in the Evil Genius folder, and I recommend that you continue this practice as you develop your own applications to keep them separate from other files in your PC so that they will be easy to find when you come back to them later.

Experiment 79
The "Hello World!" Application Explained

Parts Bin

Assembled PCB with BS2

Tool Box

PC
RS-232 cable

The language that is used by the BS2 is PBASIC and is a variation on the BASIC language that has been used to teach people how to program for more than 30 years. For the rest of the section, I will be explaining to you the basics of programming and how PBASIC code is written for the BS2.

The first program run on your BS2 was "Hello World!," which was used to demonstrate the link between your PC and the BS2 mounted on the PCB that came with the book. Chances are that by looking

at this application, you will be able to figure out what it is doing.

The program consisted of the following seven lines:

```
'   The first application
'{$STAMP BS2}
'{$PBASIC 2.5}

      debug "Hello World!"

      end
```

The program above is what is known as an application source code or just code. *Source code* is a set of human-readable instructions that are converted by a compiler into a format that a computer can process directly. The seven-line program above is converted into a set of tokens that are downloaded into the BS2 and executed. The tokens for each statement are usually one or two bytes (or characters) long. They are read and executed very quickly by the BS2. The source code only has meaning to a programmer; it is not loaded into the BS2.

The single quote at the start of the first three lines is used to indicate the start of a comment that continues to the end of the line, the comment in this case being the message: "The first application." Comments are used by programmers to document the source code and help explain what the program is doing. They can be located anywhere in the source code (including after a program instruction). It is very important for you to come up with comments that are meaningful and should try to explain something that is useful to both yourself and anybody else who is looking at the source code.

Going off on a personal rant, I hate applications that have comments that repeat or paraphrase what a line of code is doing. I could have written the first application as the following:

```
'   The first application
'{$STAMP BS2}
'{$PBASIC 2.5}

  debug "Hello World!"
'   Print greeting

  end
'   Stop the program
```

This code explains *what* the program instructions (called *statements*) are doing, but not *why* the instructions are there or why they are needed to carry out the required task. As I work through the applications in this book, try to notice how I have written the comments and how I use them to explain what is happening or why I have used the code that I have, without explaining how each instruction works.

The next two lines are not comments, but instructions to the BS2 compiler. The first of the two lines ('{$STAMP BS2}') tells the compiler that the target device is the BASIC Stamp 2 and it produces

tokens that are specific to this device. The next line specifies that the code will take advantage of PBASIC version 2.5.

The forth and sixth lines of the program are not an instruction to the BS2, but blank lines known as *whitespace*. I suggest placing blank lines around blocks of code to help space them out visually. By adding blank lines before and after a block of code, the reader's eye will be directed to them, and it will be obvious where the block begins and ends. Judiciously using whitespace in your application will make your programs easier to read and understand.

The fifth line (debug "Hello World!") is the first statement that will be compiled, converted into a token, and downloaded into the BS2. This statement will command the BS2 to send the string of characters "Hello World!" back to the PC, where they will be displayed in the "Debug Terminal" dialog box.

The operation of the "debug" statement is the primary method that you will use for returning information regarding the execution of an application to a PC. In this section and the ones that follow, I will use it to return feedback information from the BS2 to you.

Along with the "debug" statement, a number of other built-in "function" statements are available to you as you create your application source code. These will be explained in this and the following sections.

The string of characters in the double quotes ("Hello World!") can be changed within the Stamp Windows Editor simply by moving the cursor to the statement and changing the message. You may want to try changing the string to "Hello from Myke" and see what happens when you execute it again using Ctrl-R or the Run button on the Stamp Editor Window.

When you change this string, you change the program's source code. There should be a "Modified" message in the bottom line of the BASIC Stamp Windows Editor that indicates the application has been changed, but the changes haven't been saved. If you are too close the program, you would get the message indicating that the source code has changed and would you like to save it. If you click "Yes," then the Stamp Editor Window software will save the changed source code as "Hello World!"

This is why I recommend creating separate folders for each of your applications. Instead of saving the changed source code as "Hello World!," click File and then Save As; you can save it (in the "Hello World!" folder) as "Hello from Myke" or "Hello World—Updated."

The final line (and statement) in the "Hello World!" application source code is "end." This statement stops the BS2 from executing further and places it in a power down mode where it will only consume 40 uAs of current to wait for another command from the Stamp Windows Editor.

When you first start writing your own BS2 applications, you should *always* put an "end" statement at the end of your application. When a BS2 application is loaded, it is simply stored in the program memory of the BS2 without erasing any previous programs. If your application is shorter than a previous one and it doesn't have an "end" statement, then the BS2 continues executing the code from the previous application after it has executed past the end of the current one. In most robot applications, execution takes place in a continuously repeating loop, and there is no danger that execution will continue past your code—but, just to be on the safe side, remember to put in the "end" statement to make sure that the BS2 stops if its execution continues past where you think it should be.

Experiment 80
Variables and Data Types

Parts Bin

Assembled PCB with BS2

Tool Box

PC
RS-232 cable

Robots (and their controlling computer programs) are designed to work in a changing or variable environment. In some circumstances, the changes are internal to continue to the next step in the process, whereas in others the changes are due to variances in the environment the robot is working in. This changing data must be stored as well as read back when the application executes.

Variables, which are numeric or character data that is stored in memory in the computer system, are used to store and retrieve data needed by the application. As a rule, variables can be written to and their values read back. Along with explaining what variables are, I will review the different data types that are available generically as well as in the PBASIC programming language.

In the previous section, I introduced you to the concept of flip flops and how they can be used to store a single unit of data called the bit. The BS2 provides 206 bits for use in your programs. Although in some programming instances single bits (with numeric values of 0 or 1) are useful, for most of your applications you will require larger numbers. Instead of requiring you to develop program source code to combine bits together, the PBASIC language provides this function automatically.

The ways that PBASIC handles single bits and groups of bits are as shown in Table 13-1.

To define a variable, the "declaration" statement is used and is in the format:

```
VariableName var Type
```

The "VariableName" can be any string of characters up to 32 characters long, starting with the underscore character ("_") or an alphabetic character (a through z or A through Z). In the body of the vari-

Table 13-1 PBASIC data sizes and number ranges

Number of Bits	Type	Data Range
1	Bit	0 to 1
4	Nib	0 to 15
8	Byte	0 to 255 or 128 to 127
16	Word	0 to 65535 or -32,768 to 32,767

able name, along with the underscore character and alphabetic characters, numeric characters (0 through 9) can be used.

"Type" in the variable declaration is one of the four types listed in the table above. "Nib" is an abbreviation for the word "nybble," which is commonly used to describe four bits. The nybble is a very useful data type, especially when you want to display non-decimal data efficiently.

PBASIC variables can be "overlaid" on other variables or parts of the variables can be used as separate variables. To access a small part of a variable with an overlaid variable, the "modifiers" listed in the BS2 Quick Reference are used. For example, defining a single bit in a "Flag" variable would be accomplished with the code:

```
Flag var byt       '  8 Bit Variable
LED var Flag.bit1  '  Single Bit of "Flag"
```

This could be repeated for all eight bits in the "Flag" variable. Using the "bit#" modifier, you can specify the use of each bit in a byte, which is an advantage because in some cases, the PBASIC compiler will define each "bit" in its own byte. You will find that your memory space (which is only 26 bytes [206 bits divided by 8 bits/byte] to begin with) will be used up very quickly if you want to have several bit variables.

To test out the operation of variables, start up the Stamp Windows Editor and key in the following program:

```
'  Variable Display - Write to and Display
Different Variable Types
'{$STAMP BS2}
'{$PBASIC 2.5}

'  Variables
BitVar  var bit
```

```
ByteVar var byte
WordVar var word

'  Assignment Statements
    BitVar = 0      ' Bits can be 0 or 1
    ByteVar = 0     ' Bytes can be 0 to
                      255
    WordVar = 0     ' Words can be 0 to
                      65,535

'  Display Variables using debug statement
    debug ? BitVar
    debug ? ByteVar
    debug ? WordVar

    end
```

When you have finished entering the program, create a Variable Display folder for it inside the Evil Genius folder and save the program as Variable Display. Once you have done this, run it with the 0 loaded into the three variables as well as different values. You might want to try different values greater than the size of each variable type to see what kind of message comes back from the compiler (as well as what is displayed on the Debug Terminal dialog box).

When you set a variable to a value greater than the value that it can hold (try "BitVar = 3"), you will discover that the part of the number that is less than the maximum value that I listed in the table will be saved in the variable. There will be no error message. PBASIC ANDs the value to be stored in the variable with the maximum value the variable can hold. This is an important point to remember because sometimes you will discover that the value you expect to be in a variable is much less than what it should be because of this ANDing.

This application first defines three variables (a bit, byte, and word), loading different values into them and printing them out before ending the program. The three statements that make the variables equal to the numeric value are called *assignment statements*. Passing a constant value to a variable, as is done in this experiment, is the simplest form of the assignment statement that can be used for much more complex operations.

In the debug statements, the contents of the variables are displayed, but I used the "?" formatter to display the variable name along with its value. After printing the variable name and its value, a new line on the display is started. I will be explaining formatters and the different ways data can be displayed using the debug statement in the next few experiments.

Parts Bin

Assembled PCB with BS2

Tool Box

PC

RS-232 cable

When I present to you some robot applications, a big part of the work will be to quantize or convert data into numeric form so the BS2 can process it. Later in this book, I discuss the appropriate data formats for position, speed, light levels, I/O bit states, and other parameters that will make the data coming back from the robot application program easier for you to understand. Before getting to this point, I want to introduce you to the different data formats available in PBASIC for the BS2 and show how they can be used. These different formats are used to output data, either to you or another user or to another computer system.

A multidigit number is made up of several numbers, each one a multiplier times the base to the power of the position of the digit. This is exactly how decimal numbers are written with a base of 10. Figure 13-10 shows how the number "123" is really the sum of the value of each digit.

In engineering and computer science, the first digit is always zero and not 1 as you would expect. The reason for this is due to the use of binary values in which the smallest possible value is zero; if numbers were started at 1, then the first possible value would be ignored.

The method shown in Figure 13-10 can be used for expressing binary numbers (which only have two dif-ferent possible values for each digit) as decimal values and vice versa. For example, consider the decimal number 42. Using Figure 13-11, the 32 digit (bit 5) is 1, along with the 8 digit (bit 3) and the 2 digit (bit 1). All the other digits are zero. In PBASIC, binary numbers have a % prefix character in front of the number. If the % is not present, then the number is not binary. In PBASIC, 42 decimal is represented as %101010 binary.

Writing and saying multidigit binary values can take a long time; a great example of this is when Bender in *Futurama* joins a robot church and recites "Grace" in binary for several hours. To avoid this, most people use hexadecimal (base 16) numbers for binary data. Four binary digits (one nybble) represent a single hexadecimal digit. In Table 13-2, I list the binary, decimal, and the mnemonics for the 16 different hexadecimal numbers. Hexadecimal is often abbreviated to the three letters hex, just as binary is abbreviated to bin and decimal to dec. For hexadecimal values over nine, the letters A through F are used and often referred to by their phonetic mnemonics. Hexadecimal numbers have a $ prefix character placed in front of them to indicate that they use base 16.

$$
\begin{aligned}
1 \times 10^2 &= 100 \\
+\ 2 \times 10^1 &= 20 \\
+\ 3 \times 10^0 &= 3 \\
\hline
&\quad 123
\end{aligned}
$$

Figure 13-10 *The number 123 represented as a sum of powers of 10*

$$
\begin{aligned}
1 \times 2^5 &= 32 \\
+\ 0 \times 2^4 &= 0 \\
+\ 1 \times 2^3 &= 8 \\
+\ 0 \times 2^2 &= 0 \\
+\ 1 \times 2^1 &= 2 \\
+\ 0 \times 2^0 &= 0 \\
\hline
\%101010 &= 42
\end{aligned}
$$

Figure 13-11 *The number 42 represented as a sum of powers of 2*

Table 13-2 Decimal, binary, hexadecimal and mnemonic cross-reference

Dec	Bin	Hex	Dec	Bin	Hex	Mnemonic
0	0000	0	8	1000	8	
1	0001	1	9	1001	9	
2	0010	2	10	1010	A	"Able"
3	0011	3	11	1011	B	"Baker"
4	0100	4	12	1100	C	"Charlie"
5	0101	5	13	1101	D	"Dog"
6	0110	6	14	1110	E	"Easy"
7	0111	7	15	1111	F	"Fox"

PBASIC will convert data into the different numeric bases automatically for you. I recommend that you look for a cheap scientific calculator with this capability built-in. This capability is very rarely advertised, so when you are looking at calculators, look for ones that have the functions DEC, BIN, HEX, and OCT on the keypad along with the letters A through F. OCT is the representation for base 8 numbering and was quite popular 30 years ago. Except for some C programming situations, octal is rarely used because it is quite awkward to use (each digit is made up of three bits and does not fit evenly into a byte or a word).

To demonstrate the different formats available to display numbers in PBASIC, you can key in the number format application below and save it in its own folder (Number Format) in the Evil Genius folder. As in the previous application, change the value assignment statement and run the application again to see what different decimal values look like in the different formats.

```
'   Number Format - Display a Value in
Different Number Formats
'{$STAMP BS2}
'{$PBASIC 2.5}

'  Variables
Value    var byte

'  Assignment Statements
     Value = 123 'Arbitrary Value to
Display

'  Display Variables using the debug
statement
```

```
debug "Decimal ", ? Value
debug "Binary ", IBIN ? Value
debug "Hex ", IHEX ? Value

end
```

In the debug statements, I have placed the IBIN and IHEX formatters before the characters to print the variable name and contents of the Value variable. The IBIN and IHEX formatters will convert the contents of Value to binary and hexadecimal and place the % and $ data format indicators in front of the values. Along with IBIN and IHEX, a number of other formatters are available in the debug statement and they are listed in the BS2 Reference at the end of the book.

I recommend that you use just these two formatters (along with dec for decimal) for numeric data to make sure that you don't get into the situation where the number 10 is printed out and you automatically assume that it is 10 (decimal) and not 2 (binary) or 16 (hexadecimal). The other formatter options for decimal, binary, and hexadecimal numbers can be confusing if they are used when you are first starting to learn about programming and are not sure what the displayed data means.

The different number formats are selected to make reading data output and programming hardware interfaces easier. When you first start programming, I recommend that you stay with decimal (base 10) as much as possible because this is what you are most comfortable with. In the next section, I will show instances where base 2 and 16 numbers are best used when interfacing to hardware devices.

Before going on to the next experiment, I want to share with you an interesting story that shows how far we've come. As you may or may not know, *hex* is not the correct prefix for 16 (*hex* actually means 6). The correct prefix is *sex* and a base 16 number should be known as sexadecimal. This was obviously a source of mirth for early programmers. When IBM introduced their System/360 computers in the early 1960s, the documentation referred to base 16 numbers as hexadecimal because the company was uncomfortable with the idea that their computers were programmed in sex and didn't think it was appropriate for a machine that would be programmed and used by a variety of different people, some of whom could be female.

Experiment 82
ASCII Characters

Parts Bin

Assembled PCB with BS2

Tool Box

PC
RS-232 cable

In the previous experiment, I showed how groups of bits can be combined to create reasonably large values as well as be displayed as numeric values in different formats. For most computer communications and computer programming and interfacing, bytes are used to represent different English letters, numbers, and different characters. Table 13-3 shows the most common way of representing different characters is the *American Standard Code for Information Interchange* (ASCII) character set.

Special characters (located in the 32 bytes with values of $00 to $1F) that you will probably want to use (with their PBASIC equivalents) are listed in Table 13-4. Other control characters are available to control the Debug Terminal that can be reviewed in the BS2 reference.

The 128 different characters are normally loaded into a byte and passed to the receiver. You have already used ASCII characters in your experiments as the messages and data passed to the Debug Terminal from the debug statements.

Earlier in this section, I noted that a byte is 8 bits and could contain up to 256 different values (from 0 to 255). The ASCII character set was designed for a 7-bit container (and has a maximum of 128 different values). Enhancements to the ASCII character set provide an additional 128 characters to the standard for different situations to allow a byte to send up to 256 different characters at a time. The most popular of the enhancements was developed by IBM for the PC more than 20 years ago. The IBM-enhanced character set provides special characters for languages other than English, the ability to graphically box in areas of text, and common characters (such as the Greek alpha) that are not built into the standard ASCII character set.

Table 13-3 ASCII code

0	1	2	3	4	5	6	7	8	9	10	11	12	13	14	15
NUL	SOH	STX	ETX	EOT	ENQ	ACK	BEL	BS	HT	LF	VT	NP	CR	SO	SI
16	17	18	19	20	21	22	23	24	25	26	27	28	29	30	31
DLE	DC1	DC2	DC3	DC4	NAK	SYN	ETB	CAN	EM	SUB	ESC	FS	GS	RS	US
32	33	34	35	36	37	38	39	40	41	42	43	44	45	46	47
SP	!	"	#	$	%	&	'	(	)	*	+	,	-	.	/
48	49	50	51	52	53	54	55	56	57	58	59	60	61	62	63
0	1	2	3	4	5	6	7	8	9	:	;	<	=	>	?
64	65	66	67	68	69	70	71	72	73	74	75	76	77	78	79
@	A	B	C	D	E	F	G	H	I	J	K	L	M	N	O
80	81	82	83	84	85	86	87	88	89	90	91	92	93	94	95
P	Q	R	S	T	U	V	W	X	Y	Z	[	\	]	^	_
96	97	98	99	100	101	102	103	104	105	106	107	108	109	110	111
`	a	b	c	d	e	f	g	h	I	j	k	l	m	n	o
112	113	114	115	116	117	118	119	120	121	122	123	124	125	126	127
p	q	r	s	t	u	v	w	x	y	z	{	|	}	~	DEL

Table 13-4 Most commonly used ASCII and PBASIC control characters

Symbol	PBASIC Symbol	Function
NUL	CLS	ASCII—Normally it is used to terminate a string. In PBASIC, it is used to clear the Debug Terminal dialog box.
SOH	HOME	ASCII, Start of data header. In PBASIC, it is used to move the cursor to the top left-hand corner of the Debug Terminal dialog box.
BEL	BELL	Beeps the system (PC for BS2) speaker.
BS	BKSP	Backspace.
TAB	TAB	Horizontal tab.
LF	none	New line (or NL).
CR	CR	Carriage return.

For all the experiments in this book, I will only be using the common ASCII characters listed. Different PCs, as well as different fonts used within different applications, define the enhanced 128 characters differently. This means that an application that presents beautifully in one application on a certain PC will look very poor (or even been unreadable) on another PC or in another application.

Along with displaying standard characters, the ASCII character set has a number of "control characters" (the ones relevant to PBASIC are listed in Table 13-4). These codes can be used to move characters around the debug terminal's window or indicate the end of a line (the "cr" character).

Save the ASCII display application below its own folder in the Evil Genius folder:

```
'  ASCII Display - Display the ASCII Charac-
ter for a numeric Value
'{$STAMP BS2}
'{$PBASIC 2.5}

'  Variables
Value var byte

'  Assignment Statements
  Value = 123          '   Arbitrary Value to
                             Display

'  Display Variables using the debug state-
ment
```

```
debug "ASCII ", ASC ? Value

end
```

Like the previous experiments, run ASCII display after setting Value to different constant values. You might even want to assign a value greater than 127 ($7F or %1111111) to see the enhanced ASCII characters are for your PC (running Stamp Windows Editor).

In this and the previous experiment, notice that I have linked a constant string (ASCII) with the output value. This is done by placing a comma (,) between the two items to be output, and it runs them together without any spaces or new line characters; this is called *concatenation*. I'm pointing this out because data can be formatted together on a single line in many ways using a single debug statement; with a bit of creativity, you could come up with some very attractive displays on the Debug Terminal.

A string is a number of characters that are stored together in a computer system. Constant strings are identified in PBASIC as a set of ASCII characters enclosed in a pair of double quotes (""). One of the primary functions of constant strings is to provide a message for the human operator, as I have done in the programs presented in this section.

Parts Bin

Assembled PCB with BS2

Tool Box

PC

RS-232 cable

To a certain extent, you are probably familiar with the specifications given to you when you look at a personal computer. Most modern PCs have three memory specifications that you will consider when you are buying a system. The hard disk space (or memory) is measured in terms of billions of bytes (called gigabytes), and the main memory of the system is measured in millions of bytes (called megabytes). To speed up the execution of applications, the processor has local memory called a *cache* that is measured in thousands of bytes (called kilobytes). The memory in any of these three areas of the PC (shown in Figure 13-12) can be used to store any variable data that you have. I'm sure that it is a shock to you to discover that the BS2 only has one variable memory area and it can only store 26 bytes, about 30,000 times less than the smallest data area available in the *original* PC.

If you think about what I've just written here, you're probably wondering how exactly the BS2 can do anything useful; even the smallest application or game that you run requires a *lot* more space than the paltry 26 bytes available in the BS2. If you count the number of characters in each sentence in this book, you'll discover that virtually all of them require more than 26 bytes to store the ASCII characters that make them up. If you look at the previous application and count the number of string characters that are

used in the debug statements, you'll see that they require 19 bytes, leaving only 7 for code and the Value variable.

Before going on too far, I should point out that the variable memory comparison of the BS2 to the memory comparison of the PC is not an "apples to apples" comparison between the capabilities of the two devices. In the PC, the disk, memory, and cache can be used for either code or variables, whereas in the BS2 the 26 bytes can only be used for variables. Figure 13-13 shows the BS2 with the two chips were the different memory is located.

Two kilobytes of EEPROM are available for application code and constant storage in the BS2. This space is used for the quoted strings, such as the ones in the debug statements in the previous experiment using Ctrl-M (Figure 13-14). If you were to look at the previous application, you would discover that only one byte of the 26 bytes of variable storage is used by the application, whereas almost 30 bytes are used for the application's code and quoted strings.

Although the variable memory and application code are stored differently in the BS2 than in your PC,

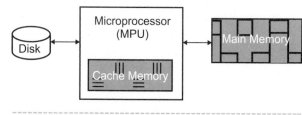

Figure 13-12   *Memory systems in a PC*

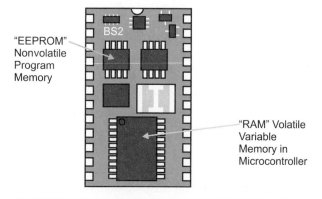

"EEPROM" Nonvolatile Program Memory

"RAM" Volatile Variable Memory in Microcontroller

Figure 13-13   *BS2 with memory locations marked*

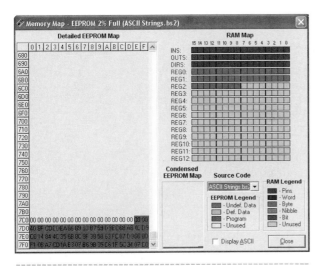

Figure 13-14 *Dialog box showing BS2 memory utilization brought up with Ctrl-M*

the variable memory behaves the same way in both devices. So far in the experiments, I have been defining and accessing single variable bytes *directly*. The variable space allows for variables to be defined with multiple bytes or words and allows individual bits, bytes, or 16-bit words to be accessed arbitrarily by use of an index. An *index* is a numeric value that is specified explicitly or mathematically and allows for a great deal of freedom to access any data within a variable array.

It is important that all variable memory in a computer system is set up as an array and can be accessed indirectly by using an index. The classic way of presenting an array is as a row of post boxes used for sorting mail as in the variable array shown in Figure 13-15. The street name is analogous to the variable name, and to access the mailbox for an address, the house's street number (which is analogous to the variable's index) has to be known.

Array variables use a slightly different format of the declaration statement I presented in the earlier experiment:

```
VariableName var type(size)
```

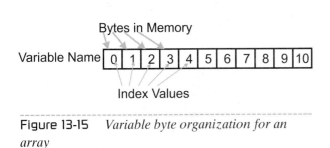

Figure 13-15 *Variable byte organization for an array*

"Size" is the number of bytes or words used by the array (up to 26 bytes or 13 words). "VariableName" and "type" are the same as I presented earlier.

So, to define a 10-byte array, the following statement is used:

```
ArrayVariable var byte(10)
```

To print out the third byte (also known as the third element in the array) to the debug terminal, use this statement:

```
debug ASC ? ArrayVariable(2)
```

The first byte in the array has an index value of 0, which is why I use an index of 2 to access the third byte in the statement above.

Arrays can be used to save numeric data as well as character data, which results in data similar to the quoted strings. In this experiment, I will show how an array is defined, loaded with data, and then printed out along with a quoted string. The code should be saved as ASCII Strings in its own folder in the Evil Genius folder:

```
'  ASCII Strings - Read and write to a byte
Array
'{$STAMP BS2}
'{$PBASIC 2.5}

'  Variables
ASCIIArray var byte(5)

'  Assignment Statements
    ASCIIArray(0) = "E"    '  Load "Evil"
                                  into
    ASCIIArray(1) = "v"    '  "ASCIIArray"
    ASCIIArray(2) = "i"
    ASCIIArray(3) = "l"
    ASCIIArray(4) = 0

'  Display Variables using the debug state-
ment
    debug STR ASCIIArray, " Genius", cr

    end
```

It is probably surprising to you that I have ended the string of characters stored in ASCIIArray with a zero (or ASCII NUL) character. The zero at the end of the string of characters in the array indicates to the STR formatter that the string has ended and no more bytes are to be printed out. The NUL character-ended string is known as an ASCIIZ string.

Experiment 84
Using Mathematical Operators in the Assignment Statement

Parts Bin

Assembled PCB with BS2

Tool Box

PC

RS-232 cable

Now that you know just about everything there is to know about variables in PBASIC (as well as most other programming languages), I wanted to take a look at what can be done with them in an application. PBASIC offers a number of different mathematical operations (most of which you are familiar with) for processing numeric data. If you are familiar with programming already, then you will find PBASIC to be very similar to other languages that you have worked with before, but there is one little twist that you will have to be aware of.

The basic form of the PBASIC mathematical assignment statement is:

```
DestinationVariable = {ValueA} Operator
ValueB {Operator ...}
```

Everything to the right of the equals sign is known as an expression. Along with being used to specify the value to be saved in the assignment statement, expressions are also used in other statements (and PBASIC functions). The braces ("{" and "}") are used to indicate the characters within them are optional. The " . . . " is used to indicate that the previous character strings are repeated in the statement. These are commonly used symbols used in defining programming statements, functions, and operations.

The "Values" can be variables, array variable elements, or constants. They can also be values with operators put into parentheses. If values have parenthesis around them, then they are evaluated first. The "Operator" can be one of the following in Table 13-5.

I have marked ValueA as an optional term because there are a number of unary operators that perform their operations on a single variable.

The twist I was talking about at the start of the experiment is how the operators are evaluated. In most high-level languages, the different operators are given an order of operations or priority. This is not the case for PBASIC; it evaluates the expression from left to right.

In a traditional high-level language, if you wanted to multiply 3 by 4 and add 5 to the result (to get 17) you would use the statement:

```
A = 5 + 3 * 4
```

But in PBASIC, this statement would be evaluated as 5 plus 3 (8) first and then multiplied by 4 for a result of 32. To get around this problem, you could break it up into two parts, the first being to save the first calculation of the expression (3 * 4) in a temporary variable and then add 5 to the product:

```
Temp = 3 * 4
A = Temp + 5
```

This method works well despite requiring an extra variable to save the temporary value.

The second method is to "force" the first operation to be evaluated by placing it and its parameters in parenthesis:

```
A = 5 + (3 * 4)
```

This format is often used for high-level languages and is probably easier to read than the first method. Remember that your ability to add small expressions within parenthesis to larger expressions is not unlimited; I would recommend that no more than two are used in each assignment statement.

Table 13-5 PBASIC mathematical operators

Operator	Function
+	Return sum of two values
-	Return difference of two values; return negative of a value
*	Return product of two values
*/	Return middle two bytes of product of two 16-bit values
**	Return upper two bytes of product of two 16-bit values
/	Return quotient of a dividend and divisor
//	Return remainder of dividing a dividend by a divisor
<<	Return first value shifted to left by other value bits
>>	Return first value shifted to right by other value bits
&	Return bitwise AND of two values
|	Return bitwise OR of two values
^	Return bitwise XOR of two values
~	Return complement of a value (same as XOR-ing with $FF)
MIN	Limits Value to specified low
MAX	Limits Value to a specified maximum
DIG	Return specified decimal digit of a value
REV	Reverse specified number of bits of a value
ABS	Return absolute value of a positive or negative value
DCD	Return value with specified bit set
NCD	Return most significant bit of value
SQR	Return square root of value
SIN	Return sine of Value assuming a 256 point circle with a radius of 127
COS	Return cosine of Value assuming a 256 point circle with a radius of 127

Try out the simple Operator Test application (in its own folder in the Evil Genius folder):

```
'   Operator Test - Look at how different
operators behave
'{$STAMP BS2}
'{$PBASIC 2.5}

'   Variables
Result var Word

'   Assignment Statement - change to test
different operators
  Result = 25 + 32

'   Display Result
  debug ? Result

  end
```

In this experiment, change the addition operator (+) with the operators listed in the table along with changing the parameters. The different operators should be quite easy to understand in most cases, except for the trigonometric operators (SIN and COS).

Experiment 85
Creating Simple Program Loops

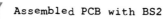

Parts Bin

Assembled PCB with BS2

Tool Box

PC

RS-232 cable

When I took my first class in computer programming, I was shown the diagram below of a typical program (Figure 13-16). With what I have shown you so far in this section, you should be able to create applications that follow this data flow. The variable initialization statements are the input, any change (mathematical operations) performed on the input is the processing, and the results on the Debug terminal are the output. This ability allows you to use the BS2 as a simple calculator, but it will not be very useful for programming a robot.

For a robot to execute indefinitely, you will have to first set up the initial conditions and then repeat a series of statements. If you think about how you would like your application to execute, it would probably look something like Figure 13-17, in which the same commands get repeated over and over again. If you look at the flowchart, the code flow going from the bottom of the program back up to the top will look like a ring or loop. This looping of the program execution will allow the processor to execute the same commands over and over again.

In PBASIC, the "do" and "loop" statements (normally referred to as just "do loop") allow for looping code exactly as shown in Figure 13-17. The two statements take the following form:

```
do  ' Start of Code to repeat
'  Statements that execute repeatedly
loop
```

To demonstrate how the do loop statements work to provide a method of repeating a series of statements, enter the following code into the Stamp Windows Editor software. Once you've done this, save it as Looping Statements in its own folder in the Evil Genius folder.

```
'  Looping Statements - Demonstrate the
operation of the "do - loop"
'{$STAMP BS2}
'{$PBASIC 2.5}

'  Variables
Counter var Word

'  Initialization
  Counter = 0

  do ' Return here for next counter value

'  Processing
    Counter = Counter + 1
' Increment counter

'  Display the Counter Value
      debug ? Counter

      pause 500
' Delay 500 msecs before repeating

  loop ' Loop around and do again
```

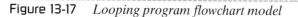

Figure 13-16   *Basic programming model*

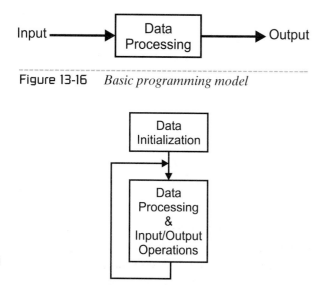

Figure 13-17   *Looping program flowchart model*

When you run this application, the counter will increment itself twice every second; the pause statement stops the BS2 from executing for the number of milliseconds specified. By specifying 500, the BS2 will stop for 500 milliseconds, or a half-second. Pause does not put the BS2 into a low-power mode.

Looking at the execution path of this program, you will see that it first initializes the Counter variable to 0 and then enters the loop. In the loop, the counter is incremented (or has 1 added to it). Next, the current value of Counter is output, and execution stops for a half-second due to the pause statement and then loops back to the do statement and starts the process over.

This program will loop indefinitely (only stopping when power is taken away from the BS2 or if it is reprogrammed). It is known as an *infinite loop* for this reason. In traditional programming, you should work at avoiding infinite loops because if execution gets into one, then it will never return or stop. When we work with the robots, you'll discover that this is not necessarily a bad thing, and you will see that many robot applications will involve an infinite loop that will only end when the robot's batteries wear out (or you execute the "Power Off" application, which is shown in a later experiment).

Experiment 86
Conditionally Looping

Parts Bin

Assembled PCB

Tool Box
PC
RS-232 cable

One of the important points in the definition of structured programming is the ability to execute iteratively, using repeated statements that perform a specific task. In the previous experiment, I showed you the do loop programming statements that allow you to loop application code repeatedly, but will not allow you to leave it. In this experiment, I want to show you how to execute looping code while a set of conditions is valid and introduce you to how conditions are tested in PBASIC (and most other languages).

PBASIC has three ways of exiting a do loop section of code, but I am only going to focus on the do-while loop, which executes the code within the loop while the tested parameters are within a valid range. The test for the parameters takes place using the expression format:

```
ValueA Condition ValueB
```

ValueA and ValueB can be variables, array elements, or constant values. Condition is the test and

can be one of the six tests listed in Table 13-6. The expression is evaluated and if it is true, then execution will continue in the loop code. If it is false, then execution will jump to the first statement following the loop statement.

If you wanted to print out numbers from 1 to 10, executing code in do-while loop code while the current number is less than or equal to 10, you could use the code:

```
  Number = 1
  do while Number <= 10
' Loop while Number is Less than 11
    debug ? Number, cr
    Number = Number + 1
  loop
  end
```

To indicate the code inside the do-while loop statements, I indented the "debug ? Number, cr" and "Number = Number + 1" statements by two columns. This makes the code inside the loop very easy to recognize and is a code-formatting technique

Table 13-6 PBASIC conditional execution tests

Condition	Operation	Complement
=	Return True if ValueA is equal to ValueB	<>
<>	Return True if ValueA does not equal ValueB	=
>	Return True if ValueA is greater than ValueB	<=
>=	Return True if ValueA is greater than or equal to ValueB	<
<	Return True if ValueA is less than ValueB	>=
<=	Return True if ValueA is less than or equal to ValueB	>

that I recommend you use in your own applications. The indentation can be whatever you feel comfortable with. Personally, I like two spaces, but you may want to use more or you may want to tab the source over. If you tab the code, then remember that different applications display the source code differently (for example, something that looks great on the BASIC Stamp Windows Editor can look very disorganized when displayed on a web page using Internet Explorer). If you are going to display your application code anywhere, then I recommend that you just use spaces to move code and comments to the right and make sure that it will be displayed as the same one regardless of the method used to display it.

To demonstrate how the do-while loop construct works, I want to expand on the simple example above and calculate the first 16 squares by the use of difference theory. Hundreds of years ago, calculations of values (like squares) were extremely difficult to do manually. Rather than rely on complex operations such as multiplication and division, the final values were calculated using addition and subtraction using a value (called a *difference*) that is predicted algorithmically. This method was to be used by Babbage's "Difference Engine" to perform complex calculations without the need for multipliers and dividers.

If you look at different square values, you will discover that they are related to the previous square value by a Delta value that is two greater than the value used to calculate that square. In Table 13-7, I

have outlined the first five squares along with the appropriate differences. The Delta is the value added to each difference.

I used the difference theory to calculate the first 16 squares in the Squares program that should be saved in the Squares folder in the Evil Genius folder:

```
'  Squares - Calculate squares based on
difference
'{$STAMP BS2}
'{$PBASIC 2.5}

'  Declare Variables
n var byte
Difference var byte
Delta var word
Square var word

'  Initialize Variables
  n = 1
  Delta = 1
  Difference = 2
  Square = 1

'  Find Squares for first 16 numbers
  do while n <= 16
    debug dec n, " squared = ", dec
Square, cr

'  Calculate New Square and Delta value
    Square = Square + Delta + Difference
    Delta = Delta + 2
    n = n + 1
  loop

  end
```

In this program, I used the do-while loop statements to repeat the square calculating code until n is greater than 16. When n is greater than 16, the do-while condition is no longer true and execution jumps out of the do-while loop. In this case, the next statement is the end statement, which stops the application.

Table 13-7 Illustrating how squares can be calculated using differences

"n"	Square	Difference	Delta	Square of "n + 1"
1	1	2	1	4
2	4	2	3	9
3	9	2	5	16
4	16	2	7	25
5	25	2	9	36

Experiment 87
"Power Off" Application

Parts Bin

Assembled PCB with BS2

Tool Box

PC

RS-232 cable

After finishing with the looping application, you may have shut down the Stamp Windows Editor, unplugged the PCB from the PC, and assumed that the BS2 had stopped running. Unfortunately, this is not the case; the BS2 will run the current application as long as it has it in memory. The pause instruction does not place the BS2 into a low-power mode, it will continue to drain the 9-volt battery as long as it is installed in the PCB. To make matters worse, the PCB that you have built to run the BS2 applications does not have an off switch; the application will continue to run as long as the battery is inserted into the socket.

During normal operation, the BS2 will draw 8 milli-amperes of current. With a fresh 9-volt alkaline battery, you can expect something less than one day of life. With this information in hand, you might be wondering if a mistake was made when the PCB was designed.

When I designed the PCB to be included with this book, I wanted to take advantage of the low-power modes available in the BS2 and avoid a power switch for it altogether. This does not violate my "rules of robotics." As in all the applications where motors are driven, the batteries powering the motors do have switches to make sure the robot can be positively turned off and prevented from moving unexpectedly.

By executing the end statement, the BS2 is put in a low-power state, consuming only 40 micro-amperes of current. When the BS2 is in the low-power state, a fresh 9-volt alkaline could be expected to run for over a year! With this ability built into the BS2, I decided to forego putting in a power switch on the PCB and take advantage of the low-power mode that could be entered by simply executing an "end" state-

ment as I do in the application that I call "Power Off" shown here:

```
'  Put BS2 to sleep with to save power
(and not remove
'   battery).
'{$STAMP BS2}
'{$PBASIC 2.5}

'  Constant Declaration
AllInput con 0

 outs = AllInput
' All Pins now I/Ps to avoid
'   current source/sink

 debug "Goodbye...", cr

 end
```

This application can be run with any of the applications or circuits presented later in the book, and it will put the BS2 into low-power mode by simply connecting the PCB to a PC and running (Ctrl-R) this application. Looking at the code, there is probably at least one statement that you will be unsure about.

In this application, I introduce the concept of constants. These are labels, similar to variable labels in which a constant value is assigned to. Unlike variables, constant labels cannot be changed by the application. For example, if you were to put in the statement:

```
AllInput = 42
```

The Stamp Windows Editor would return the message "Expected a Variable, Label, or Instruction," indicating that the constant label cannot be changed.

The "outs = AllInput" statement forces all the I/O pins in the BS2 to be "Inputs," and they are unable to source current to or sink current from other devices. By doing this, I will make sure that there isn't any additional current drain on the BS2 when it is turned "off" by this application.

Once all the I/O pins are put into input mode, I send a "Goodbye" message to indicate that the application is running correctly and then put the BS2 into low-power mode using the end statement.

If you read through the BS2 documentation, you'll probably come to the conclusion that an application that just consisted of an end statement would perform the same task as what I have done here. The

Goodbye message is certainly not required. If you read the BS2 documentation, you will discover that when a new application is downloaded, the BS2 is reset and all the I/O pins are put into input mode automatically (which means the "outs = AllInput" statement is redundant). But if I presented you with an application that just consisted of an end statement, then I wouldn't have had the opportunity to introduce you to the variable-like register interface of the I/O pins or show you how to define constants for your own applications.

Experiment 88
Conditionally Executing Code

Parts Bin

Assembled PCB with BS2

Tool Box

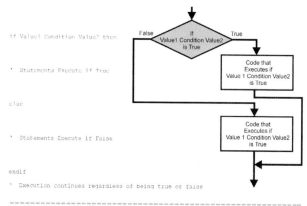

PC
RS-232 cable

Being able to loop an application is a huge advancement in your ability to create meaningful programs, and it will allow you to create robot applications that repeat the same operations indefinitely. This capability, however, will not allow you to respond to different situations and inputs while the application code is executing. PBASIC has a couple of different built-in statements to allow you to change execution if specified conditions are met. The most basic one is the if statement that causes execution to jump to a different location in an application if a specified condition is true. The problem with just showing the if decision is that you won't know how to arrange your PBASIC code. To avoid this confusion, I prefer to characterize conditionally executing code not just as one statement, but three—showing what executes if the condition is true, as well as if it is false, and then execution from the two blocks merging together as I have shown in Figure 13-18.

On the left of the flowchart in Figure 13-18, I have included the PBASIC statements that make up the conditionally executing code that starts with the if

statement. The first statement is the if-then statement that tests the conditions needed to determine which block of code is to execute. If the condition is true, then the code following the if-then statement executes until it encounters an else or endif statement. Along with executing the code if the condition is true, you can also execute code if the condition is false by placing this code after an else statement. The endif statement indicates that the conditionally exe-

Figure 13-18   *If flowchart 2*

cuting code is finished and execution should resume whether or not the condition is true or false.

The else statement as part of conditionally executing code is optional. If you wanted to just execute code if the condition is true, you could eliminate the else statement and the statements that follow and just key in the following:

```
if Value1 Condition Value2
' Execute if Value1 Condition Value2 is
true
 endif
' Execute if regardless of being true or
false
```

In some applications, you will see that the programmer only wants to execute code if the condition is false. The code he or she will come up with will look something like this:

```
if Value1 Condition Value2
else
'  Executes if test result is false
endif
```

I would like to discourage this type of programming because it is confusing to read. In the table listing the different conditions earlier in the section, I listed the "complement" conditions to the basic six conditions available to PBASIC.

Instead of using the complement, you can also place the word "not" in front of the Value1 Condition Value2 expression, but this can make the code confusing to read (although in some cases, it will make the function of the code more obvious). Examples of the two different options are shown next:

```
if A <= B then

if not A > B then
```

Personally, I will always write out my applications using the first example's form because I feel that it is the clearest when I am reading over the source code. You may feel differently and want to use the second example's form.

It is possible to put in expressions for the "values" such as

```
if A + 4 < 37 then
```

But I would discourage you from doing this because it can be difficult to read when you are first starting to program. Although you will want to use this form when you become more comfortable with programming, and it will make your programs more efficient, when you are just starting keep the statements as simple as possible. If you do this, you might want to put parentheses around the arithmetic expression to make the statement more readable.

Along with the not operator, the if statement also has the "and" and "or" operators that will allow more complex if statements. The "and" and "or" operators work just like the digital logic "and" and "or" I discussed earlier in the book. Like the complex arithmetic statements, the conditions are read from left to right and code only executes when they all have been executed. Some examples of these operators are as follows:

```
if A > B and A < C then
'  Execute if "A" is between "B" and "C"

if A = CR or A = 10 then
'  Execute if "A" is a line end character
(10) or Carriage Return
```

To test out the operation of the if statement, try out "If Test" (that should be stored in the If Test folder in the Evil Genius folder). This application will print out the numbers from 1 to 30 that are evenly divisible by 3:

```
'  If Test - Print Numbers evenly
divisible by 3
'{$STAMP BS2}
'{$PBASIC 2.5}

'  Declarations
i var byte
j var byte

 i = 1

 do while (i < 31)

   j = i // 3
' Store the remainder of i/3
   if j = 0 then
     debug dec i, " is evenly divisible
by 3", cr
   endif

   i = i + 1

' Try the next variable

 loop

 end
```

Experiment 89
Advanced Conditional Execution

Parts Bin

Assembled PCB with BS2

Tool Box

PC
RS-232 cable

When I first wrote the experiment dealing with conditional execution, I was doing it with the version of PBASIC previous to 2.5, which did not have the structured features of the current version. In this version, the if statement was a lot simpler and I could explain both it and another statement type in just one experiment. In this experiment, I would like to expand upon some of the features of the "if-else-endif" statement that will make it easier for you to develop complex applications as well as another statement type that will allow you to execute multiple if statements without having to write them repeatedly.

When I described the if-else-endif statement structure in the previous experiment, I presented it in the format:

```
if Value1 Condition Value2 then
'  Execute if Value1 Condition Value2 is
True
   else
'  Execute if Value2 Condition Value2 is
False
   endif
```

Along with this statement structure, you can combine these multiple statements into a single line:

```
If A > B then C = A else C = B
```

This single-line format will avoid the extra lines of the multiline format, but it will not allow you to easily explain what the line is doing. The example above of loading the greater example is perfect for this format because a single operation is carried out. If you are going to carry out multiple statements depending on whether or not the condition is true, then I recommend using the multiline format because what is happening should be easier to understand than if

everything is crammed into a single line. Ideally, you should be able to explain what is happening in the line with a single comment.

When I key in my programs, I indent each level of conditional statements by two spaces (with the starting code two spaces in). I suggest that you follow this convention so you do not find your indented code squeezed over the right of the program, making it very difficult to read. You can nest your if-else-endif statements (or any other conditional or looping statements):

```
if A = "A" then
   if B = "B" then
'  Execute if variables have first two
letters of the alphabet
     else           ' A = "A" and B <> "B"
'  Execute if A has "A" but B is different
   endif
else                ' a <> "A"...
```

Instead of resorting to using multiple if statements to carry out different operations depending on a single value, you can use the select-case statement. This statement will compare a single value with a case value, as I have shown in the following example:

```
select RobotCommand
   case 1
'  "if RobotCommand = 1 then"
     '  Move the robot forward
   case 2
'  "if RobotCommand = 2 then"
     '  Move the robot backwards
   case else
'  Execute if the "RobotCommand" is
anything else
     debug "Invalid Command Received", cr
   endselect
```

The values after "case" could have a condition operator ($=$, $<>$, $>$, and so on) before the test value if you want to capture a number of values. I would recommend that you do not take advantage of this feature until you are very comfortable with programming. Adding the condition operators is quite difficult to do because they do not allow for testing a range of values (other than from the lowest possible value to a point or from a point to the highest possible value).

To look at how the different if statements work, I wanted to do something kind of interesting and print out a sine wave with a period of 20 and an amplitude of 7 so it can be displayed effectively on the debug terminal (Figure 13-19). This application treats the debug terminal as a raster display and prints a sine wave value if its column is the same as the currently printed row. After keying in the application below, save it as "Sinewave" in its own folder in the Evil Genius folder:

```
'  Sinewave - Output a sine wave on the
Debug Terminal
'{$STAMP BS2}
'{$PBASIC 2.5}

'  Variables
Row        var byte
Col        var byte
SineValue var word

'  Initialization
 Row = 1

  do while (Row < 8)
    Col = 0
    do while (Col < 40)
      SineValue = Col * 14
```

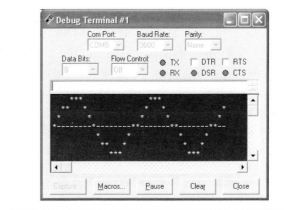

Figure 13-19   *BS2 sinewave*

```
      SineValue = sin SineValue
      if SineValue > 32767 then
        SineValue = SineValue ^ $FFFF + 1
/ 37
        SineValue = 4 - SineValue
      else
        SineValue = SineValue / 37 + 4
      endif
      select Row
        case 4
          if (SineValue = 4) then debug
"*" else debug "-"
        case else
          if (SineValue = 8 - Row) then
debug "*" else debug " "
      endselect
      Col = Col + 1
    loop
    Row = Row + 1
    debug cr
  loop

  end
```

Experiment 90
Using the "for" Loop in Your Application

Parts Bin

Assembled PCB with BS2

Tool Box

PC
RS-232 cable

With what I have shown you, you now have the capability to develop programs for just about any requirements that you are given. This is probably a shock to you because you probably look at a favorite PC application or game and think that you will never be

able to develop a program like that. Maybe today you wouldn't be able to develop these applications, but you are further along the road to understanding how to program than you might think. Over time, as you get more experience programming different applications, you will gain the skills, confidence, and background necessary to develop very complex and comprehensive applications.

The first version of the experiment's software will sort six random integers and print them out in order. Start up the Stamp Windows Editor, enter in this application, and save it (as Bubble Sort) in the For Statement folder located in the Evil Genius folder:

```
'  Bubble Sort - Sort a list of numbers
'{$STAMP BS2}
'{$PBASIC 2.5}

'  Variables
ArraySize con 6
SortArray var byte(ArraySize)
i var byte
j var byte
Temp var byte

'  Load Initial values into the Array
SortArray(0) = 55: SortArra(1) = 5
SortArray(2) = 100: SortArray(3) = 2
SortArray(4) = 65: SortArray(5) = 4

  i = 0  '  Print out the Initial Order
  debug "Initial Number Order: "
  do while i < ArraySize
    debug dec SortArray(i)
    i = i + 1
    if i < ArraySize then debug ", " else
debug cr
  loop

i = 0  '  Perform Bubble Sort on Numbers
  do while i < ArraySize - 1
    j = 0
    do while j < ArraySize - 1
      if SortArray(j) > SortArray(j + 1)
then
        Temp = SortArray(j + 1)
        SortArray(j + 1) = SortArray(j)
        SortArray(j) = Temp
      endif
      j = j + 1
    loop
    i = i + 1
  loop

  i = 0
  debug "Sorted Number Order: "
  do while i < ArraySize
    debug dec SortArray(i)
    i = i + 1
    if i < ArraySize then debug ", " else
debug cr
  loop

  end
```

Running this application will create the debug terminal display in Figure 13-20 and sort the six numbers in "SortArray" one element at a time. It will compare each element in the array with the one following it. If the current element is greater than the

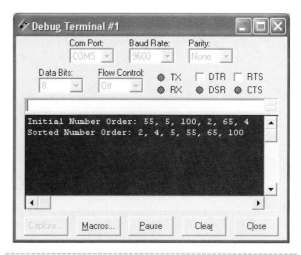

Figure 13-20 *Sort output*

one that is after it, then it will swap their positions. The tests are repeated once for each element in the array to make sure that even if the largest number in the array was at the start, it would be moved to the rear of the array at the end of the program. This type of sort is known as a "bubble sort" and is probably the most inefficient sort known to man for large arrays—I won't go into details why because for small lists of numbers like the one used in this application, it is fine, but if you have to sort a large array then it will take longer than virtually any other sorting algorithm.

In the bubble sort, you'll see that I have to read through SortArray several times, and to do this, I initialize a variable and run it through a do-while loop while it is less than the final value (while incrementing it inside the loop). Although the do-while loop works well for this application, it is not the most efficient method of looping a set number of times; the best way to do this is to use the for-next statements that take the form:

```
for variable = InitialValue to EndValue
{step StepValue}
'  Code to be repeated
next
```

In the for statement, the count variable is defined, along with its initial value as well as its ending value. When the for statement ends, you'll find that the variable will be greater than the end value. How much depends on the step value. This value defaults to 1 if no value is specified, or it can be any specified positive or negative value. You can change the variable in the loop, but this is not recommended. Changing the loop variable could cause execution to leave the for loop unexpectedly. For this reason, be very careful about how you use the loop variable in your pro-

gram, because you could inadvertently change it, resulting in the program behaving unexpectedly.

To show how the for-next statements can improve an application, consider the Bubble Sort program rewritten to For Sort. It is quite a bit shorter and, I think, much easier to read and follow:

```
'  For Sort - Sort a list of numbers using "for" statement
'{$STAMP BS2}
'{$PBASIC 2.5}

'  Variables
ArraySize con 6
SortArray var byte(ArraySize)
i var byte
j var byte
Temp var byte

'  Load Initial values into the Array
 SortArray(0) = 55:  SortArray(1) = 5:  SortArray(2) = 100
 SortArray(3) = 2:   SortArray(4) = 65: SortArray(5) = 4

 debug "Initial Number Order: "
 for i = 0 to ArraySize - 1
   debug dec SortArray(i)
   if i < (ArraySize - 1) then debug ", " else debug cr
 next

 for i = 0 to ArraySize - 2     '  Perform Bubble Sort on Numbers
   for j = 0 to ArraySize - 2  '  Find highest in List
     if SortArray(j) > SortArray(j + 1) then
       Temp = SortArray(j + 1) '  Swap the two Values
       SortArray(j + 1) = SortArray(j)
       SortArray(j) = Temp
     endif
   next                        '  Repeat through the list
 next                          '  Re-Start List to find next highest

 debug "Sorted Number Order: "
 for i = 0 to ArraySize - 1
   debug dec SortArray(i)
   if i < (ArraySize - 1) then debug ", " else debug cr
 next

 end
```

Parts Bin

Assembled PCB with BS2

Tool Box

PC
RS-232 cable

As you write more and more programs, you will discover that you are writing the same things over and over again. This is something that you will have to get used to (at least until you learn how to copy code from previous applications using the cut and paste features of Windows), but it will be a problem when you end up putting in the same pieces of code in the same application.

To show you what I mean, consider an application that uses a countdown timer. In the Timer Demonstration application, I first count down from 5 and then from 10:

```
'   Timer Demonstration 1 - Count down
from the specified value
'{$STAMP BS2}
'{$PBASIC 2.5}

'   Variables
j var byte

'   Mainline
  debug rep "."\5
' Count down from 5 seconds
  for j = 1 to 5: pause 1000: debug rep
bksp\1: next
  debug "5 Second Delay Finished", cr, cr

  debug rep "."\10
' Count down from 10 seconds
  for j = 1 to 10: pause 1000: debug rep
bksp\1: next
  debug "10 Second Delay Finished", cr, cr

  end
```

In this application, a set number of periods are written to the Debug Terminal using the rep debug statement formatter. Once this is done, I count down, waiting one second each loop (the pause 1000 statement) before sending a backspace (bksp) character to the Debug Terminal to erase the period on the right of the line. Once I have erased the periods, I print a message along with two carriage returns so

the next text written to the display will be two lines below the message.

The colon (:) character used in the code allows me to place multiple statements on the same line. I do not recommend this practice for most situations because it can make it much more difficult to understand what the code is doing. Cases occur, however, like this one, when the colon is used to bring together a number of statements so they perform a single function (in this case, backspace once per second for a set number of times). Grouping a series of statements like this into a single function can enhance the readability of the code.

This application works very well and is kind of interesting to watch and try to understand. When a problem occurs, I repeat almost exactly the same three lines that are used to display a number of dots, then erase them, and put in a message saying that the delay has finished.

PBASIC, like most languages, provides the ability to define and call subroutines that will allow code to be reused in an application. The gosub statement saves the address of the statement following it so once the subroutine code (starting at the subroutine "label") has finished executing, a return statement will allow it to continue executing where it left off. Figure 13-21 shows how the gosub and return statements are used to jump and come back from a subroutine.

To show how the Timer Demonstration 1 could take advantage of using the subroutine capability of PBASIC to eliminate one of the sets of code used to display the periods and then erase them, key in the following application and save it as Timer Demonstration 2 in your Timer Demonstration folder.

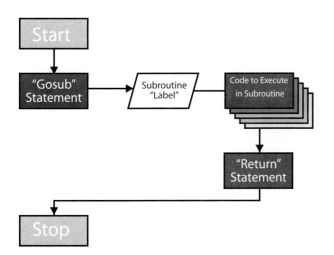

Figure 13-21 *Subroutine flowchart*

```
'   Timer Demonstration 2 - Use a Count
down Subroutine
'{$STAMP BS2}
'{$PBASIC 2.5}

'   Variables
i var byte
j var byte

'   Mainline
  i = 5
'   Count down from 5 Seconds
  gosub CountDown

  i = 10
'   Count down from 10 Seconds
  gosub CountDown

  end

CountDown:
'   Count down subroutine
  debug rep "."\i
'   Count down "i" seconds
  for j = 1 to i: pause 1000: debug bksp:
next
  debug dec i, " Second Delay Finished",
cr, cr
  return
'   Return to caller
```

This program executes exactly the same way as the previous one, but where I hardcoded the delays in the first version, I use the variable i (which can be called a subroutine *parameter*) to indicate to the subroutine how many seconds it must delay for. The first time the CountDown subroutine label is called (using the gosub statement), i has a value of 5 and the second time, it has the value of 10. This value is used to specify how many periods are written to the Debug Terminal as well as how many times the for loop executes to erase the different periods.

Using Ctrl-M to show the program data that is stored in the BS2's EEPROM, I can see that Timer Demonstration 1 requires 160 bytes of EEPROM storage and Timer Demonstration 2 requires an even smaller 100 bytes. The use of the subroutine in this application results in a more than 35 percent reduction in program memory space required. You will find that this reduction in space is quite reasonable for applications that use subroutines over ones that don't. Along with the very tangible benefit of the amount of program memory saved by use of a subroutine, you will find that the elimination of the need to repeat the same code in your program repeatedly to be quite an important intangible benefit of using subroutines.

For Consideration

If you have worked with the Parallax BS2 before, you'll know that I have not said anything about three of the most popular basic statements in PBASIC:

```
goto Label
if Condition then Label
branch value, Label0, Label2,
Label3{, ...}
```

Along with ignoring these statements, I seem to have minimized the importance of Label in PBASIC programming and only promoted its use as an indicator for the start of subroutines. If you take a look around the Internet (at sites such as www.hth.com/filelibrary/txtfiles/losa.txt) for sample BS2 applications, you may be confused at my reticence to use them because after looking at them, it seems impossible to write PBASIC applications without these three statements.

As I have explained programming in this section, I have taken advantage of the structured programming features built into the latest version of the BASIC Stamp compiler.

Structured code is built from very definite blocks of code that execute under specific conditions. This is an important aspect of the structured programming philosophy. Each block of code should perform a specific function—ideally after each block of code, you should leave a line of whitespace before and after it so somebody reading the code can see where the block begins and ends.

Each function of the program is broken into a single block of code, including an if statement indicating why they are executing, separated by a blank line. The resulting snippet of code is very easy to understand and specific functions are easy to find. The goals of structured programming are that it is easy to write and, once it is written, easy to understand and change. An important aspect of being easy to understand and change is the need for somebody other than the original author to fix the program at some later date.

Previous versions of PBASIC did allow for sequential statement execution, as this is a feature of most traditional programming languages, but it did not allow for what I call "positive" conditional programming. To execute a block of code conditionally (after a test), you would have to write it in the format:

```
if Not_Condition then goto Skip

'  Code executes if "Condition" is true

Skip:
```

instead of the format presented in this section:

```
if Condition then

'  Code executes if "Condition" is true

endif
```

In the first case, the if statement causes the execution to jump over the block of code when the negative condition needed for it to execute is encountered. For example, if you wanted to execute the block of code if A is less than B (A < B), then you would have to convert that to jump over it if A is greater than or equal to B (A >= B). This is why when you see list of the different if statement conditions you often also see a list of the complementary (or negative) conditions as well.

I call the first example of the if statement *negative programming*. It adds an extra layer of complexity to learning how to program and the need for some definite thinking skills to recognize that the code that follows the if statement is executed because the condition is false and not true. Negative program-

ming is just not intuitively obvious and should be avoided.

The "else" option of the if statement allows your code to execute one or another different path if the condition is not true. For example, using the original PBASIC statements, to implement an if-else-end section of code, you could write it as something like

```
if Condition then True_Skip
goto False_Skip

True_Skip:

'  Code executes if "Condition" is true

   goto If_End

False_Skip:

'  Code executes if "Condition" is false

If_End:                               '
Finished
```

or, eliminating the goto statement after the if statement, as

```
if Not_Condition then False_Skip

'  Code executes if "Condition" is true

   goto If_End

False_Skip:

'  Code executes if "Condition" is false

If_End:                               '
Finished
```

What happens in these two blocks of code is quite confusing, especially when compared to what the same block of code looks like when it is written to take advantage of the if-else-endif statement available with the latest version of PBASIC:

```
if Condition then

'  Code executes if "Condition" is true

   else

'  Code executes if "Condition" is false

   endif
```

Another reason why I prefer the structured if-else-endif statements is that it cuts down on the number of labels I have to name. In the example above, True_Skip, False_Skip, and If_End are quite clear, but their clarity in the application goes down geometrically the more you use them. It is much harder to try and think up meaningful names for the different labels, and if you cut and paste code to save on typing, you will probably make the mistake of using the same label or goto statement somewhere (which can be difficult to find). On the surface, this may seem trivial, but I always seem to run out of good descriptors for skip, else, and end after using them three or more times.

This should explain why I don't like the "if Condition then Label" statement in programming, but you might want to use the goto statement for loops as something like

```
Loop:
'  Code executes within the loop
   goto Loop
```

instead of the do loop statements presented in this section. The difficulty in coming up with different labels probably seems lessened, and the readability between the two methods is insignificant. I would recommend sticking with the do loop statements because their purpose is immediately clear (the statements make up a loop), and you can add the conditional "while" to the "do" statement easily.

Along with a do-while condition loop, PBASIC also has a "do loop until Condition" that I would recommend you stay away from, because it is another example of negative programming. You stay in the loop while something is not true, rather than the while condition in which the execution stays in the loop while the condition is true.

While in a do loop, you can get out of it by executing an exit statement (like "if Condition then exit"):

```
   do
'  Code executes in Loop
      if Condition then exit
'  Code executes after test
   loop
```

This is similar to a goto statement such as

```
   do
'  Code executes in Loop
      if Condition then Loop_Exit
'  Code executes after test
```

```
   loop
Loop_Exit:
```

and should be avoided for the same reasons as I stated for the "if Condition then Label" statement, as well as the very important reason that it can be considered to be unstructured; it will cause execution to stop in the middle of a block of code. You could argue that a block of code exists before and after the if statement, but this adds to the complexity of the code; remember you want to make your code as easy to read as possible.

I could go on about why I do not recommend the use of the branch statement (the need for differently named labels, the unstructured appearance of the code, and so on), but the real reason why I don't recommend it is that it is simpler to use the select-case statement. The branch statement is quite simple to work with and can handle a variety of different values, but this results in complex, difficult-to-read code, whereas the select-case statement handles different values as a matter of course.

When I have talked about structured programming, you may have noticed that I have not used any graphics to explain the different concepts. This was done to demonstrate one of the most important aspects of structured programming; when a program is written with structured programming concepts in mind, the result can be expressed quite easily verbally and does not require any kind of diagrams. Nonstructured programs cannot be explained very easily verbally, and flow charts and other visual aids are often required.

Being able to express your structured programs verbally allows you to very simply document them. In fact, if you have taken care in naming your variables and subroutines and have explained the operation of the I/O pins using the pin declaration (explained in the next section) and comments, you may not have to document your application at all. Properly documenting programs is difficult to do and often gets left to the end of a program (where it doesn't get done at all). To minimize the documenting hassle, make sure that your programs are well written and follow the structured programming philosophy that I have set out here. It's much easier to document a well-written and structured program than one that has gotos all over the place with labels like "loop_end_4."

Interfacing Hardware to the BASIC Stamp 2

In the previous section, I introduced you to the basics of programming the BS2. The information provided so far is quite generic and can be applied to just about any programming situation. These skills can be applied to other microcontrollers, desktop computers, personal digital assistants, and sophisticated tools. By making this information generic, you should be able to read other introductions to programming and be able to follow along and apply what is being presented to the BS2. You now have a reasonably good grouding in programming and are ready to learn how to create robot applications using the BS2 because I have not presented the hardware interfaces that make it unique to other devices.

The BS2 that you are using is based on the Microchip PIC16C57 PICmicro® microcontroller. This microcontroller contains program code that decodes and executes the tokens provided by the Stamp Windows Editor software and interfaces to the PC that downloads the application into the BS2. Along with providing these functions, the PICmicro microcontroller also provides the BS2's *input/output* (I/O) pin hardware.

The PICmicro *microcontroller* (MCU) I/O pins are designed to be in either output or input mode by enabling or disabling the *TRI-State* (TRIS) enable register controlled tri-state driver shown in the diagram of the pin's internal circuitry shown in Figure 14-1.

When Parallax architected the BASIC Stamp and PBASIC, they wanted to avoid the complexities of the PICmicro MCU I/O pins, so they specified the I/O pins to be accessible in a manner similar to that of variables. They added three labels (along with a number of sublabels) that allow software to interface directly with the BS2's PICmicro MCU I/O pins. These registers (and subregisters, allowing you to access smaller groups of pins) are listed in Table 14-1.

PBASIC provides the pin type that allows you to define a pin using the statement:

```
Label pin #
```

where "#" is the pin number (0 through 15). To define "outputpin," make the BS2's "P0" pin an output, and drive out a low voltage, you could use the code

```
outputpin pin 0
  Output outputpin
' P0 is put into output mode
  outputpin = 0
' Set pin to "low" or zero output
```

A more direct way of making the P0 I/O pin a "low" output is to use the "low" built-in function statement. I find this single statement to be a lot more intuitive than the statements above—to perform the same function the simple statement that follows is used:

```
    low outputpin
' Make P0 output, drive out "low" voltage
```

Along with "low," you can use a number of other built-in statements to set the I/O mode of the pins as well as the output state of the pins. These functions are listed in Table 14-2.

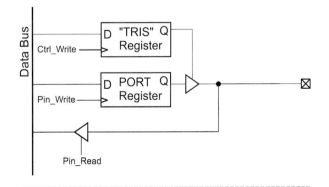

Figure 14-1   *I/O pin*

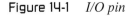

Table 14-1 Labels built into PBASIC to access the BS2 input/output (I/O) pins

Word Name	Byte Names	Nybble Names	Bit Names	Function
INS	INL, INH	INA, INB, INC, IND	IN0–IN15	Read state of input pins
OUTS	OUTL, OUTH	OUTA, OUTB, OUTC, OUTD	OUT0–OUT15	Save new output state to I/O pins
DIRS	DIRL, DIRH	DIRA, DIRB, DIRC, DIRD	DIR0–DIR15	Change I/O pin mode (1 for output, 0 for input)

Table 14-2 PBASIC I/O pin statements

Statement	Functional Description
low #	Make I/O pin # an output and set to low voltage
high #	Make I/O pin # an output and set to high voltage
input #	Make I/O pin # an input
output #	Make I/O pin # an output
reverse #	Reverse I/O pin # from an input to an output
toggle #	Make I/O pin # an output and toggle its state

The function statements listed make controlling the I/O pins quite simple and eliminate the need to write to individual I/O pins with the assignment statements showed previously. Defining a pin allows you to access the data as if it were a variable while allowing you to use the label as a constant in the function statements. If you have multiple I/O pins that you want to access or you are checking the state of a bit (or group of bits), you will have to use the I/O pin register names that are listed. These names are used in exactly the same way as variable names.

Along with providing simple, digital I/O, the I/O pins on the BS2 can be used in a number of different PBASIC function statements. In this section, I will introduce you to a number of these functions and how they provide different capabilities to the BS2. These built-in functions are one of the major strengths of the BASIC Stamp family, and they allow you to create very complex applications very easily.

When you read about the serial I/O functions that are available to the BS2, you will discover that 17 I/Os can be used and not the 16 that you are probably expecting. The seventeenth (number "16") I/O that you can use for serial operations is the programming interface. This can be a convenient way to communicate with the BS2 in situations where a serial port interface is already built in (such as in the PCB provided with this book).

Experiment 92
Controlling an LED

Parts Bin

Assembled PCB with breadboard, battery and BS2 installed

LED, any color

Tool Box

PC

RS-232 cable

Wiring kit

In the previous section, when I wanted to display data from the BS2, I used the "debug" statement. This statement works quite well at this task, but it isn't practical in most robot applications—for example, if you had a wall-following robot and wanted to understand exactly how the program was executing, chances are you could not reasonably pass an RS-232 cable from the robot to a PC and have the robot work correctly. The ability to output execution information from the BS2 to the user without using the "debug" statement is an important capability for robot applications. In this experiment, and the others that follow, I will present you with different ways in which data can be output from the BS2 without requiring a direct interface to a computer.

The most basic output device for the BS2 is the *light-emitting diode* (LED). I have discussed the LED at length earlier in the book, and in this experiment it will be connected to the BS2 and controlled to turn on and off.

The circuit that you will be working with consists of wiring an LED to one of the BS2 I/O pins and its regulated power supply output as shown in Figure 14-2. Figure 14-3 shows how the LED will be wired into the PCB's breadboard.

I would like to point out that if you look on the Internet, you will find BS2 and PICmicro microcontroller application circuits that do not use current-limiting resistors at all (in this circuit I take advantage of the built-in 220 Ω resistors built onto the PCB). In these cases, the designers are counting on the PICmicro MCU's I/O pins only being able to supply a maximum of 20 mA to an external device or to sink a maximum of 25 mA from an external

device. I would recommend that you do not follow these examples as they are wasteful of current and do not provide any protection for the BS2/PICmicro

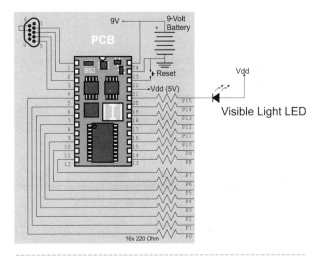

Figure 14-2   *LED output*

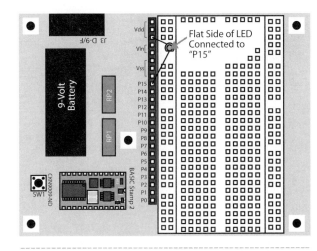

Figure 14-3   *Adding a single LED to the PCB*

microcontroller I/O pins. Eliminating the current-limiting resistor could result in a circuit that could burn out the I/O pin, leaving you with a damaged BS2.

An aspect of the design that you will find surprising is that I connected the LED's anode (positive connection) to one of the PCB's power supply connections and the cathode (negative connection) to the BS2's I/O pin. This connection follows a convention that is commonly used for microcontrollers because some early chips (the Intel 8051 is a prime example of this) could not source current; they could only sink it. To allow LEDs to be connected directly to the I/O pin, the anode would have to be connected to the power source and the cathode to the I/O pin, just as I have done here.

Flashing the LED twice per second by accessing the I/O pin like a variable, using assignment statements, could be accomplished using the application code:

```
' LED Flash Demonstration 1 - Flash LED
on P0 2x per second
'{$STAMP BS2}

LED pin 15               ' Define the I/O Pin

' Mainline
    dir15 = 1            ' P15 is an output
do
```

```
    LED = 0              ' LED On
    pause 250            ' Delay 1/4 second
    LED = 1              ' LED off
    pause 250
Loop                     ' Repeat
```

Save this as "LED Flash 1" in the LED Flash folder that you have created in the Evil Genius folder. Once you have done this and tested it, you can create the LED Flash 2 application (saving it in the same folder), which uses the low and high built-in PBASIC function statements:

```
' LED Flash Demonstration 2 - Flash LED
on P0 2x per second
'{$STAMP BS2}

LED pin 15               ' Define the I/O Pin

' Mainline
do
    low LED              ' LED On
    pause 250            ' Delay 1/4 second
    high LED             ' LED off
    pause 250
    Loop                 ' Repeat
```

LED Flash 2 works identically to LED Flash 1 because they are essentially the same application. I did not need a "dir0 = 1" or "output 0" statement in LED Flash 2 because the first "low 0" statement places the I/O pin in output mode before driving it low.

Experiment 93
Cylon Eye

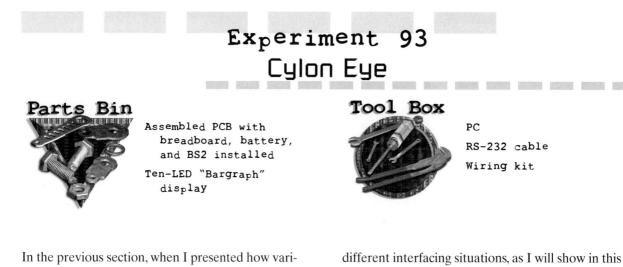

Parts Bin

Assembled PCB with
    breadboard, battery,
    and BS2 installed
Ten-LED "Bargraph"
    display

Tool Box

PC
RS-232 cable
Wiring kit

In the previous section, when I presented how variables were declared, I showed how other variables could be based on them, either as renamed variables or as a smaller part of them. I rarely use this capability for straight programming, but it is very useful for

different interfacing situations, as I will show in this experiment.

In the 1970s, a moderately successful show called *Battlestar Galactica* was on television and was pretty forgettable except for the robots. The robots weren't great conversationalists (their stock phrase was "By your command"), but they did have a pretty cool eye. The robot's eye consisted of a red light that swung back and forth, scanning the area in front of it. Although the series is largely forgotten, this scanning eye lives on as an indicator in a variety of different computer systems to output that the system is "alive and functioning."

The circuit consists of wiring the 10-LED bargraph display to the BS2 as I've shown in the schematic diagram (Figure 14-4). Like the other experiments that use LEDs, I take advantage of the current-limiting resistors built in to the PCB.

When you are wiring in the 10-LED bargraph display, install the LED bargraph with the anode orientation notch cut into the corner of the display on the bottom right-hand corner (Figure 14-4).

Save the code used for the experiment following in the Cylon Eye folder, located in the Evil Genius folder on your PC:

```
' Cylon Eye - Scan an LED across the
display
'{$STAMP BS2}
'{$PBASIC 2.50}

' Variables
Direction var byte
LSB pin 0
' Least Significant LED bit
MSB pin 9
' Most Significant LED bit

' Initialization
 outs = %1111111110
' Make all the LED pins outputs
 dirs = %1111111111
' With LED at P0 on
 Direction = 0
' Start Running Up

 do

  if (Direction = 0) then
   outs = outs << 1 + 1
' Shift the Lighted LED up
   if (MSB = 0) then
    Direction = 1
' Change the direction of the Movement
   endif
  else
   outs = outs >> 1
' Shift the Lighted LED down
   MSB = 1
' Make sure the MSB bit is set
   If (LSB = 0) then    ' At the bottom
    Direction = 0       ' Start going up
   endif
  endif

  pause 100
```

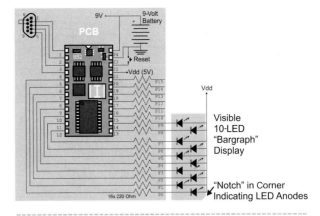

Figure 14-4   *Cylon eye*

```
' Take 1 second to run across 10 LEDs

loop
```

The application should be quite simple to understand. When you get it running, it really is quite attractive (and could be the basis for a Christmas or other holiday decoration).

Instead of declaring the pins LSB and MSB from the outs directory, I could have simply polled the individual bits (that is, out0 and out9 for LSB and MSB, respectively) or used mathematical values to find if the numbers were at either extreme. For example, I could have written the first test ("if (MSB = 0) then") to

```
if (out9 = 0) then
```

or put in the entire bit string that is expected for the bits. Another way the "if (MSB = 0) then" statement could be written out is

```
if (outs = %0111111111) then
```

This method of testing the data has a certain appeal because the expected state of each bit is displayed in the source code. In this application, writing out each bit as you expect it to be is in character with the application because of the way I initialized the "dirs" and "outs" registers.

It might be surprising to see that I used two statements for the shift down operation. The first statement shifts the bits to the right (moving the lighted LED down), and the second explicitly sets the most significant bit of the display (the "MSB = 1" statement). I could have combined the two statements into one of the three below:

```
   outs = outs >> 2 + 512
' Shift down and set the MSB
   outs = outs >> 2 + $200
' Shift down and set the MSB
   outs = outs >> 2 + %1000000000
' Shift down and set the MSB
```

I refrained because these statements do not seem as easy to understand as the other two statements when the application is to be shared with somebody who is not familiar with programming. These three statements will shift down the value in the outs register and then set the most significant bit. This is just like I did with the shifting up, adding 1 to the bit that would no longer be set after the data is shifted up. The problem with adding, 512, $200 or %1000000000 to the shifted-down value is that I don't believe what the code is doing is as obvious to you as setting the most significant bit explicitly.

Experiment 94
Hitachi 44780-Controlled Liquid Crystal Display

Parts Bin

Assembled PCB with
   breadboard, battery,
   and BS2 installed

16x2 Character LCD

10k pot

Tool Box

PC

RS-232 cable

Wiring kit

So far I have concentrated on using LEDs for output devices, but you can consider other methods of passing information to the user. I am sure that you are familiar with *liquid crystal displays* (LCDs); they are used in virtually everything from watches to appliances to computer displays. Their primary attraction is the low power they consume.

You will probably be very surprised to discover that there actually is a liquid in an LCD. Nematic (based on the Greek word *nema* or string) crystals, which are very long, thin molecule that react to electrical fields, are suspended in a liquid (usually water). As shown in Figure 14-5, when an electrical field is applied to them in the liquid, these crystals arrange themselves so they are parallel with the electrical field and block light. When no electrical field is applied, the crystals are aligned randomly and light can pass through them. LCD displays operate with very low current drains because no current is flowing through the liquid.

The actual interface to the LCD controller hardware is quite complex and must be very accurately timed. To make interfacing easier to the LCD, a num-

ber of different interfaces have been developed for them. The most popular of these is the Hitachi 44780 chip that is bonded to the LCD carrier. The LCD, chip, and carrier are usually referred to as *modules* and have the 14 connection holes listed in Table 14-3. The LCD module works like a teletype or a single-line TV display—as you write characters to it, a cursor will move to the right to prepare for the next character.

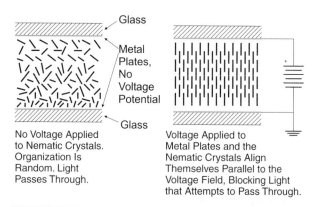

Figure 14-5   *LCD operation*

No Voltage Applied to Nematic Crystals. Organization Is Random. Light Passes Through.

Voltage Applied to Metal Plates and the Nematic Crystals Align Themselves Parallel to the Voltage Field, Blocking Light that Attempts to Pass Through.

Table 14-3 Module connection pins

Pins	Description/Function
1	Ground
2	Vcc
3	Contrast voltage
4	"RS" — _Instruction/register select
5	"RW" — _Write/read select
6	"E" clock
7 – 14	Data I/O pins (Do on Pin 7/D7 on Pin 14)

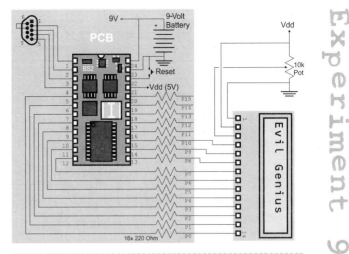

Figure 14-6 *LCD interface*

I typically attach a series of pins to the 14 connector pins so that the LCD can be easily mounted on a breadboard. In some LCDs, you may discover that there are 16 connector holes, with the extra 2 holes used for backlighting. Some other LCD modules have two rows of seven or eight pins. For the ease of creating the experiments in this book and wiring them to the breadboard, you should just use LCD modules that have a single row of pins.

Finding an LCD module with the Hitachi 44780 chip on it is quite easy, and most electronic stores have a number of different ones in stock. I recommend that you buy a 16x2 display because it is very common and usually reasonably cheap. Looking around a surplus store, you should be able to find one either loose or as part of another product for a dollar or so.

Wiring the LCD to the BS2 is quite straightforward as you will see in Figure 14-6. The only unexpected aspect of the circuit is the potentiometer used to set the contrast voltage used by the LCD. The voltage produced by this voltage divider is used to specify the darkness of the characters on the LCD. Depending on the type of LCD that you are using, you will find that this voltage will either be high or low.

To simplify the wiring, I have simply passed the data signals directly between the BS2 connector and the LCD. By doing this, I have to "rotate" the data bits (using the "rev" operator) so the ASCII characters read from the "lookup" statement can be passed directly to the LCD module.

The application code will print out "Evil Genius" on the top line of the LCD and can be used as a basis for other applications.

```
' LCD Test - Display a simple message on
an LCD module
'{$STAMP BS2}
'{$PBASIC 2.50}

' Variables
i var byte
Character var byte      ' Character to
Display
LCDData var outl        ' Define LCD Pins
on BS2
LCDE pin 8
LCDRW pin 9
LCDRS pin 10

' Initialization
  dirs = %11111111111
' Make Least Significant 11 Bits Output
LCDRW = 0: LCDRS = 0: LCDE = 0
' Initialize LCD interface
pause 20
' Wait for LCD to reset itself
LCDData = $0C: pulsout LCDE, 300: pause 5
' Initialize LCD Module
pulsout LCDE, 300: pulsout LCDE, 300
' Force reset in LCD
LCDData = $1C: pulsout LCDE, 300: pause 5
' Initialize/Set 8 Bit
LCDData = $08: pulsout LCDE, 300
' No Shifting
LCDData = $80: pulsout LCDE, 300: pause 5
' Clear LCD
LCDData = $60: pulsout LCDE, 300
' Specify Cursor Move
LCDData = $70: pulsout LCDE, 300
' Enable Display & Cursor

Character = 1: i = 0
LCDRS = 1            ' Print Characters
do while (Character <> 0)
  lookup i, ["Evil Genius", 0], Character
  if (Character <> 0) then
   LCDData = Character rev 8: pulsout
LCDE, 300
   i = i + 1
  endif
loop

end
```

Experiment 95
Musical Tone Output

Parts Bin

Assembled PCB with breadboard, battery, and BS2 installed

LM386, 6-volt LM386 audio amplifier in 8-pin "DIP" package

Two 1k resistors

0.01 μF capacitor, any type

0.1 μF capacitor, any type

330 μF 16-volt electrolytic capacitor

10k pot

Eight-ohm SPKR

Tool Box

PC

RS-232 cable

Wiring kit

So far I have introduced you to three ways in which the BS2 can be used for feeding back information to you during its operation. The "debug" function is useful for returning detailed information to either you or the user, but it requires a PC connection. LEDs can be seen from a fair distance away, but do not do well with providing a lot of information or variances of information (changing output intensity). Finally, LCDs can display a lot of data and indicate variances, but they are difficult to read from far away (it isn't unusual to see robot scientists crawling right behind their robots to read LCD output information). If you have an application that requires different levels of information and the BS2 may be some distance away, you might consider adding a speaker that can provide status information from some distance away.

PBASIC provides the "freqout" function that will drive out a musical tone from a specified pin. The format for the freqout function is given below. The Duration is the length of time that the note will play. Frequency1 is the primary frequency output from the BS2 specified in Hertz and, if you desire, you can have a second frequency (Frequency2) to make a two-note chord.

```
freqout Pin, Duration, Frequency1{,
Frequency2}
```

Chances are you would think that the freqout output would be a square wave at the frequency specified—this is not the case. The output consists of a series of pulses that are designed to be "filtered" into a smooth sine wave as I've shown in Figure 14-7, which is an oscilloscope picture of the two lines during operation. In Figure 14-7, the top wave is the output from the BS2 I/O pin and the lower line is the resulting sine wave.

The filter circuit that is used consists of the resistors and capacitors between the BS2 output pin and the LM386 input in Figure 14-8.

When you have built the circuit, key in the following application code.

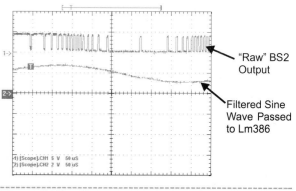

"Raw" BS2 Output

Filtered Sine Wave Passed to Lm386

Figure 14-7 *Freqout wave*

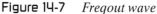

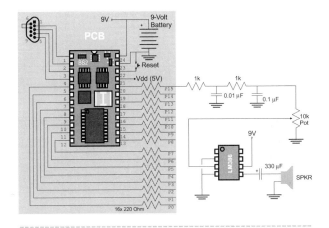

Figure 14-8   *Note output*

```
' Hawaii 5-0 - Play the Theme
'{$STAMP BS2}
'{$PBASIC 2.50}

' Variables
i var byte
Note var word
Duration var word
SoundOut pin 15

' Note Definition
A   con 880
AS  con 932
B   con 988
C   con 1046         ' Middle "C"
CS  con 1108
D   con 1174
DS  con 1249
E   con 1318
F   con 1396
FS  con 1480
G   con 1568
GS  con 1662
hA  con 1760
hAS con 1873
hB  con 1976
hC  con 2094
hCS con 2218
hD  con 2350

' Application
 Note = 1: i = 0
```

```
' Start reading through
 do while (Note <> 0)
 ' Loop through the tune
   lookup i,[G,G,hAS,hD,hC, G,-1,G,G,F,
hAS,G,0],Note
   if (Note <> 0) then
     lookup i,[1,1,1,2,4, 5,2,1,1,1,
2,7,0],Duration
     Duration = Duration * 208
 ' Start with 72 beats per minute
     if (Note <> -1) then
       freqout SoundOut, Duration, Note
     else
       pause Duration
 ' -1 means space, Just Delay
     endif
   endif
   i = i + 1
 end
```

This application plays the opening few bars of the theme of the TV show *Hawaii 5-0* and sounds surprisingly good considering how simple the circuit is. In the Note Definition section of the application, I have listed out the basic frequencies for the notes from A below middle C to D an octave above. When -1 is encountered, a space is put into the tune. The Duration is in the units of quarter notes, and I have multiplied it by the number of msecs that a single quarter note should take (playing at 72 beats per minute) to get the actual delay for the tune. As an exercise for you, you may want to try to program a different tune into the application. To do this, you will have to find some sheet music and transpose the notes and delays into the lookup table that I created. It's not very hard and actually quite fun.

When you run the application, you will find that the 10k potentiometer must be set quite low or else the LM386 audio amplifier will be overloaded and nothing will be output. Even with a low potentiometer setting, you will find that the sound output is quite loud.

Experiment 96
Electronic Dice

Parts Bin

Assembled PCB with breadboard, battery, and BS2 installed

Seven LEDs, any color

10k resistor

Push button

Tool Box

PC

RS-232 cable

Wiring kit

When I have looked at robots that are controlled by the BS2, I have discovered that one of the most useful PBASIC built-in functions is never used, even though it is very well suited for robot applications. This function is the "button" function, and it provides you with a very simple framework for debouncing button input. Actually, if you look at all the different BS2 applications that are available on the Internet, you will discover that the button function is almost never used. This is a mystery to me because the button function is probably one of the most unique and useful functions defined in the PBASIC language.

```
button Pin, DownState, Delay, Rate,
Workspace, TargetState, Address
```

where the button statement's parameters are explained in the PBASIC Reference.

The trick to the button function is that you have to remember that it executes each time the button is pressed and either increments or clears the Workspace variable. The function does not provide timing for you—you will have to place the statement inside a timed loop. This is not as hard as it sounds if you consider that you can count on each PBASIC statement taking 250 µs (using Parallax's specification that the BS2 runs at 4,000 statements per second), and you can take advantage of the "pause" function built into PBASIC.

Let's have a bit of fun and for this experiment create some "digital dice" using the BS2 and some LEDs started by a press of a button. The circuit schematic is shown in Figure 14-9.

The wiring for the experiment is a bit more complex because of the desire to make the LEDs look

like dice. The code should be saved as "Digital Dice," in a Digital Dice folder in the Evil Genius folder:

```
' Digital Dice - Create Digital Dice with
a push button
'{$STAMP BS2}
'{$PBASIC 2.50}

' Variables
ButtonPin pin 15
ButtonCount var byte
Dice var byte
i var byte
j var byte

' Initialization
outs = $ffff
' Make all the LEDs high/off
dirl = %1111111
' Make P0 through P6 outputs

do

ButtonDownWait:
' Wait for Button to be pressed
Dice = Dice + 1
' Randomize the Dice Value
pause 4
' Loop takes approximately 5.5 ms
button ButtonPin, 0, 10, 180,
```

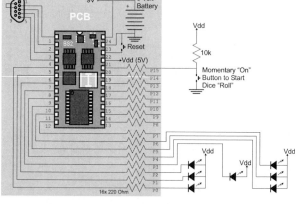

Figure 14-9   *Digital dice*

```
ButtonCount, 1, ButtonDown
  goto ButtonDownWait
' Debounce after 55 msecs, repeat 1x sec
ButtonDown:
  for i = 1 to 5        ' Button Pressed
   for j = i to 5
' Display as running down
    pause i * 125
' Increasing Delay for value displays
    Dice = Dice + 1
    select (Dice // 6)
' Display a Dice Value using "Select"
     case 0: out1 = %1110111
' Start with "1"
     case 1: out1 = %0100010 ' Display "5"
     case 2: out1 = %0111110 ' Display "2"
     case 3: out1 = %1100011 ' Display "3"
```

```
     case 4: out1 = %0001000 ' Display "6"
     case 5: out1 = %0101010 ' Display "4"
    endselect
   next
  next
 loop
```

Looking at the code, I'm sure you're thinking that it's nice, but how is it useful for robotics? I find the button function to be very useful for implementing "whiskers" on a robot. Multiple whiskers can be polled with multiple button statements, each one with its own Workspace variable.

Experiment 97
Keypad Input

Parts Bin

Assembled PCB with breadboard, battery, and BS2 installed

Nine 4.7k "SIP" resistor

Keypad

Tool Box

PC
RS-232 cable
Wiring kit

In the previous experiment, I showed how individual buttons are polled and debounced using the PBASIC button function. Many applications have more than one button, and using something like the button function would not be very efficient both in terms of the number of lines that have to be polled as well as how the code polls the buttons. In most applications, where multiple buttons are required, a matrix of buttons is used rather than individual ones. The matrix of buttons is wired together as in Figure 14-10, with a set of row and column connections that allow individual switches to be "addressed" and polled.

Although using four wires to access four switches does not seem better than wiring individual switches, when many switches are involved, the power of the matrix becomes obvious. For example, a 16-button matrix could be addressed using eight wires (four row and four column). If you were to find a surplus telephone keypad (don't take apart one at home), you would discover that it is wired as a four-by-three matrix, with seven wires coming out.

I put the word "addressed" above in quotations because how a matrix of switches (often called a switch matrix keyboard or switch matrix keypad) is read is different from how memory is addressed. The rows are normally pulled up and the columns are tied

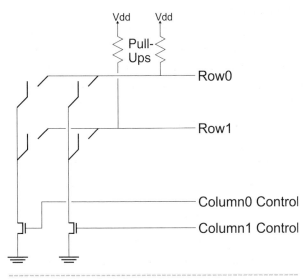

Figure 14-10   *Switch matrix operation*

to ground using switches, as I have shown in Figure 14-10. To read a column of switches, the column's transistor is turned on (pulling the column to ground), and then the individual switches are polled. If any return a 0 (instead of the 1, which results from being pulled up), then the switch at the intersection of the row and column is identified as closed and the button pressed.

You will probably run into three difficulties when you are working with switch matrix keypads. The first is what happens if two keys are pressed—you may find that along with the two keys, other keys that are not pressed also return 0, wrongly indicating that they have been pressed. In most keypads and keyboards, this is not a serious problem because they are often arranged with the Shift, Ctrl, Alt, and other keys that can be pressed with others during an operation that are wired to different rows and columns from the letter and number keys.

You must use open-collector (or open-drain as I have shown in Figure 14-10) drivers for pulling the columns to ground. In the application code created for this experiment, I use the pins connected to the keypad in either input mode or 0 output to simulate the open collector driver.

Last, you will find that the wiring of the keypad (or keyboard) will not be laid out in a logical order. The reason for this is expediency on the part of the keypad manufacturer; keypads are generally designed to use a single-sided PCB, and to avoid wiring the other side, the connections will seem to be randomized. In the experiment code that I wrote, I treat *all* the connections as potential columns and pull them to ground while polling the other pins, which avoids this problem.

In other books, I explain how to decode the connections coming from a keypad and document them as a matrix to allow the controller to be wired efficiently. This experiment avoids this step and works through each line as a "column" polling the other lines as "rows" and returns a unique hex value for each switch. This method is probably the most efficient way of decoding a keypad; once you have run it and recorded the hex values for the different switches, the code and hex values can be used in a "select" statement in an application to perform a unique button function.

The schematic for this experiment is very simple, as shown in Figure 14-11. I used a 10-pin *Single Inline Package* (SIP) resistor for the pull-ups. The SIP consists of nine resistors with a common pin (marked by a dot).

The application "Switch Matrix" will return a unique hex value for each button. The indicated keypad that I used was from a local surplus store, and has 20 buttons and 9 pins to interface to.

```
' Switch Matrix - Return Switch Matrix
Keypad keys as Hex
'{$STAMP BS2}
'{$PBASIC 2.50}

' Variables
i var byte
j var byte
Flag var byte
LastButton  var word
CurrentButton var word
ButtonCount var byte

' Mainline
 ButtonCount = 0: LastButton = -1 ' No
Button Pressed Yet
 do
  Flag = 0
' Note if button Pressed
  for i = 0 to 7
   dirs = DCD i: outs = 0
' Make I/O "i" "0" Output
   for j = i + 1 to 8
    if (ins.lowbit(j) = 0) then
     Flag = 1          ' Button Pressed
     CurrentButton = (j * 8) + i
' Record Address
    if (LastButton = CurrentButton) then
     ButtonCount = ButtonCount + 1
' Previous Button
    else
' New Button Reset the Counter
     LastButton = CurrentButton:
    ButtonCount = 0
   endif
  endif
 next
next
```

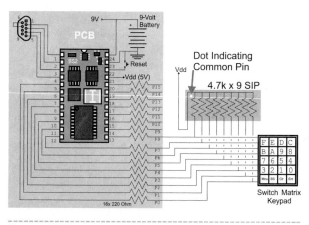

Figure 14-11 *Switch matrix keypad*

```
  if (Flag = 0) then                          else
' No Button Pressed?                    ' Held down, no autorepeat
    ButtonCount = 0: LastButton = -1 ' No        if (ButtonCount = 3) then ButtonCount
  else              ' Else Button Pressed   = 2
    if (ButtonCount = 2) then                  endif
' Button Held Down?                         endif
      Debug "Button = ", shex LastButton, cr   loop
```

Experiment 98
Resistance Measurement

Parts Bin

Assembled PCB with
   breadboard, battery,
   and BS2 installed

Ten-LED bargraph display

10k CDS cell (LDR)

Three 0.01 µF capacitor,
   any type

330 µF, 16-volt elec-
   trolytic capacitor

10k pot

Eight-ohm speaker

LM386, 6-volt, 8-pin DIP
   package

Two 1k resistor

Tool Box

PC

RS-232 cable

Wiring kit

When I created the simple light-seeking robot project earlier in the book, I took advantage of the variable CDS cell resistance to work with a 555 timer. CDS cell resistances can used in the same way for the "rctime" function built into the BS2. Figure 14-12 shows how you can easily wire a resistor with a capacitor to the BS2 to measure its time delay. The following circuit can also be used for reading the position of a potentiometer.

The recommended format for the rctime function is the following:

```
high Pin            ' Set Pin State
before rctime
rctime Pin, 1, Variable
```

(The parameters are explained in the PBASIC Reference at the end of the book.)

If you read through the explanation of the rctime function in the PBASIC programming manual, you

will see how the expected returned value is derived. I'm going to skip over this and go directly to the general case where the count returned is defined by the equation:

$$Count = 600,000 \times R \times C$$

"R" is the resistance in ohms and "C" is the capacitance in farads. If you were expecting to use a 10k

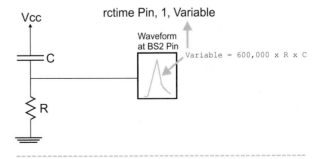

Figure 14-12   *rctime function*

(dark resistance) with a 0.01 µF capacitor, the expected "Count" would be

$$Count = 600{,}000 \times 10k \times 0.01\ \mu F$$

$$= 600{,}000 \times 10(10^3) \times 0.01(10^{-6})$$

$$= 60$$

Using a 10k CDS cell that decreases in resistance as light is applied to it, the expected range of values is from 0 to 60. When I tested my circuit, I found that the maximum value with the CDS cell and 0.01 µF capacitor that I used was 67, so this is quite an accurate approximation of the expected output.

This experiment's circuit will output the (base two) order of magnitude display of the CDS resistance (which is dependent on the light striking it) using an LED bargraph display and a speaker. The circuit to be used for this experiment is shown in Figure 14-13.

The source code for this experiment (called "Light and Sound") is listed here.

```
' Light and Sound - Measure resistance of
a CDS cell/Display Data
'{$STAMP BS2}
'{$PBASIC 2.50}

' Variables
CDSValue var word
' CDS cell Value Returned
Translate var word
' CDS cell Value Order of Magnitude

' Initialize
dirs = %0001001111111111
' P15 - CDS Cell Input, P12 - Sound
outs = $ffff

do
 high 12
 rctime 12, 1, CDSValue
' Read the CDS Cell
 Translate = NCD CDSValue
' Get the high order bit
 outs = DCD Translate - 1 ^ $3ff
 freqout 15, 100, (Translate * (440 / 7))
+ 440
 loop
```

I would like you to notice two things about this application. The first is how I converted the light delay value to an order of magnitude for the LED display and the speaker output. The speaker and

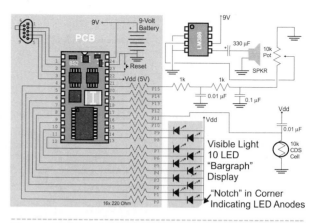

Figure 14-13   *Light theramin*

LED outputs are good methods to use because you can see/hear them across the room without having to hover over your robot. Second, notice that I was able to write the entire application without using any "if" statements—instead I calculated the values to be output all in a single statement, taking advantage of the left-to-right order of operation in the BS2.

When I built this application for the first time, I ran into three problems. The first was that I did not put in the "high 12" statement that sets the state of the I/O pin used for the rctime function. If this is not in, rctime will always return "1," indicating the proper state of the pin before operation. Next, I got confused wiring the circuit—this is the most complex BS2 circuit so far in the book, but I can say that it is the most rewarding and shows how light and sound can be used for outputting variable data information. Finally, I was not paying attention to the 9-volt battery in the PCB and the operation of the circuit became erratic. This was a good way for me to remember to check the battery (using a battery tester) if it does not seem to work according to the software that was written for it.

After working through this experiment, look at the "outs =" and "freqout" statements. To demonstrate that you have figured out how the code works, why don't you change it so that more LEDs light and the tone goes higher as more light is applied to the CDS cell?

Experiment 99
PWM Analog Voltage Output

Parts Bin

Assembled PCB with breadboard, battery, and BS2 installed

Two 10k resistors

100 Ω resistor

0.47 µF capacitor, any type

Push button

Tool Box

PC

RS-232 cable

Wiring kit

DMM

Creating true analog voltages in an all-digital device like the BS2 is surprisingly easy. In this and the next experiment, I will show how you can create analog voltages that can be used as reference voltages for power supplies, LCD displays, and other circuits. Earlier in the book, I presented some discrete circuits that can produce a pulse width modulated analog signal, and in this experiment I will use the built-in PBASIC PWM function to produce analog voltages.

```
PWM Pin, Duty, Cycles
```

The typical circuit used with the PWM function in the BS2 is the resistor/capacitor filter network presented in Figure 14-18. The voltage across the capacitor is the duty cycle divided by 255 and multiplied by the BS2's power source (nominally 5 volts).

The PBASIC PWM command executes over a 1 msec period (1,000 Hz). And when it finishes, it sets the pin to "input" mode to allow any resistor/capacitor networks to stay at a constant value instead of draining or being sourced by the BS2. To get a stable and accurate voltage output, Parallax recommends that you use the formula:

Charge Time = 4 x R x C

For a resistance of 10k and a capacitance of 0.47 µF, this time is as follows:

Charge Time = 4 x 10k x 0.47µF

$$= 4 \times 10(10_3) \times 0.47(10_{-6}) \; seconds$$

= 0.0188 seconds = 19 msecs

The resulting PWM function statement that would be used in the application is

```
PWM Pin, Duty, 19
```

To demonstrate the operation of the PWM, I have created the application circuit (Figure 14-14) that will allow you to change the output voltage of the circuit by a "notch" each time you press the button and measure it using your *digital multimeter* (DMM).

The application code, which I would like you to name "PWM Test" and store in the PWM Test folder, located in the Evil Genius folder, is the following:

```
' PWM Test - Output PWM Value on Button
Press
'{$STAMP BS2}
'{$PBASIC 2.50}

' Variable/Pin Declarations
PWMDuty var byte
PWMOut pin 15
ButtonPin pin 3

' Initialization
PWMDuty = 0              ' Start at 0% Duty
Cycle PWM

 do              ' Loop forever
  debug dec PWMDuty, "/255=", DEC1 PWMDuty
/ 51, "."
  debug DEC2 ((PWMDuty // 51) * 100) / 51,
cr
  do while (ButtonPin = 0)
' Wait for Button High
    PWM PWMOut, PWMDuty, 19
' Output the PWM Value
  loop
  do while (ButtonPin = 1)
' Wait for Button Low
    PWM PWMOut, PWMDuty, 19
' Output the PWM Value
```

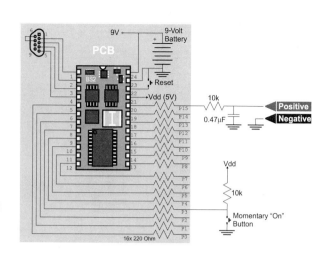

Figure 14-14   *PWM circuit*

```
loop
  PWMDuty = PWMDuty + 1
loop
```

In this application, each time the button is pressed, the PWM Duty cycle is increased until it reaches 255. The purpose of the two debug statements is to predict what value you should be reading on the DMM (assuming the high value is 5 volts and the low value is zero). As you press the button, you should see the analog voltage outputs increase by 20 millivolts or so.

Experiment 100
R-2R Digital-to-Analog Converter

Parts Bin

Assembled PCB with
  breadboard, battery,
  and BS2 installed

10k resistor

220 Ω resistor

Seven 100 Ω resistors

Push button

Tool Box

PC

RS-232 cable

Wiring kit

DMM

In this experiment, I want to go back to the basic electrical theory presented earlier in the book and show you a very clever digital-to-analog converter that doesn't need to be refreshed, has no time delay before the output is valid, and can have any number of bits of resolution that you would like.

The circuit is known as the R-2R digital-to-analog controller and takes the form of Figure 14-15 using switched 2R resistors connected to Vcc or Ground.

In Table 14-4, I have listed the different switch values (0 for Ground and 1 for Vcc) and the outputs from the DAC.

If you look at the circuit, I think you can easily solve it for the cases where SW2, SW1, and SW0 are set to %000, %001, %010, %100, and %111. The cases where two switches are placed at Vcc and one at ground like Figure 14-16 are more difficult.

Table 14-4   Three-bit R-2R DAC output values

SW2	SW1	SW0	Output
0	0	0	0 Volts
0	0	1	1/8 Vcc
0	1	0	1/4 Vcc
0	1	1	3/8 Vcc
1	0	0	1/2 Vcc
1	0	1	5/8 Vcc
1	1	0	3/4 Vcc
1	1	1	7/8 Vcc

Looking at this ugly circuit, you are probably at a loss as to how to solve it and find Vout. To help you, I

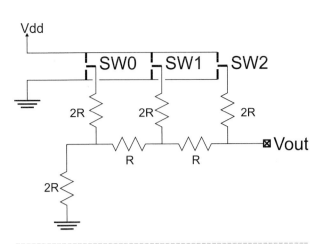

Figure 14-15   *Three-bit R-2R DAC*

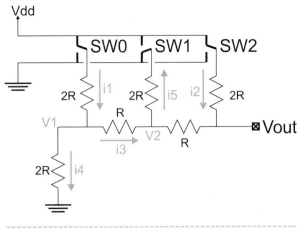

Figure 14-16   *Three-bit R-2R DAC 5-8 out*

have labeled the currents through each of the resistors and labeled the voltages at the two nodes that have three resistors connected to them. Using the electronics rules I presented earlier, we can use Ohm's and Kirchoff's Laws to find the voltages at V1, V2, and Vout.

To do this, we start by writing out everything we know about the circuit. Some of the formulas I found useful are

```
V1 = 2R x i4 = Vdd - (2R x i1)

i3 = (V1 - V2)/R

i1 + i2 = i4 + i5
```

By substituting different values in the equations, you will discover that you can rewrite the first and last equations as

```
4V1 = Vdd + 2V2

6V1 = 5Vdd - 5V2
```

By multiplying the second equation by two and replacing the value of "12V1" with three times the value of "4V1" from the first, you will discover that V2 is equal to 7/16 Vdd. When you work through the circuit, you will find that Vout is equal to (R/3R (Vcc – V2)) + V2, which simplifies to 10/16 Vdd or 5/8 Vdd (which is what is listed in Table 14-4 for SW2 = 1, SW1 = 0 and SW0 = 0).

Chances are that if you are like me, you would burn through a lot of paper coming up with this answer if you were to calculate it yourself. It's a good exercise to go through all eight possible outputs in this circuit to help cement the basic principles in your mind—it is unreasonable to go through all 256 possible outputs of the 8-bit DAC circuit in Figure 14-17 created for this experiment.

```
' R-R2 DAC Test - Output PWM Value on
Button Press
'{$STAMP BS2}
'{$PBASIC 2.50}

ButtonPin pin 15

' Initialization
  dir1 = %11111111        ' All 8 Low Bits
Outputs
  out1 = $ff              ' All High + 1 = 0

  do                ' Loop forever
    out1 = out1 + 1        ' Increment the
Output Value
    debug dec out1, "/255=", DEC1 out1 / 51,
```

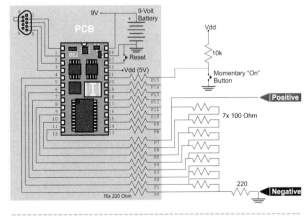

Figure 14-17   *R-2R circuit*

```
"."
  debug DEC2 ((out1 // 51) * 100) / 51, cr
  do while (ButtonPin = 0) ' Wait for
Button High
  loop
  do while (ButtonPin = 1) ' Wait for
Button Low
  loop
 loop
```

When you run this experiment, you will probably find that the actual value isn't as accurate as the PWM DAC. The reason for this is the use of 100-ohm resistors for the R values—this isn't a "true" R-2R digital analog converter. If you are able to substitute in 110-ohm resistors, you will find that the accuracy of the circuit is very good and matches that of the PWM DAC in the previous experiment.

Sensors

When I wrote *Programming Robot Controllers*, I described robot sensors as having exactly the same purpose as the sensors used on the *U.S.S. Enterprise* in the *Star Trek* TV show. Sensors should be able to look at the environment and report if there is something within the sensor's detection envelope. If an object is detected, even if it could damage the robot, the sensor should not do anything about it; that is the job of the robot's controlling software, just as it is the captain of the *Enterprise*'s responsibility to do something about an oncoming "Bird of Prey."

Figure 15-1 shows a number of the different objects that a robot should be required to be aware of. These different objects could be something that the robot must avoid or a place where it is supposed to go. This is the reason why I want the robot's "smarts" to be in the central control code, not in the sensor routines. A sensor may detect and object, only to turn inappropriately because additional sensor data is not being included in the decision-making process.

One important point to note in Figure 15-1 is that when I drew the robot, I included a light gray cone indicating the field of view of the sensors. This cone is trying to indicate that a given sensor cannot sense objects all the way around the robot and that it can only sense objects a specific distance away. When you are looking at different sensors, you will have to make sure you understand the field of view and the depth of view (distance) of each sensor that you are considering.

I have not yet talked much about the central control software, but its integration with sensors and drives is critical to the success of the robot. This does not mean that you have to design your robot to include every possible sensor and make sure that it can handle absolutely any contingency. As I will explain, this means defining what the robot is going to do. You must select sensors and drives that will allow it to navigate its environment and perform the required task under the control of the central control software.

Sensors have to be designed for a dizzying array of different situations. In Table 15-1, I have listed just a few of the different things that a robot may have to be aware of and the sensors used to detect them. In this section, I will introduce you to many of these different sensors and comment on their effectiveness as well as situations where they will be used to best advantage.

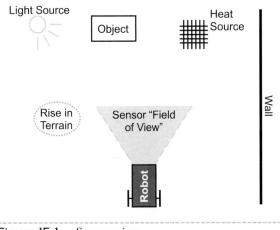

Figure 15-1 *Sensor view*

Table 15-1 Different robot sensors and their characteristics

Parameter	Sensor	Comments
Light	CDS cells	Multiple CDS cells placed about a robot, each given its own unique point of view to help identify bright and dark parts of the robot's environment. Easy to implement.
	Video camera	Capture scene in front of robot. Capable of identifying light/dark areas in environment fairly easily. Very difficult to identify objects, their orientation, and position from the robot.
Objects	Wire wiskers	Simple to implement. Drawbacks include being easily damaged and a potential source of ESD to electronics if allowed to rub along a surface.
	Ultrasonic ranging	Good directional measurement of objects in beam path. Requires large amounts of power and the sensing angle is very narrow.
	IR reflections	Good detection of objects at relatively low cost and low power consumption. Very wide angle sensing. Difficult to determine distance to object.
	Microphone	Detects sound made when robot collides with an object. Can be difficult to implement due to other sounds found the robot.
	Radar	Provides accuracy in distance and direction is dependent on device.
	Video camera	Allows for plotting of object around robot. Can be difficult to implement hardware/software and provide proper lighting for objects to be easily detected.
Sound	Microphone	Allows robot to be controlled with a shout or a clap. Low cost and quite easy to implement. May have some problems with noise coming from robot.
Surface	Skid sensors	Detect if wheel or leg is slipping on surface. Skip and skid information is sensened when tracking robot's movement and position.
	Tilt sensors	Return information if robot is going up or down a hill and requires a change in power.
Location	Odometry	Stores the movement of the robot so that its current position can be determined arithmetically. Fairly easy to implement, but it can be very difficult determining the exact position of the robot.
	Compass	Returns current direction of robot. Useful for odometry to determine if robot is not going straight and to set the initial direction for movement. Very useful for helping a robot find a specific location or follow a pre-programmed route.
	GPS	Use GPS constellation to determine position (and heading of the robot) to within several meters. GPS satellites can be difficult for the receiver to find if it is located within a building.
	Ultrasonic ranging	Used to triangulate the position of robot from objects around it. Can be difficult to implement and requires mechanism for moving ultrasonic transmitter and receiver around.
	Radar	Can be used to identify positions of objects around the robot like ultrasonic ranging.
	Video camera	Used to locate and plot objects around the robot. This is very difficult to implement.
Heat (including identifying humans and animals)	Pyrometer	Used to detect humans. Very sensitive with a very wide field of view.
	Infrared photodiodes	Difficult to set up for detecting heat. Infrared photodiodes are designed for optical interrupters and may not be sensitive to the frequencies given off by people. Can be used to detect fire quite well.
	Video camera	Most video cameras will detect light in the near infrared. May require a chilled lens/camera implementation to ensure heat is visible.

Experiment 101
bLiza, the Snarky Computer

Parts Bin

Assembled PCB with BS2

Tool Box

PC
RS-232 cable

As you start working with robot sensors, you may want to simulate the data returned from them so that you can test out the software that handles the data. This is actually a good idea because when you start working with intelligent sensors, including many of the different sensors listed in the introduction to this section, you will find that handling the data is just as difficult as wiring the sensor to the robot. When I introduced you to programming, I had you explicitly change or hard code the new values into the application's source code. Changing your source code every time you want to try something new is not very efficient and allows for the possibility that when you change the value, you end up changing something else as well.

If you are working on a PC application, you might create a data file that is read in and the data treated as if it came from the sensors. The advantage of doing this is that you can change the data easily without affecting the application. The problem with this method is that you can't apply it to a BS2, which doesn't have a file system from which the data could be read. A potential solution is to use the BS2's capability to store data in its program memory EEPROM using the "data" statements, but technically this involves changing the source code, which is exactly what we didn't want to do. What is needed is some way of passing arbitrary data to the BS2.

The PBASIC "debugin" command fits the bill almost perfectly. This command is the opposite of the "debug" command; instead of passing data from the BS2 to you, you can pass data to the BS2 using this command, which has the following format:

```
debugin formatter Variable
```

The formatter used for debugin is the same formatter command used for "debug" and "serout." So, to input a number from the user console into the variable "i," the following statement would be put in the application:

```
debugin dec i
```

Using the debugin statement for entering in decimal data is quite simple, and if you have worked through all the programming experiments, I would expect that you are able to do this quite easily by yourself. Rather than go through an example that you could create on your own quite easily, I wanted to stretch a bit and use the debugin statement as the input method for an experiment into artificial intelligence.

When I use the term *artificial intelligence*, I am mostly interested in trying to come up with a programming algorithm that mimics simple animals (such as ants). This is contrary to what most people think of and are interested when the term comes up. They would like to see a computer (or robot) behave exactly as if it were a human being. This notion of what artificial intelligence is has been rooted in many people's minds because of the English computer scientist, Alan Turing, who suggested that a computer would be "intelligent" if a person could communicate with it as if it were another person and not be able to tell the difference.

One of the first experiments in developing a program that passed the Turing test was called Eliza. Eliza was developed by Joseph Weizenbaum at the Massachusetts Institute of Technology in 1965 as a demonstration of how a computer could demonstrate what appears to be artificial intelligence by looking for keywords in a statement and responding to them. Eliza was an amazing program for its time. It could generate remarkable responses to statements passed to it (via an RS-232 console) from ordinary people. What was truly amazing was that this program was implemented in just 8k of memory.

The challenge I took on was to try and replicate this work on a BS2 (which has only 2k of memory)

using the debug and debugin statements. Surprisingly enough, I was quite successful with my "bLiza" program, listed here. I realize that the application is a bit long, but seeing it work makes the process quite worthwhile.

```
'  bLiza - Trying to implement "Eliza" A.I. Demonstrator in a BS2
'{$STAMP BS2}
'{$PBASIC 2.5}

'  Variable Declarations
InputString  var byte(20)         '  20 Character String
Temp         var byte
i            var byte
k            var byte
RandomWord   var word
j            var RandomWord.HIGHBYTE
RandomCount  var RandomWord.LOWBYTE
LastCharFlag var bit
FoundFlag    var bit

'  Initialization
  debug "Hello.  I'm bLiza.", cr

'  RandomCount = 0                  '  Random continually updated/no initial
                                    '   value required

'  Main Loop
  do                                '  Loop Forever
    i = 0
    InputString(i) = 0             '  Start with Null string
    LastCharFlag = 0               '  Want to execute at least once
    do while (LastCharFlag = 0)    '  Wait for Carriage Return
      debugin str Temp\1           '  Wait for a Character to be input
      if (Temp = cr) then
        LastCharFlag = 1           '   String Ended
      else
        if (Temp = bksp) and (i <> 0) then
          debug " ", bksp          '  Backup one space
          i = i - 1                '  Move the String Back
          InputString(i) = 0
        else
          if (i >= 18) or ((Temp <> " ") and ((Temp < "a") or (Temp > "z")) and ((Temp < "A")
  or (Temp > "Z"))) then
            debug bksp , 11        '  At end of Line or NON-Character
          else                     '  Can Store the Character
            if ((Temp >= "a") AND (Temp <= "z")) then Temp = Temp - "a" + "A"
            debug bksp, str Temp\1
            InputString(i) = Temp
            i = i + 1
            InputString(i) = 0     '  Put in new String End
            RandomCount = RandomCount + Temp
          endif
        endif
      endif
    loop

    if (i = 0) then                '  Blank String
      debug "Type something!", cr
    else                           '  Respond to Comment

      i = 0: k = 1: FoundFlag = 0  '  Search for input match
      do while (FoundFlag = 0)     '  Look through the Data tables
        j = 0                      '  Keep Track of the Characters
        read i, Temp               '  Read the Current Character
        if (Temp = 0) then         '  If First is Zero, then no match
          FoundFlag = 1
```

```
      else
        do while ((Temp = InputString(j)) and (InputString(j) <> 0))
          j = j + 1
          i = i + 1
          read i, Temp
        loop
        if (Temp = 0) then      '  String Match
          FoundFlag = 1
        else                    '  No Match
          do while (Temp <> 0)
            i = i + 1
            read i, Temp
          loop
          i = i + 1             '  Point to Start of Next string
          k = k + 1             '  Indicate next string type
        endif
      endif
    loop                        '  On Exit, "j" points to mismatch

    i = 0                       '  Remove everything in InputString
    do while (InputString(j) <> 0)  '    before "j"
      InputString(i) = InputString(j)
      i = i + 1: j = j + 1      '  After copying byte, point to
    loop                        '    the next one
    InputString(i) = 0          '  Put in Null at end of new string

    RANDOM RandomWord           '  Get Random Response Value
    i = (RandomWord / 8) & 3
    if (i = 0) then i = 3       '  Result 1, 2 or 3

    if (k = 1) then
      j = 11                    '  Only one response to being hated
    else
      j = ((k - 2) * 3) + i
    endif

'  Produce Response to Input
    if (j <= 9) then
      select(j)
'  Responses to "I AM ..."
        case  1:  debug "Do you like being ", str InputString, "?"
        case  2:  debug "And you're happy?"
        case  3:  debug "That explains your friends."
'  Responses to "I HAVE A ..."
        case  4:  debug "Do you like it?"
        case  5:  debug "Knowing you, it's cheap."
        case  6:  debug "Bite me ", str InputString, "-boy."
'  Responses to "I WANT ..."
        case  7:  debug "All the young dudes want a ", str InputString, "."
        case  8:  debug "Well, if you want it..."
        case  9:  debug "Ask the police."
      endselect
    else
      if (j <= 18) then
        select(j)
'  Responses to "I HATE ..."
          case 10:  debug "That's too bad."
          case 11:  debug "Should I be scared?"
          case 12:  debug "Are you a psycho?"
'  Responses to "MY ..."
          case 13:  debug "Honest?"
          case 14:  debug "You must be proud."
          case 15:  debug "No way, Jose!"
'  Response to "BECAUSE ..."
          case 16:  debug "Wrong."
          case 17:  debug "Sure...I believe you."
          case 18:  debug "That's dumb."
        endselect
      else
        select(j)
'  Responses to "I Like ..."
          case 19:  debug "You should stay off the Internet."
          case 20:  debug "Good for you!"
          case 21:  debug "Don't tell anybody."
```

```
'   Responses to "IT IS ..."
            case 22:  debug "That and four dollars will buy a cup of double-latte."
            case 23:  debug "Check your facts."
            case 24:  debug "You believe that?"
'   Responses to "WHY [...]"
            case 25:  debug "Live with it."
            case 26:  debug "That's life."
            case 27:  debug "Nobody knows why."
'   Responses to anything else
            case 28:  debug "Are you gaining weight?"
            case 29:  debug "I'm impressed...NOT!"
            case 30:  debug "Tell me more."
          endselect
        endif
      endif
      debug cr
    endif
  loop

'   Potential Starts to answers (conversation keys)
RT01  data "I HATE YOU", 0
RT02  data "I AM ", 0
RT03  data "I HAVE A ", 0
RT04  data "I WANT ", 0
RT05  data "I HATE ", 0
RT06  data "MY ", 0
RT07  data "BECAUSE ", 0
RT08  data "I LIKE ", 0
RT09  data "IT IS ", 0
RT10  data "WHY ", 0
RTEnd data 0
```

In the source code, I have shaded three areas that I would like to bring to your attention. The first is the data entry code in which I wait for the user to enter data, and then check to see if it is a carriage return, backspace, blank, or alphabetic character. This code loops until a carriage return (Enter key on your PC) character is received. As it is waiting for the enter key, it parses all the other ASCII characters that come in, adding characters and blanks to the "Input-String" array and taking them away when the backspace key is pressed. As characters are added to InputString, you should notice that they are converted to uppercase characters. This is done to allow easier comparisons in the second shaded block of code. When you run bLiza, note that the text being input (on the top input bar) is displayed as uppercase in the main window of the Debug Terminal.

In the second shaded block of code, I am comparing the input string to each of the 10 sets of keys stored in the EEPROM using the data statements. This code compares the statements to the keys and stops if there is a match. When I work with text strings, I always end them with a Null ("0") character to indicate when they end. The string compare code compares each character in InputString to the current data value, and if the two strings match, it returns the number of the data statement.

The Null character at the end of a string of characters is used by PBASIC as an indicator for when to stop. The traditional term for this type of string is ASCIIZ, meaning that it is a string of ASCII characters terminated by a zero, and it is used in many different programming languages.

After comparing the strings, I move everything after the matching part to the front of the string (this is the third shaded part of the program). I did this because some responses use the object of the sentence and by removing the noun and the verb, the object can be easily accessed.

These three shaded areas consist of code that is normally built into programming libraries. If you are developing your own text input programs using the BS2, you should remember these routines and use them as a base for your own code. For example, in the first shaded block of code, you may want to add the ability of entering in numerics (0 through 9) or other characters. Changing characters from lowercase to uppercase is quite simple. I first check to see if the character is in the range of "a" to "z" and if it is, I then subtract the ASCII value for "a" and then add the ASCII value for "A." This is a trick you might want to keep in the back of your mind in case you ever have to change characters to all upper- or all lowercase.

Experiment 102
Multiple Seven-segment Displays

Parts Bin

Assembled PCB with breadboard and BS2

Two dual seven-segment, common cathode displays

Four ZTX649 NPN transistors

Tool Box

Wiring kit

The previous section devoted to interfacing different hardware devices to the BS2 probably seems to be more appropriate for this experiment, but I wanted to introduce you to how advanced robot programming is done. Advanced robot programming consists of applications that incorporate motor control, sensor interfacing, output operation, and task programming. Creating applications for each of these interfaces is quite easy—putting them together is quite hard. Implementing a multiple seven-segment display requires very similar skills to that of developing a robot application. The code for creating the data must be integrated with formatting the data, as well as driving it to the displays. This corresponds to the tasks in a robot of polling the sensor hardware, interpreting the data, coming up with a response to the input, and, finally, controlling the motors. These operations can be drawn out as in the flowchart shown in Figure 15-2.

Quite a bit of care must be taken into account to make sure that these functions work together. In the case of controlling DC motors, part of the output

device control is properly timing the motor's PWM. This can be performed by specialized hardware or, with a bit of thought beforehand, by using the robot's controller. Using the robot's controller is preferable because it will minimize the cost and complexity of the overall robot. I choose to use four seven-segment displays to demonstrate how the robot code is architected, because if there are problems with it, they will be quite literally visible. The most traditional way of implementing an application that displays data on multiple seven-segment displays is to cycle through each of the displays very quickly, flashing the data on each display faster than the human eye can perceive (Figure 15-3).

As a rule of thumb, each display should be active 50 or more times per second. The slower each display is flashed on and off, the more likely the human eye will pick up the flashing. A flashing multicharacter display is not attractive and could cause headaches in some people (especially if the displays are very bright). The time each display is turned on must be as

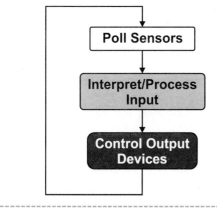

Figure 15-2   *Robot code flow*

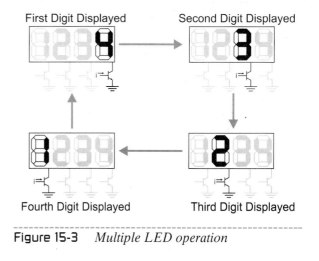

Figure 15-3   *Multiple LED operation*

equal as possible. If one display is on for a longer period of time than the others, then it will appear brighter, and conversely, a display active for a shorter period of time will appear dimmer. The frequency and duration of a display being active are analogous to the period and duty cycle of a motor's PWM, which cannot be as easily observed as the LED display. When working with multiple displays, in order to meet the 50-times-per second-guideline, you are actually going to have to loop through your display code 50 times per second multiplied by the number of displays. So, for a four-digit display, you will have to loop 200 times per second, and you will have a total application execution time of 5 msecs in which you will have to poll your sensors, respond to the sensor data, and output it. If you plan for this timing when you create your application, you will find that it is not as difficult as it may seem when you first think about it.

For this experiment, you will have to wire in four seven-segment displays digits. I used two dual com-

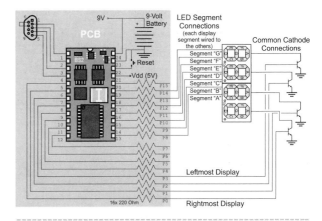

Figure 15-4 *Multiple LED circuit*

mon cathode displays, which are available in 18- (with a decimal point) or 16-pin packages. These displays were connected to the PCB's BS2 as shown in Figure 15-4. To test the application, I created the "Counter Display" application.

```
'   Counter Display - Display seconds on Four 7 Segment LED Displays
'{$STAMP BS2}
'{$PBASIC 2.5}

'   Variables and I/O Port Pin Declarations
Counter var word               '  Second counter
CurLED  var byte               '  Currently Displayed LED
Display var byte(4)            '  Display Variable
Dlay    var word               '  Delay Count
DispOut var OUTS.HIGHBYTE       '  Output Bits for LED Displays
DispDir var DIRS.HIGHBYTE
LEDCtrl var OUTS.NIB0           '  Four Transistor Control Bits
LEDDir  var DIRS.NIB0

  CurLED = 0: Counter = 0      '  Initialization
  LEDDir = %1111: LEDCtrl = 0  '  Transistor Pins O/P & Off
  DispDir = %01111111: DispOut = 0 '  Display Pins O/P
  for Dlay = 0 to 3: Display(Dlay) = 0: next

  do                           '  Loop to Display Number on LED
    Dlay = Dlay + 1            '    Displays Incrementing (4.25 ms)
    if (Dlay >= 235) then      '  One Second Passed?
      Dlay = 0: Counter = Counter + 1 '  Reset Dlay and Inc Time
      Display(3) = (Counter / 1000) // 10
      Display(2) = (Counter / 100) // 10
      Display(1) = (Counter / 10) // 10:  Display(0) = Counter // 10
    endif                      '  Finished updating seconds

    CurLED = (CurLED + 1) // 4: LEDCtrl = 0  '  Roll to Next Display
    lookup Display(CurLED), [$7F, $06, $5B, $4F, $66, $6D, $7D, $07,
          $7F, $6F, $00], DispOut '  Setup Character Output
    lookup CurLED, [%0001, %0010, %0100, %1000], LEDCtrl
  loop                                       '  Repeat forever
```

Each do-loop iteration will execute between 2.75 and 4.25 ms, which meets the 5 ms requirement for four seven-segment displays and the "Dlay" count of 235 was found using the worst case (4.25 ms) do-loop timing. I found that with a Dlay count of 235, the counter incremented significantly faster than once per second, and sometimes a character would appear to flash to counter these problems. You may want to repeat the code after the if as an else condition with-

out incrementing "Count." By repeating the code in the else, you will have more constant timing that is important for an application such as a motor PWM. Finally, if you want to display leading blanks, then you would have to change the "Display" variable update code to the following:

```
if (Count > 100) then Display(2) =
(Count / 100) // 10 else Display(2) = 0
```

Experiment 103
RCtime Light Sensor

Parts Bin

Assembled PCB with BS2

Two 10k CDS cells

Two 0.01 μF capacitors, any type

Two dual seven-segment LED displays

Four ZTX649 NPN transistors

Tool Box

PC

RS-232 cable

It will probably be surprising for you, but I consider the most basic robot sensor to be vision. This is not the same vision as you are provided by your eyes, but the ability to detect what direction in front of the robot is brightest. When I introduced you to the first robot (the 555-based light-seeking robot), I used a *Light-Dependent Resistor* (LDR) in conjunction with a capacitor to produce a timed signal that was used to control a robot. The LDR and capacitor combination can be used to provide a rough light measurement using the RCtime built-in function (Figure 15-5).

When using the state "1" circuit, the code to read the value of the LDR (or a potentiometer) is the following:

```
high Pin            ' Discharge Cap
pause 1
'  Wait for Cap charge to Discharge
rctime Pin, 1, PinValue
'  Wait for Cap to discharge
```

The time required to measure the resistance can be minimized by using a small resistor or capacitor. For this experiment, I have specified a 10k (maxi-

mum resistance) LDR and a 0.01 μF capacitor that will result in a maximum value of 60 and require 1.5 ms to perform the operation. This delay to poll an LDR will be a problem for the basic four seven-segment LED display circuit, as it will cause one display to be active for an abnormally long time, causing it to flash.

To help even out the time sensing of the LDR and minimize the impact of converting the value and passing it to the displays, I broke up the three statements needed to read the LDR (plus two others to display the result). These statements were then executed one at a time sequentially per loop in a software state machine. The software state machine executes one function per loop (simulating the output of the state machine) and updates the state for the next loop iteration based on different inputs and the current state. I like to use the software state machine in applications like this, where I can spread out a task over several loop iterations.

To poll two LDRs and present the results on seven-segment displays, add the two LDRs and 0.01

```
rctime Pin, 0, Variable                    rctime Pin, 1, Variable
```

 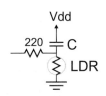

$$\text{Value} = 600 \times R \text{ (in k)} \times C \text{ (in } \mu F)$$

Figure 15-5   *RCtime wiring*

µF capacitors to the four seven-segment LED display circuit used in the previous experiment (Figure 15-6). When this is done, enter in the following "RCtime

Display" application and put it either into its own folder or the one you created for the "Seven-Segment Display" in the Evil Genius folder.

```
'  RCtime Display - Display the values from the LDRs
'{$STAMP BS2}
'{$PBASIC 2.5}

'  Variables and I/O Port Pin Declarations
RightLDR        pin 4
LeftLDR         pin 5
RLDRVal         var word            '  Saved LDR Values
LLDRVal         var word
CurLED          var byte            '  Currently Displayed LED
Dsplay          var byte(4)         '  Display Variable
Dlay            var byte            '  Delay Count
DispOut         var OUTS.HIGHBYTE   '  Output Bits for LED Displays
DispDir         var DIRS.HIGHBYTE
LEDCtrl         var OUTS.NIB0       '  Four Transistor Control Bits
LEDDir          var DIRS.NIB0
State           var byte            '  Read State Variable

  CurLED = 0:                       '  Initialization
  State = 0                         '  Start from Beginning
  LEDDir = %1111: LEDCtrl = 0       '  Transistor Pins O/P & Off
  DispDir = %01111111: DispOut = 0 '  Display Pins O/P
  for Dlay = 0 to 3: Dsplay(Dlay) = 10: next  '  All Blank

  do                                '  Loop to Display Number on LED

    select(State)         '  State  Major          Minor
      case 0:             '    0    Read Right LDR  Start Read
        high RightLDR
        State = 1
      case 1:  State = 2  '    1    Read Right LDR  Cap Charge
      case 2:             '    2    Read Right LDR  Read Cap Charge
        rctime RightLDR, 1, RLDRVal
        State = 3
      case 3:             '    3    Read Right LDR  Format MS Char
        if (RLDRVal <= 9) then Dsplay(1)=10 else Dsplay(1)=RLDRVal/10
        State = 4
      case 4:             '    4    Read Right LDR  Format LS Char
        Dsplay(0) = RLDRVal // 10
        State = 10
      case 10:            '   10    Read Left LDR   Start Read
        high LeftLDR
        State = 11
      case 11:  State = 12 '  11    Read Left LDR   Cap Charge
      case 12:            '   12    Read Left LDR   Read Cap Charge
        rctime LeftLDR, 1, LLDRVal
        State = 13
      case 13:            '   13    Read Left LDR   Format MS Char
```

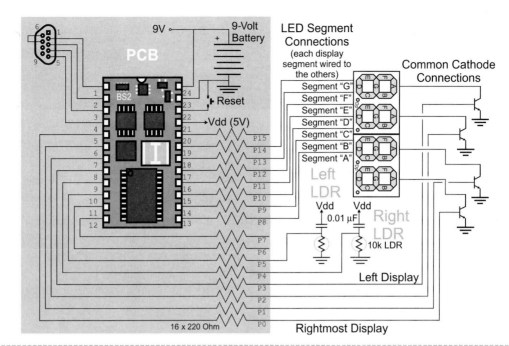

Figure 15-6   *LDR light sensor circuit*

```
        if (LLDRVal <= 9) then Dsplay(3)=10 else Dsplay(3)=LLDRVal/10
        State = 14
      case 14:                '    14    Read Left LDR    Format LS Char
        Dsplay(2) = LLDRVal //10
        State = 0               '  Start Over again...
  endselect

  CurLED = (CurLED + 1) // 4: LEDCtrl = 0  '  Roll to Next Display
  lookup Dsplay(CurLED), [$3F, $06, $5B, $4F, $66, $6D, $7D, $07,
          $7F, $6F, $00], DispOut '  Setup Character Output
  lookup CurLED, [%0001, %0010, %0100, %1000], LEDCtrl
loop                            '  Repeat forever
```

Experiment 104
Differential Light Sensors

Parts Bin

Assembled PCB with
    breadboard and BS2

LM339 quad comparator

Two 10k CDS cells

Three 10k resistors

0.01 μF capacitor, any
    type

Tool Box

Wiring kit

DMM

The PBASIC RCtime statement allows you to very simply and quantitatively measure the light falling on a CDS cell. Unfortunately, when you are using many other microcontrollers, a function or statement like

RCtime will not be built into the chip or the code development tools (the compiler and assembler that you are using). If you are programming your robot in assembler, then you could probably develop the code that provides the same function as RCtime, but if you are new to assembler programming for this chip, it may be outside your ability to program efficiently. A simple way of finding out which light sensor is being exposed to the brightest light is to wire them together as a voltage divider. By measuring the voltage at the connection between the two CDS cells, you can tell which one is getting the most light. As I show in Figure 15-7, when an equal amount of light is falling on the two CDS cells of the voltage divider, the voltage output will be one-half the voltage applied to the voltage divider (I assumed "Vdd" for this example).

In this circuit, when more light falls on one CDS cell, you will find that the voltage across the other CDS cell will be greater, resulting in a voltage output that will not seem to make sense when you first think about it. To test out the operation of the two-CDS-cell voltage divider, you could wire up this circuit quickly and look at the voltage output when you put your hand over one CDS cell and then the other. If you are going to perform this simple experiment, I recommend that you follow the conventions for "left" and "right" along with how they are wired so your results will match the results in Figure 15-7.

Using the BS2, you do not have the ability to measure the voltage coming from the CDS cell voltage divider, but you can easily compare it against a known value. By setting up another fixed value resistor, or a voltage divider to provide a nominal half-voltage, a comparator will tell you very simply whether or not one CDS cell or the other is exposed to more light.

You might want to use a potentiometer instead of two equal value resistors. This rational is based on most resistors having a tolerance of 5 percent, which can result in a voltage that is quite a bit off from the ideal one-half of the input voltage. This is not a major consideration because you will find that when the light input to the CDS cells is even slightly different, the voltage divider output will be considerably outside the possible error voltage caused by the resistor tolerance.

The second point is that if the CDS cell voltage divider output is equal to the fixed resistor voltage divider, the circuit will behave as if the light input is brighter on one side than the other. If this circuit were used in a robot, you might find that the robot turned in one direction more than you would like. You could fix this problem by adding a second comparator and modifying the fixed resistor voltage divider as shown in Figure 15-8. When this circuit is used in an application, if the microcontroller reads 0 from both comparators, then the light falling on both CDS cells is approximately equal and it should just go forward. The 1/10 R resistor provides a gap between the voltage for situations where there is a clear difference between left and right. When the light level on one CDS cell changes, it will cause the output of one of the comparators to change to 1, and in the case of a light-seeking robot, it can be commanded to turn toward the light.

In practical robot applications, you will find that you do not need to add the extra comparator and the modified fixed resistor voltage divider. If you pro-

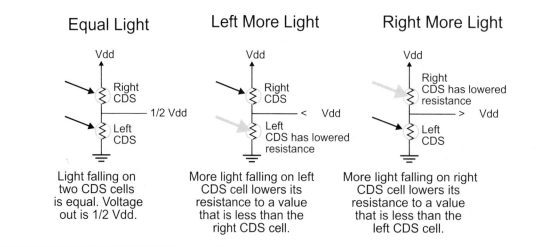

Equal Light

Light falling on two CDS cells is equal. Voltage out is 1/2 Vdd.

Left More Light

More light falling on left CDS cell lowers its resistance to a value that is less than the right CDS cell.

Right More Light

More light falling on right CDS cell lowers its resistance to a value that is less than the left CDS cell.

Figure 15-7 *Differential light sensor operation*

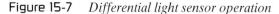

gram the robot to move forward while it is turning (which, in the case of a two-wheel differentially driven robot, is best accomplished by "pulsing" one wheel forward while the other is stationary), then you will not have any problem with the robot turning away from the light.

One problem with comparing light levels using a two-CDS-cell voltage divider as I have shown here is that this circuit cannot detect the case when it is in a completely dark room, which is a feature you get for "free" with the RCtime light-measuring circuits.

To demonstrate the operation of the two-CDS-cell differential light-sensor circuit, you can wire it to a BS2 according to the circuit shown in Figure 15-9. For this circuit, I use the regulated 5 volts provided by the BS2 for the voltage dividers and the comparator (the LM339 chip). I am only using one of the four comparators in the LM339, but I still have to provide

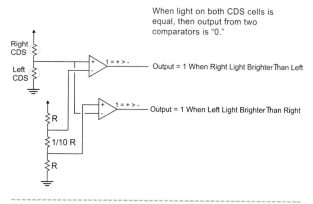

Figure 15-8   *Fixed differential light sensor*

power to it (including a decoupling capacitor). The LM339 has open collector outputs, which is why I added the 10k pull-up resistor on its output. The application code will indicate when the light input level shifts from one CDS cell to the other.

```
'   Differential Light Sensors - Indicate which CDS Cell is Getting Most Light
'{$STAMP BS2}
'{$PBASIC 2.50}

'   Variables
CDSCells          pin 15        '  CDS Input Cells
LeftMsgFlag       var bit       '  Message Out Indicator Bits
RightMsgFlag      var bit

'   Initialization/Mainline
  input CDSCells
  LeftMsgFlag = 0: RightMsgFlag = 0
  do                            '  Repeat forever
    if (CDSCells = 0) then      '  Brightest to Left
      if (LeftMsgFlag = 0) then
```

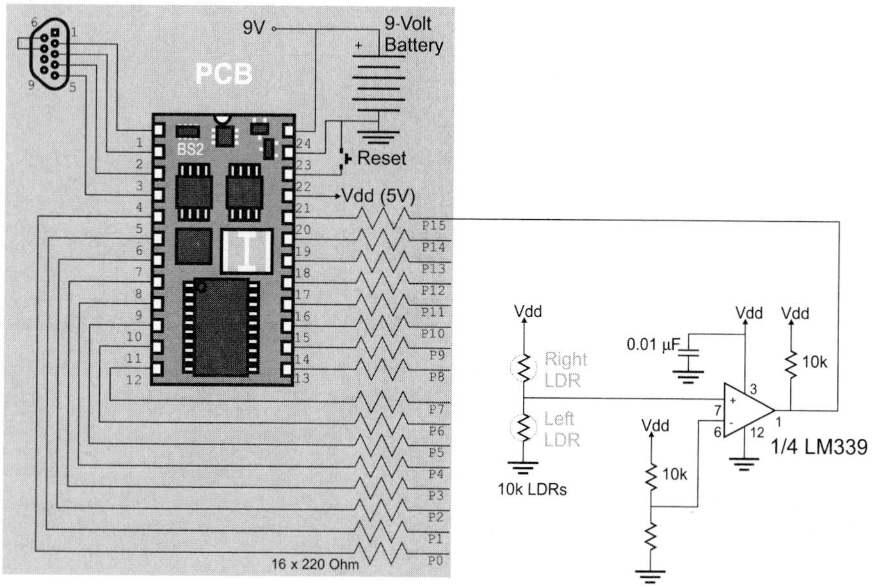

Figure 15-9   *Differential light circuit*

```
          debug "Brighter Light to the Left", cr
        endif
      LeftMsgFlag = 1: RightMsgFlag = 0
    else                           '  Brightest to Right
      if (RightMsgFlag = 0) then
        debug "Brighter Light to the Right", cr
      endif
      LeftMsgFlag = 0: RightMsgFlag = 1
    endif
  loop                            '  Loop forever, end of application
```

Experiment 105
Sound Control

Parts Bin

Assembled PCB with
   breadboard and BS2

LM324 quad op-amp

74LS74 dual D flip flop

Electret microphone

Four 2.3M resistors

10k resistor

Two 470 Ω resistors

Two 220 Ω resistors

Six 0.1 µF capacitors, any type

Tool Box

Wiring kit

Chances are, when you first saw the title for this experiment, you thought that it would involve being able to command the robot using sound (and ideally your voice). You may have thought that I was going to give you the secret to voice recognition and allow your robot to respond as if it were a dog. Unfortunately, this is not the case, but in this experiment, I will show you how to create an application that responds to loud sound input quite effectively, which will allow you to at least shout at the robot to stop it before it rolls into Aunt Martha.

The circuit shown in Figure 15-10 filters out high frequencies (the ones most likely to be found in a robot), leaving the low-frequency sounds that are most likely to be caused by external sources. The two op-amps behave as filters and very high gain amplifiers of the signal.

Figure 15-11 shows the circuit in operation. To test it out, I simply said "D'oh!" in a reasonably loud voice. The various high-frequency harmonics have been filtered out from the signal, leaving just a few "wide" changes to the input signal that are passed to

the D flip flop's clock. The clock changes the state of the flip flop, and it can be polled at any time by the BS2.

As the BS2 cannot continuously poll the output of the op-amps, I have passed its output to a D flip flop's clock, which will change the output state of the flip flop from "0" (which is loaded into the 74LS74 by the BS2) to "1." By doing this, the BS2 just has to poll the output of the D flip flop periodically to see if there have been any loud sounds passed to the circuit that the BS2 must respond to. After noting the sound by the change in the flip flop, the BS2 can reset the flip flop and wait for the next sound to be received.

The wiring is a bit challenging because of all the discrete resistors and capacitors, but it does fit on the breadboard with some space to spare. Once you have built the circuit, key in the following application:

```
'  Too Loud - Indicate when sound is
received by the BS2
'{$STAMP BS2}
'{$PBASIC 2.50}

'  Variables
```

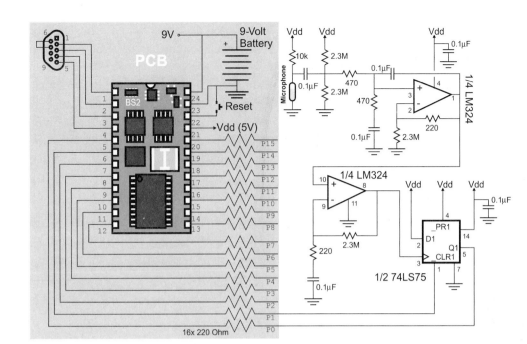

Figure 15-10   *Sound circuit*

```
ResetPin      pin 1
SoundInput    pin 0

'  Initialization/Mainline
   high ResetPin
   input SoundInput
   do
     pulsout ResetPin, 10
'  Reset the '74
     pause 1000
     if (SoundInput = 1) then
'  Poll Sound Input Pin
        debug "Keep it Quiet - that was TOO
LOUD!", cr
     endif
   loop
```

This application should be very straightforward for you to build and should not present any surprises, except if a radically different microphone than the one I had is used. You may find that you have to experiment with the 10k microphone pull-up in the circuit until the input to the circuit is in the 50 to 100 mV range shown in Figure 15-11.

Depending on the microphone used, you may find that the circuit is more sensitive than you would like. The first op-amp circuit is a double-pole Butterworth low-pass filter with a gain of approximately 1 and a cutoff point around 340 Hz. This circuit will very effectively filter out all higher-frequency components of sound, just resulting in a square wave of approximately a 3 ms period. I find that this circuit works quite well, and there is very little reason to modify it.

The second op-amp circuit is a noninverting amplifier, and its gain is defined by the formula:

$$Gain = 1 = R2/R1$$

Here "R2" is the 2.3M resistor and "R1" is the 220 Ω resistor. Using these values, you will get a gain of about 10,000 times. By changing the value of "R2" (the 2.3M resistor), you can change the gain of the amplifier, making the entire circuit less sensitive to surrounding noise.

As well as using a microphone to pick up loud sounds, such as you shouting "Stop" when the robot

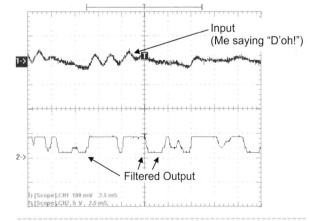

Figure 15-11   *Sound operation*

is about to hit an object, it can be used to detect when the robot collides with another object by mounting it to bumpers or the chassis of the robot. I've found this to be a reliable method of detecting collisions, although others have found the opposite to be true because of motor, gear, and wheel noise conducted through the robot—I believe the high frequency filtering that I use minimizes the pickup of ambient noise. Using this circuit for detecting object collisions is something to think about if you have a robot that may collide when turning or running in reverse. Rather than putting object sensors around the perimeter of the robot, you could attach a microphone to the chassis, simplifying the mechanical design of your robot.

Experiment 106
Robot "Whiskers"

Parts Bin

Assembled PCB with breadboard and BS2

Two microswitches with long actuator arms

Two 10k resistors

Tool Box

Wiring kit

Twenty-four-gauge solid core wire (red)

Soldering iron

Solder

If you have looked at my designs for mobile robots, you'll see that I very rarely add "whiskers" or other physical object sensors to the design of the robot. I don't like to implement whiskers for two reasons. The first is that they are difficult to implement practically. Secondly, I would like to detect objects several inches (10 cm) or more from the robot to allow the robot to stop or turn away and not risk colliding with the object. You may want to add whiskers to your robot as they are very easy to understand and are relatively simple to build into the robot.

I used to think that a cat's whiskers were completely redundant when the cat matured to the point where its eyes opened. Actually, a cat's whiskers are an important source of sensory input throughout the animal's life. The whiskers help protect the cat from obstacles when it is in a completely dark environment (cats can see better than humans in very low light conditions, but they are just as blind as we are when there is no light). They also help direct the cat's mouth towards its prey. Having three times the diameter of regular hair and set into the skin much deeper, a cat's whiskers are also very sensitive to sub-audible sounds and vibrations.

As I show in Figure 15-12, whiskers have three primary uses in robots. When an object in front of the robot is detected, it results in a command to the robot's motors to stop or it initiates a sequence to turn or back away from the object. A whisker can also be used to detect a wall beside the robot, which allows it to keep one "hand" on the side of a maze, allowing it to find its way out. When implementing a wall sensor, make sure that the whisker is connected to ground and ground is passed to the operating surface. You do not want a static electrical charge building up on the robot as you run it. Finally, whiskers running across the floor can be used to detect

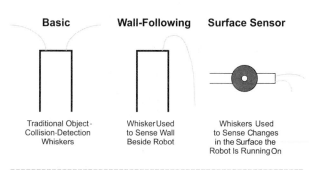

Figure 15-12   *Whisker types*

rougher surfaces or changes in the running surface (from a smooth floor to a raised carpet).

Whiskers can be implemented in a number of different ways; you'll see ones that look like large loops, others like a cockroach's antennae, and ones that act like skids to help balance the robot. Whiskers change an electrical signal when they are in contact with an object (Figure 15-13). A pulled-up circuit that the whisker pulls down when it makes contact is the most basic method of interfacing the whisker with the robot (the left drawing of Figure 15-13). This signal needs to be debounced just as if it were a momentary on push button. The amount of force on a whisker can be measured using multiple contacts, each one closed with a varying amount of force, or by connecting the whisker to the wiper of a potentiometer.

A few practical issues must be considered when adding whiskers to your robot. Many people use a fairly small diameter wire, such as a guitar string or piano wire. These small diameter wires produce a whisker that is quite sensitive and potentially very easily deformed. The ease with which the whisker can be deformed can be a major headache and will require you to check them before running the robot. Instead of a wire, I usually use a "microswitch" with an actuating arm (Figure 15-14), which cannot be easily deformed.

To demonstrate the operation of whiskers in a BS2-based robot, I want you to wire two micro-switches as shown in Figure 15-15. When you have the circuit built, then you can key the following code. This code uses the "Button" PBASIC statement to

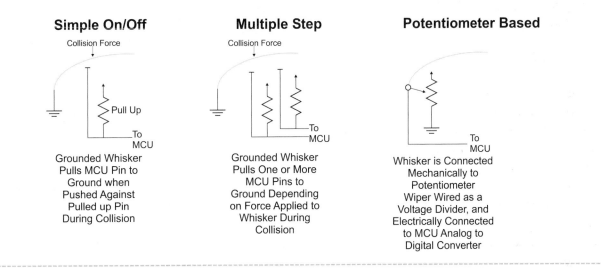

Figure 15-13 *Whisker implementation*

Figure 15-14 *A small microswitch with a built-in actuator arm is ideal for use as a robot's whisker.*

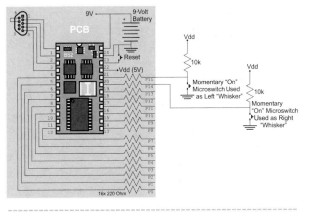

Figure 15-15 *Whisker circuit*

debounce the two whiskers' signals, which is important because you may find that as your robot runs, a homemade whisker will periodically bounce, returning a false collision. The code appears to be somewhat cumbersome because of the need for debouncing when the whisker is in contact with an object, as well as debouncing when the whisker has stopped being in contact with the object.

```
'   Whisker Sensors - Monitor two Whiskers on front of Robot
'{$STAMP BS2}
'{$PBASIC 2.50}

'   Variables
LeftWhisker      pin 15          '   Define the Whiskers
RightWhisker     pin 14
LeftFlag         var bit         '   Touched Flags
RightFlag        var bit
LeftCount        var byte        '   Debounce Counter Variables
RightCount       var byte

'   Initialization
  LeftFlag = 0: RightFlag = 0 '  Clear Flags/No Hits

  do                              '   Repeat forever
    if (LeftWhisker = 0) and (LeftFlag = 0) then '  Wait for Left
      button LeftWhisker, 0, 10, 180, LeftCount, 1, LeftButtonDown
      goto LeftButtonDownSkip
LeftButtonDown:
      LeftFlag = 1: debug "Ouch, collision on Left Side", cr
LeftButtonDownSkip:
    else                         '   When Whisker Released, Reset
      if (LeftWhisker = 1) then Leftflag = 0
      LeftCount = 0              '   Reset the Left Whisker Count
    endif

    if (RightWhisker = 0) and (RightFlag = 0) then '  Wait for Right
      button RightWhisker, 0, 10, 180, RightCount, 1, RightButtonDown
      goto RightButtonDownSkip
RightButtonDown:
      RightFlag = 1: debug "Ouch, collision on Right Side", cr
RightButtonDownSkip:
    else                          '   When Whisker Released, Reset
      if (RightWhisker = 1) then Rightflag = 0
      RightCount = 0              '   Reset the Left Whisker Count
    endif
  loop                            '   Loop forever, end of application
```

Experiment 107
IR Object Sensors

Parts Bin

Assembled PCB with breadboard and BS2

Sharp GD2D120 (Digi-Key part number 425-1162-ND)

LM339

10k breadboard-mountable potentiometer

10k resistor

Tool Box

Wiring kit

Twenty-four-gauge solid core wire (red, white, and black)

Soldering iron

Solder

Five-minute epoxy

Along with using physical whiskers, some commonly used noncontact methods of detecting objects around your robot also exist. These methods require a bit more electronic expertise than the simple wire whiskers, but I find them generally superior to wire whiskers because they do not have to be bent back into shape, and they can often be calibrated for new environments very easily. The most commonly used method is infrared light pulses, although ultrasonic sound (frequencies above human hearing) is commonly used as well. These methods work on the same theory as "SONAR" in submarines; energy is output, and then a receiver waits for the pulses to return. Based on the time required for the return, or even the presence of an echo, it is determined that an object is by the robot. In this experiment, I am going to work with commonly available infrared object (or proximity) detection modules that can be purchased for a few dollars. These modules work according to the SONAR model; an I/R LED outputs a series of pulses that may be reflected by an object back to a receiver (Figure 15-16). The speed of light generally precludes the ability to measure the "flight time" (and distance to the object), which is why this type of object detector is called a *proximity* detector; it can just detect an object at a specific distance from the robot.

These I/R proximity detector modules generally have a commonly used part that you've never thought about: the infrared TV remote control receiver. This small module, which costs about $0.25, allows for the reception of simple commands (less than 16 bits long) to a TV, DVD player, or other electronic device. The data bits (or *packet*) are sent using

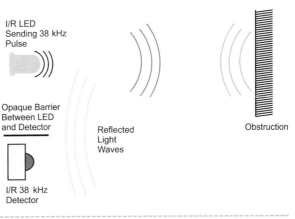

I/R LED
Sending 38 kHz
Pulse

Opaque Barrier
Between LED
and Detector

I/R 38 kHz
Detector

Reflected
Light
Waves

Obstruction

Figure 15-16 *IR detection*

the Manchester encoding scheme in which, when a signal is received, the receiver pulls its output line low. The start of the packet is indicated using a long indicator pulse, and 0's and 1's of the data are produced by varying lengths of "lows." This signal is modulated by a 38 kHz sine wave. When a "low" is to be output, an IR LED sends a 38 kHz signal that is turned off for the output to be "high."

I have used I/R TV remote controls for providing a direct control to a robot, as well as for making a simple proximity detector using a circuit like the one shown in Figure 15-17. Despite appearing very simple and being quite cheap, an IR TV remote control receiver is actually a very complex circuit, consisting of very high-gain amplifiers and filters to recognize and output the IR signals from an environment that is awash in light signals from a variety of different sources. This complexity means that if you had a constant 38 kHz signal being transmitted, the IR TV remote control would recognize it for about 50 ms

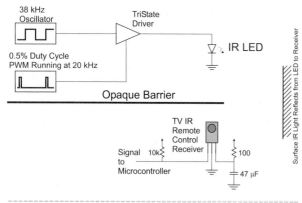

Figure 15-17 *Possible IR detector circuit*

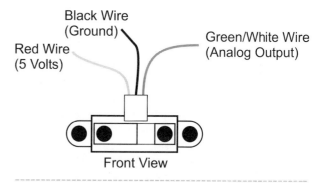

Figure 15-18 *Sharp wiring*

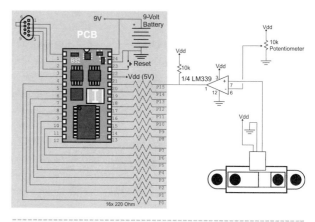

Figure 15-19 *Sharp test circuit*

and then ignore it because it is assumed to be part of the background noise. This is why after producing a 38 kHz modulating signal, I use a PWM signal to switch it on and off.

For best operation, the duty cycle of the 38 kHz modulating signal must be 50 percent. This requirement means that you cannot use an a 555 chip for generating the 38 kHz signal because you cannot produce a genuine 50 percent duty cycle signal. Coupled with this, the PWM control, although it can be easily built, will require a number of chips that will add to the complexity and cost of your implementation. If you have a microcontroller with a built-in PWM generator, you could probably very easily produce this signal, but it will take quite a few discrete chips, which makes it impractical for use with the PCB included with this book.

Instead, Sharp has a line of different IR proximity detectors that encompasses the functions shown in Figure 15-17 in a simple package that only requires you to provide power and poll the output signal. When you buy a Sharp GD2D120, you will discover that it has a small white connector that you will have to solder wires on as I show in Figure 15-18. These wires should be color-coded to make sure you don't get confused as to which wire is which. Along with color-coding the wires, it is a good idea to put five-minute epoxy over the solder joint to strengthen the mechanical joint and make sure that nothing shorts against them.

After soldering the wires to the GD2D120, you can now wire it to the breadboard (see Figure 15-19). The output from the GD2D120 ranges from 0 to 2.25 volts (the voltage level is a rough approximation of

how far away an object is), and to detect objects at a specific distance away, I used a potentiometer wired as a voltage divider to provide a comparator voltage. The potentiometer and comparator will cause the circuit to recognize an object from a set distance away. Once you have built the circuit, key in the following program to test it and save it as "IR Test" in the IR Proximity Detector folder in the Evil Genius Folder.

```
'  IR Proximity Detector - Monitor Sharp
GD2D120 Proximity Detector Output
'{$STAMP BS2}
'{$PBASIC 2.50}

'  Variables
Whisker          pin 15
Flag             var bit

'  Initialization
  Flag = 0

  do
    if (Whisker = 0) and (Flag = 0) then
      Flag = 1: debug "Ouch, collision",
cr
    else
'  When Whisker Released, Reset
      if (Whisker = 1) then Flag = 0
```

```
        endif
        loop
```

You should probably notice that the code was taken from the previous experiment and cut down to demonstrate the operation of the IR proximity sensors. It should be obvious that a big advantage to the GD2D120 is that it debounces the object detection so that you don't have to.

When the software is running, lift up the GD2D120 by the corners and point it about the room and put your hand in front of it. You should see that it detects objects six inches to a foot (15 cm to 30 cm) away, but *not* all objects; black objects and some fabrics do not reflect IR light. In the cases where you are going to run a robot with IR proximity sensors, it might be a necessary for you to include physical whiskers as well as sound-range finding to make sure the robot doesn't run into any objects.

Section Sixteen

Mobile Robots

Before starting to design your own mobile robot, I want to take a look at one that I am sure that anyone reading this book is very familiar with, both in terms of its design as well as its capabilities. This robot has demonstrated the ability to traverse a wide range of terrain, fix an impressive array of different technologies, access very sophisticated computer systems, and finally record and replay data messages. Despite having these qualities, I feel the robot was designed quite poorly, and quite a few different features should be changed for me to consider the design successful.

In case you haven't guessed, the robot I'm talking about is R2-D2 from the *Star Wars* movie saga.

The areas in which I would consider R2-D2 (and probably the whole R2 series) deficient consist of the robot's drivetrain, sensors, robot grippers, input/output, and basic programming. Despite these limitations, the robot design works well in a variety of specialized situations (most notably the maintenance and operation of different spacecraft during combat).

Before you dismiss this introduction as whimsical, I would like to point out that there were significant problems with operating R2-D2 in the first *Star Wars* movie (Chapter IV, "A New Hope"), which threatened to put the film behind schedule and over cost. Despite the apparent sophistication of C-3P0 and other very spectacular effects, one of the biggest challenges faced by the filmmakers was getting R2-D2 to work as required. I suspect that many of these problems were due to the issues that I will bring up here.

The design of R2-D2's drivetrain seems to be somewhat schizophrenic; most of the time it seems to be running around on three wheels, but occasionally, the wheels seem to retract into the robot's legs and body and it walks or wobbles around. Two major problems with this movement are understanding where the robot's center of mass should be and the difficulty the mechanism would have traversing uneven terrain (see Figure 16-1).

The center of mass issue is probably not one of the first things that comes to mind when you are looking at deficiencies in a robot, but it should be. The center of mass is the position of the robot (or a body) about which the robot would hang without tilting if a nail were driven directly through it. Pilots and aircraft designers call the center of mass the center of gravity. As I show in Figure 16-2, the best position for a robot's center of mass is as close to the center of the robot with the robot's weight spread over a wide area. When the center of mass is high up, the robot can fall over when it starts moving or stops. If the center of mass is at one end of the robot or the other, then the robot is in danger of continuously falling over or "popping a wheelie" when it starts moving in one direction. If you are familiar with racing cars, you'll know that the designers want to keep the car's center of mass as low as possible and in the middle of the car to ensure all the wheels are on the ground during acceleration, breaking, and turning.

Going back to R2-D2's center of mass (Figure 16-1), the robot's designers would have trouble figuring out what is the correct point for the center of

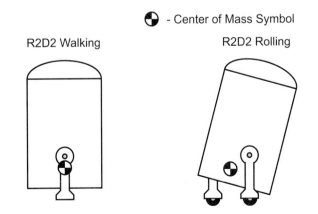

Figure 16-1 *To ensure the stability of R2-D2, its center of mass should be as low to the ground as possible and in the middle of its wheels/legs.*

mass because of the two different methods of loco-motion. When the robot is riding on wheels, it should have a center of mass somewhere between the two wheels for maximum stability, but for walking, the robot must have a center of mass that is along the axis of the robot that the legs are on or it will fall over. One of the big problems in filming the movie was R2's insistence on falling over when starting, stopping, or operating on uneven terrain. These problems should have been expected based on looking at where the center of mass of the robot was.

R2-D2 has remarkably small wheels and stumpy legs for a robot that seems to have been able to cross deserts, traverse rocky terrain, navigate swamps, run over snow, and climb up stairs. Looking at the robot layout, I would expect that it would only be able to run (or walk) on smooth, polished floors. The ability to move over different surfaces is one of the toughest assignments required of a robot—most will get stuck,

fall over on, or damage their drivetrains (which the real R2-D2 did repeatedly while filming the movies).

Most robots that are expected to run over uneven terrain are built from an articulated body such as the one shown in Figure 16-3. This example robot has three sets of wheels that are joined to the chassis and can move relatively independently. The Mars Pathfinder Sojourner robot used this type of body to run over the surface of Mars.

A big advantage of this type of mobile robot is that it keeps the chassis relatively level, which keeps the sensors at a relatively constant point of view. R2-D2's sensor's point of view will change with the angle it is set at, as well as when the dome on the top (where I presume the sensors are located) turns. A robot that maintains a reasonably flat altitude will be able to sense what is happening in its environment in a much more consistent manner than R2-D2.

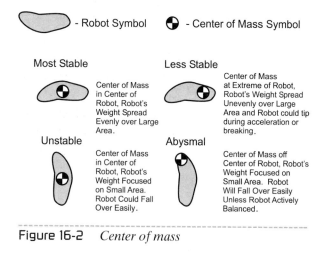

Figure 16-2   *Center of mass*

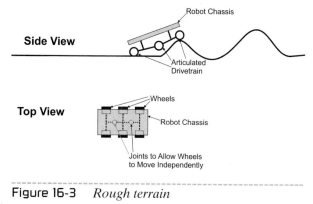

Figure 16-3   *Rough terrain*

Experiment 108
DC Motor Control Base with H-Bridge Drivers

Parts Bin

Assembled PCB with breadboard and BS2

Assembled plywood base with DC motors and four-AA-battery pack and power switch

754410 motor driver

555 timer chip

Eight 1N914 (1N4148) silicon diodes

Two microswitches with arms

Two 10k resistors

Two 1k resistors

10k potentiometer

Three 0.01 µF capacitors

Tool Box

Wiring kit

Five-minute epoxy

Screwdriver set

24-gauge solid sore wire (see text)

Soldering iron and solder (see text)

If you take a look around the Internet at different home-built mobile robots, you will probably discover that many of them use the LM293 motor driver chip. This chip is very popular because it can be used as an H-Bridge control of two DC motors with up to 1 amp of current. Officially, the LM293 is obsolete, but many sources still have quite a few in stock, or you can use the TI 754410, which is an update to the LM293 with some improvements (including thermal shutdown) and is widely available at low cost.

Wiring a 754410 to a microcontroller to control a robot's motors is quite simple as shown in Figure 16-4. It has two power inputs, one being for the logic control inputs (which are *Transistor-to-Transistor Logic* [TTL]) and the other being the power supplied to the motors. The chip does not have kickback diodes on the outputs, so you will have to add these to make sure the voltage transients produced by the motors being switched on and off do not damage the 754410 (or other chips in the robot).

Along with four inputs that control four H-Bridge halves and that output signal polarity (two halves have to be put together to create a full H-Bridge), two lines are used to enable the motor drivers. These two lines are typically used to provide PWM control to the robot, and they will allow for simplified pro-

gramming of the robot motors. To demonstrate how the 754410 works, I would like you to go back and add the finished plywood chassis with the DC motors and a four-AA-battery holder wired to it to make a simple programmable robot (your first). Before bolting the plywood chassis to the book PCB, you should five-minute epoxy two microswitches with long actuator arms to the base as the robot's whiskers (Figure 16-5). You may have to solder wires to connect these microswitches to the breadboard before gluing them to the plywood chassis.

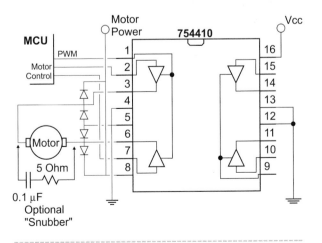

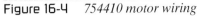

Figure 16-4 *754410 motor wiring*

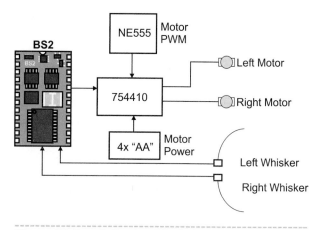

Figure 16-5 *Whisker microswitches five-minute epoxied to the finished plywood chassis*

Figure 16-6 *DC robot block diagram*

With the whiskers in place, wires for the breadboard, and the finished plywood base bolted to the book PCB, you are now ready to build your first programmable robot. Figure 16-6 is the block diagram of a robot that will move randomly about a room until it hits something, at which time it will back up and resume moving randomly. Note that I have marked the directions that control signals/power are moving in. I find it to be a good idea to always show the direction where things are going; you will have conflicts that could have been avoided by simply mapping things out when you first started your design.

The circuit shown in Figure 16-6 shouldn't be very surprising except for the use of the 555 timer. I added an adjustable monostable timer to act as a PWM generator for the motors. Depending on the motors that you are using, you may find that the robot moves too quickly for the BS2 to control it adequately; the 555 timer (with potentiometer) is a speed control for the robot (with a duty cycle ranging from 66 to 91 percent).

It should be quite easy for you to wire the robot circuit in Figure 16-7. The completed robot should be ready to roll after loading in the following application. Don't disassemble this robot when you are fin-

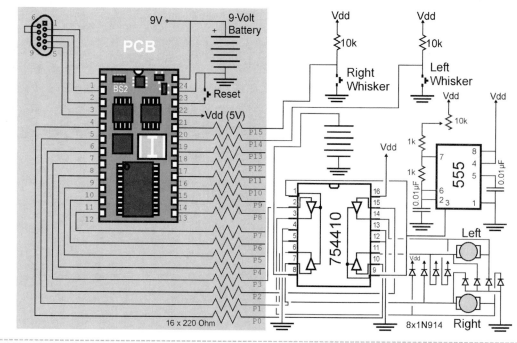

Figure 16-7 *DC robot circuit*

ished with this experiment; it is needed in the following experiments.

```
'  DC Robot 1 - First Program for the DC
Robot/Run Randomly
'{$STAMP BS2}
'{$PBASIC 2.5}

'  Variables and I/O Port Pin Declarations
i               var word
LeftCount       var byte
RightCount      var byte
RandomValue     var word
MotorValue      var RandomValue.NIB0
RandomActive    var RandomValue.NIB1
LeftWhisker     pin 14
RightWhisker    pin 15
MotorCtrl       var OUTS.NIB0
MotorDir        var DIRS.NIB0

  MotorCtrl = %0000: MotorDir = %1111
  RandomValue = 6
  i = 1
  do
    i = i - 1
    if (i = 0) then
      MotorCtrl = %0000: pause 100
      random RandomValue
'  Get New Random Value
      MotorCtrl = MotorValue
      i = ((RandomActive & 3) + 1) * 100
    endif
    if (LeftWhisker = 0) then
      Button LeftWhisker, 0, 100, 0,
LeftCount, 0, LeftWhiskerDown
      Goto LeftWhiskerSkip
LeftWhiskerDown:
'  Collision
      MotorCtrl = %0110: pause 3000
LeftWhiskerSkip:
    else
      LeftCount = 0
    endif
    if (RightWhisker = 0) then
      Button RightWhisker, 0, 100, 0,
RightCount, 0, RightWhiskerDown
      Goto RightWhiskerSkip
RightWhiskerDown:
'  Collision
      MotorCtrl = %0110: pause 3000
RightWhiskerSkip:
    else
      RightCount = 0
    endif
  loop
```

You may have noticed a few things when you have run your robot in this experiment. I lengthened one set of motor wires and the four AA battery clip wires and wired them directly to the 754410. This was done to avoid the relatively high (on the order of ohms) resistance of the breadboard pins—when you are building your own robots, you may want to create low-resistance motor wiring to minimize this parasitic resistance.

After the robot has been running for a few seconds, you may find that the 754410 becomes quite warm; this is due to the current draw through the motors and the voltage drop through the transistors in the 754410. You will not experience any heating of the 754410 if you use low current draw motors. As the 754410 heats up, you will find that its efficiency will drop (the motors will not turn as quickly and they may whine). The ultimate solution to these problems is to work at matching the motors to the 754410 or using a discrete transistor-based H-Bridge with high-gain transistors, such as the one presented in Section six. In any case, your wiring should use heavy-gauge copper and ideally be built on a PCB. The reason for using the 754410 is convenience. The 754410 provides you with a fairly cheap and easy-to-use a package that will allow you to experiment quickly with motors and robots.

Finally, you will discover that although the robot's whiskers will detect objects in front of it and back away, it will just as often run into objects going backwards and sit there with the motors stalled. In this application, I really should have provided whiskers all the way around the robot. In real applications, I would tend to use just a couple of object detectors at the front of the robot and never go backward unless I was backing away from an object.

Experiment 109
State Machine Programming

Parts Bin

Assembled PCB with breadboard and BS2

Assembled plywood base with DC motors, four-AA-battery pack and power switch, and microswitch whiskers

754410 motor driver

555 timer chip

Eight 1N914 (1N4148) silicon diodes

Two 10k resistors

Two 1k resistors

10k potentiometer

Three 0.01 µF capacitors

Tool Box

Wiring kit

Screwdriver set

Looking over the code used for the previous experiment, you'll probably see that it is quite difficult to understand. I should point out that the code works quite well; it is just very difficult to understand what is happening when the robot is moving. There are two reasons why I wrote the code the way I did in the previous experiment. The first was that I was trying to demonstrate the operation of the 754410 motor driver in a real robot, and the second was that I didn't have much space to do it in. By using the random statement and mapping bits from it directly to the motors, I was able to create code that is very small and functional. When you start creating your own robot software, you should be working at making the code as simple and as readable as possible.

This probably seems like it is very difficult to do, but by using a software state machine you will find it very easy to develop robot code. This tool is the *state machine* and is used to periodically update the state of a number of output bits based on the previous state as well as any input bit changes. The state machine is used in a variety of different applications, including microprocessors, and is usually very easy to create and program with specific values. Programming a software state machine is not very difficult; it works almost exactly the same way as the hardware state machine. The differences are quite subtle; the

hardware state register is replaced with a state variable and the ROM becomes a series of if statements (or a single select statement as I show in Figure 16-8).

The advantage of the software state machine over traditional programming methodologies is not only how easy it is to see what is happening, but also how easily it can be changed as the requirements for the application change. These advantages are only true if you follow two simple rules when making up your software state machine:

1. When you are numbering your states, make sure that they are separated by 10 instead of 1. If they are separated by 1, you will have to renumber the entire application if you have to add a new state.

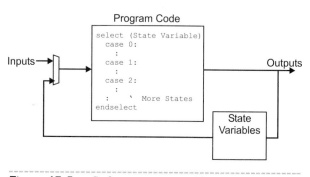

Figure 16-8 *Software state machine*

2. Do not execute any if statements within the ROM of the state machine. Instead, use external inputs to change the state if an action is required. This is an important point and one that can cause confusion for new programmers. If you are polling a bit, and a state being monitored is true, then increment the state variable before it is checked and the state code is executed.

To demonstrate the state machine, I have modified the previous application's code to use a state machine, and after the robot has backed up after a collision, it will then turn away from the whisker that had the collision. The code should be saved in the same directory as the previous experiment:

```
'  DC Robot State Machine - State Machine
Version of CD Robot 1st Pgm
'{$STAMP BS2}
'{$PBASIC 2.5}

'  Variables and I/O Port Pin Declarations
State          var byte
TurnAway       var byte
i              var word
WhiskerCount   var byte
RandomValue    var word
MotorValue     var RandomValue.NIB0
RandomActive   var RandomValue.NIB1
LeftWhisker    pin 14
RightWhisker   pin 15
MotorCtrl      var OUTS.NIB0
MotorDir       var DIRS.NIB0
```

```
State = 0
MotorCtrl = %0000: MotorDir = %1111
RandomValue = 5
i = 1
do
  i = i - 1
  if (i = 0) then State = State + 1
  if (RightWhisker = 0) and (State <>
30) then State = 10: MotorCtrl = %0010
  if (LeftWhisker = 0) and (State <> 30)
then State = 10: MotorCtrl = %0100
  if (WhiskerCount = 20) then State = 20
  select (State)
    case 0:
'   Keep Doing what you're doing
      WhiskerCount = 0
    case 1:
'   Timeout
      random RandomValue
      MotorCtrl = MotorValue
      i = ((RandomActive & 3) + 1) * 100
      WhiskerCount = 0
      State = 0
    case 10:
'   Collision
      WhiskerCount = WhiskerCount + 1
      i = 40
      State = 0
    case 20:
'   Debounced Collision
      WhiskerCount = 0
      MotorCtrl = %0110: i = 100
      State = 30
    case 30:
'   Going Backwards
    case 31:
'   Finished, Resume Operation
      i = 1
      State = 0
  endselect
loop
```

Experiment 110
Robot Moth Example

Parts Bin

Assembled PCB with breadboard and BS2

Assembled plywood base with DC motors, four-AA-battery pack and power switch, and microswitch whiskers

754410 motor driver

555 timer chip

Eight 1N914 (1N4148) silicon diodes

Two 10k CDS cells

Two 10k resistors

Two 1k resistors

10k potentiometer

Five 0.01 µF capacitors

Tool Box

Wiring kit

Screwdriver set

Flashlight

Two bricks

With the ability to create or modify an application easily using a software state machine, I wanted to create a simple robot application that demonstrates the preprogrammed behavior of searching out the brightest spot in the room. When the robot collides with the light (or wood blocks protecting the light), the robot will then back up and turn away from the collision (as it did in the previous experiment), move randomly for 30 seconds, and then resume its search for light (Figure 16-9 outlines this movement). A robot that approaches light and backs off randomly, such as this one, is often known as a *moth*.

Adding two CDS cells with capacitors to the DC robot circuit is quite simple (Figure 16-10). If you followed the wiring diagrams provided for the original DC robot, there should be space at the front left of the breadboard for these additional parts. As in the earlier experiment on wiring in the CDS cells, I bent their leads at right angles and positioned them so that they would have different fields of view.

The code for the moth application is listed here (and should be saved in the same folder as the previous two DC robot applications). I have rearranged some of the states in the software, which again demonstrates how easy a state machine is to modify.

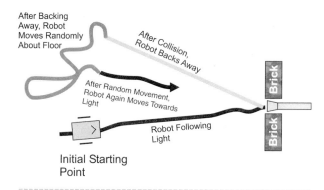

After Backing Away, Robot Moves Randomly About Floor

After Collision, Robot Backs Away

After Random Movement, Robot Again Moves Towards Light

Robot Following Light

Initial Starting Point

Figure 16-9   *Moth path*

```
'  DC Robot Moth - State Machine code to
implement a "Moth"
'{$STAMP BS2}
'{$PBASIC 2.5}

'  Variables and I/O Port Pin Declarations
State          var byte
TAway          var byte
LeftValue      var word
RightValue     var word
i              var word
j              var byte
WhiskerCount   var byte
RandomValue    var word
MotorValue     var RandomValue.NIB0
RandomActive   var RandomValue.NIB1
LeftLight      pin 13
RightLight     pin 12
LeftWhisker    pin 15
RightWhisker   pin 14
MotorCtrl      var OUTS.NIB0
MotorDir       var DIRS.NIB0
```

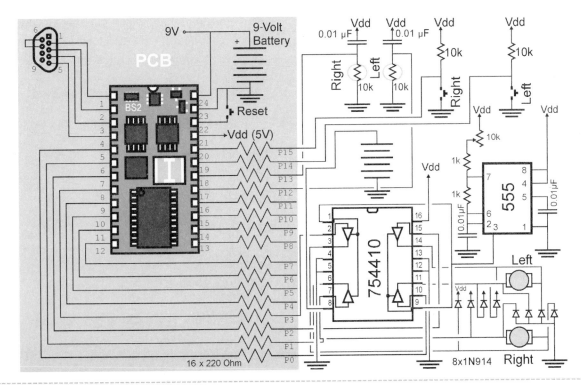

Figure 16-10 *Moth circuit*

```
State = 0
MotorCtrl = %0000: MotorDir = %1111
RandomValue = 5
i = 1: j = 1
do
  i = i - 1
  if (i = 0) then State = State + 1  '
Delay Finished
  if (j = 0) then State = 0: MotorCtrl =
%0000
  if (LeftWhisker = 0) and (State < 30
)then State = 10: TAway = %0001
  if (RightWhisker = 0) and (State < 30
)then State = 10: TAway = %0100
  if (WhiskerCount = 20) then State = 20
  select (State)
    case 0:
'  Charge CDS Cell Capacitors
        high LeftLight: high RightLight
        WhiskerCount = 0: j = 1
    case 1:
'  Read Left CDS Cell
        rctime LeftLight, 1, LeftValue:
State = 2
    case 2:
'  Read Right CDS Cell
        rctime RightLight, 1, RightValue:
State = 3
    case 3:
'  Wobble Towards the Light
        if (RightValue < LeftValue) then
          MotorCtrl = %1000
        else
          MotorCtrl = %0001
        endif
        i = 70: State = 0
    case 10:
'  Collision
```

```
        WhiskerCount = WhiskerCount + 1
        State = 0
    case 20:
        MotorCtrl = %0110: i = 100
        WhiskerCount = 0
        State = 30
    case 30:
'  Going Backwards
    case 31:
'  Finished Going Backwards
        MotorCtrl = TAway: i = 50: State =
40
    case 40:
'  Turning Away
    case 41:
'  Finished, Move Randomly
        j = 8
        i = 1
        State = 50
    case 50:
'  Random Wait
    case 51:
        random RandomValue
        MotorCtrl = MotorValue
        i = ((RandomActive & 3) + 1) * 100
        WhiskerCount = 0
        j = j - 1
        State = 50
  endselect
loop
```

When testing the robot, place a flashlight between two bricks as I have done in Figure 16-10. This will provide the robot with a light source to home in on, as well as a barrier that will cause the robot to reverse course, move about randomly, and start over.

When the robot is moving toward the light, it will waddle toward it with a curious gait. This is a result of one wheel being turned on and followed by the other. Depending on your motors and the drivers that you use, you may find that the time the robot is turning is too long and you will have to change the value given to "i" after "MotorCtrl" is loaded with the turning value.

Experiment 111
Random Movement Explained

Parts Bin

Assembled PCB with breadboard and BS2

Assembled plywood base with DC motors, four-AA-battery pack and power switch, and microswitch whiskers

754410 motor driver

555 timer chip

Eight 1N914 (1N4148) silicon diodes

Two 10k CDS cells

Two 10k resistors

Two 1k resistors

10k potentiometer

Five 0.01 µF capacitors

Tool Box

Wiring kit

Screwdriver set

Looking over the previous three experiments, I can see that the three code lines in the software for each experiment that are used to move the robot randomly are not well explained and what is happening in them is probably pretty confusing. There is a lot of synergy (to use a word I hate) in how these three lines work and when you try to visualize what is happening between the software and the robot you will discover that they are very tightly integrated, as shown in Figure 16-11. In this experiment, I would like to explain the thinking behind this code and how to see opportunities to use code like this in your own robot applications.

The "random" statement is a software linear feedback shift register operation on the value passed to it, and it returns a pseudorandom value out. The value returned cannot be easily determined from the values passed to it. It can be anywhere in the range of 1 to 65,535, and it does not repeat until all the values have been displayed. If you were displaying the different values from the random statement once per second, you would find that it would take 18 hours, 12 minutes, and 15 seconds before there was a repeating value. Each of the bits in the returned value can be set or reset, apparently randomly, and can be passed directly to the motor control bits as I have done in these experiments. When two bits are passed to two 754410 drivers that are connected to a motor, the motor reacts in one of three ways as defined by Table 16-1.

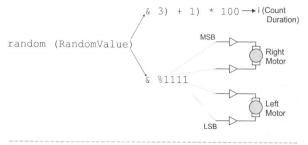

Figure 16-11   *Apparent random action*

So, by simply passing the least significant bits from the random statement, I am randomly specifying whether or not the motor is to work and in which direction it will turn. I could have used the PBASIC select statement to perform a similar function:

```
random RandomValue
MotorValue = RandomValue & %1111
select (MotorValue)
  case 0:
' Left Motor Forward, Right Motor Stopped
    MotorCtrl = %1000
  case 1:
' Left Motor Forward, Right Motor Forward
    MotorCtrl = %1010
  case 2:
' Left Motor Forward, Right Motor Reverse
    MotorCtrl = %1001
  case 3:
' Left Motor Stopped, Right Motor Forward
    MotorCtrl = %0010
  case 4:
' Left Motor Stopped, Right Motor Reverse
    MotorCtrl = %0001
  case 5:
' Left Motor Reverse, Right Motor Stopped
    MotorCtrl = %0100
  case 6:
' Left Motor Reverse, Right Motor Forward
    MotorCtrl = %0110
  case 7:
' Left Motor Reverse, Right Motor Reverse
    MotorCtrl = %0101
  case else
' Anything Else, Motors Stopped
    MotorCtrl = %0000
endselect
```

It is somewhat easier to read and understand what is happening with this code, but it has a number of concerns, including that it takes more effort to write and requires more space in the BS2 memory. Of course, looking through the statements, somebody new to programming and robots will probably have a couple of questions, such as why isn't there an explicit "Left Motor Stopped, Right Motor Stopped?" The person may note that in 8 cases (out of 16) the "case else" code will execute and both motors will be stopped. Why is that allowed? In the practical terms of readability, this modification of the code isn't any better than the original.

You may have noticed that when I have specified the motor control bits for this application, I arranged them to simplify software development. You might be thinking that in a real-world application, the motor control bits and their pins would be specified in order to simplify the wiring of the application; software can be changed easily. In this case, you would use something like the previous select statement or the following branch statement code where Lred, Lblk, Rred, and Rblk, defined using the pin statement, are the BS2 I/Os that interface to the motor driver.

Table 16-1 Motor response to different wire connections

Red Wire Bit	Black Wire Bit	Motor Response
0	0	Motor stopped (Both inputs tied to ground so no current flow)
0	1	Motor running backward (Current from negative to positive)
1	0	Motor running forward (Current from positive to negative)
1	1	Motor stopped (Both inputs tied to power so no current flow)

```
random RandomValue
MotorValue = RandomValue & %1111
branch MotorValue, [M0, M1, M2, M3, M4, M5, M6, M7, M8]
  Lred = 0: Lblk = 0: Rred = 0: Rblk = 0: goto Mend  ' Stopped
M0: Lred = 1: Lblk = 0: Rred = 0: Rblk = 0: goto Mend
M1: Lred = 1: Lblk = 0: Rred = 1: Rblk = 0: goto Mend
M2: Lred = 1: Lblk = 0: Rred = 0: Rblk = 1: goto Mend
M3: Lred = 0: Lblk = 0: Rred = 1: Rblk = 0: goto Mend
M4: Lred = 0: Lblk = 0: Rred = 0: Rblk = 1: goto Mend
M5: Lred = 0: Lblk = 1: Rred = 0: Rblk = 0: goto Mend
M6: Lred = 0: Lblk = 1: Rred = 1: Rblk = 0: goto Mend
M7: Lred = 0: Lblk = 1: Rred = 0: Rblk = 1: goto Mend
Mend:
```

Experiment 112
Remote-Control Car Robot Base

Parts Bin

Toy remote-control car

Tool Box

Everything you've got

In this book, I have tried to maintain the brave front that there is no such thing as a "failed" experiment. No matter what happens, you can always learn from your experiences and go on. Sometimes, in cases like this experiment, what you learn is the difference between hubris and humility.

The original idea for this experiment was to build a robot out of an inexpensive toy car with the ability to move on its own and with the ability to turn by the use of some kind of electrical control. In this book, you will spend a fair amount of effort building two robot chassis (DC-motor- and R/C-servo-driven bases). I thought it would be interesting to start with a premanufactured product that already had the drivetrain and steering gear in place so the effort required to convert it into a robot would be minimal. I was expecting to just add an H-Bridge for the motor and a servo for the steering gear. As an added bonus, this robot would give you experience in working with car robot bases and show why they were generally perceived as being suboptimal when compared to differential drive robots. I have never seen a toy car converted into a robot in any books or magazine articles and I was pretty sure I was onto something that everyone else had overlooked.

At a local toy shop, I thought I found the perfect chassis to build on, a replica Chevrolet Blazer police vehicle that had a wire remote control that allowed the user to command it to move forward or backward and control the steering (Figure 16-12). The size of the car is 14 inches (35.6 cm) by 5.5 inches (14 cm), which lead me to believe that I could mount one of the book's PCBs on it easily with a servo to control the angle of the front wheels. The price was certainly right ($10 Canadian). Unfortunately, there was only

one in the store; otherwise I would have bought at least two or more.

When we got the toy home, my daughter had fun with it. Along with being remote controlled, it has some flashing lights, prerecorded sounds (my favorite being "Stop, this is the police—you're under arrest!" followed by gunfire; I'm sure a real police officer would cringe upon hearing this), and a microphone in the remote-control unit that will output a person's voice through the car. Actually, she had a lot of fun with it until the batteries that came with it died; upon opening the battery door on the underside of the car, I discovered that it was empty with no contacts for replacement batteries (see Figure 16-13). This was the last time the car was seen in one piece.

Looking at the underside of the car, there were two Phillips screws holding the front of the chassis to

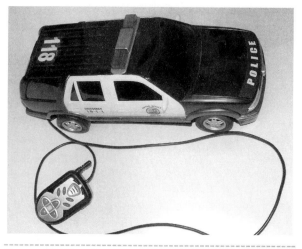

Figure 16-12   *Inexpensive wire-remote-control car intended to be used as a simple and cheap robot base.*

Figure 16-13   *Battery compartment*

the body. I thought that it would be a simple matter of removing these two screws, lifting the body off the chassis, and replacing the batteries.

No such luck, the body was glued to the chassis; I presume that the screws were there to hold the car together while the glue cured. Not to be deterred, I very carefully cut the chassis from the body along the underside with the expectation that after replacing the batteries, the body could be put back, and the toy could continue to be used as originally intended.

After 30 minutes with a rotary cutting tool and a carbide wheel, I had separated the bottom of the body from the chassis. Unfortunately, this wasn't the only place the body was glued to the chassis. In trying to pull apart the body and the chassis, I discovered that the roof was glued to the chassis and in my efforts to pull the body from the chassis, I collapsed the roof. At this point, it was no longer a toy (at the very best it was a model of the police car used by

Jackie Gleason at the end of *Smokey and the Bandit*).

With the body off, I discovered that two AA batteries had been soldered directly into the car's circuitry and then glued to the chassis.

Moving on to learn more about how the car was built, the steering mechanism was a sealed box with two solenoids inside, and you can probably guess what happened when I tried to cut this open.

The long and the short of this story is, after spending two hours carefully disassembling the toy car, all I had was a pile of deformed black plastic pieces, some with different colored wires coming from them. My initial thought was that my cheapness caught up with me—maybe if I looked at a more expensive toy car, I would have had a better experience. After spending another two hours looking at different toy cars of varying price points (including some very expensive radio-control cars), I discovered that the methodology used for assembling the original police car is the one used for all toy vehicles. They are designed for out-of-the-box use and no thought is given toward modifying them for other purposes.

It's up to you if you want to repeat this experiment. By all means you should at least investigate different products and see if you can find a toy car that can be easily converted into a robot and controlled by the PCB that comes in this book. If you do find a product that meets this criteria and you have successfully installed the PCB, by all means drop me a line to let me know.

Just remember my experience and don't spend a lot of money unless you are absolutely *sure*.

Experiment 113
R/C Servo Setup

Parts Bin

Assembled PCB with breadboard and BS2

Four-AA-battery pack with power switch connected

Radio-control (R/C) servo

Three-pin R/C servo connectors (see text)

Two-inch to 3-inch (5-cm to 7.5-cm) model aircraft wheel or servo shaft robot wheel

Two 2.2k resistors

Tool Box

Wiring kit

Five-minute epoxy

Screwdriver set

Soldering iron and solder

Knife

Rotary cutting tool (see text)

Although I introduced you to robot bases, I used the basic DC motors driving two wheels in a differential drive layout. The advantage of this type of method for controlling the robot is the relatively simple programming interface that is easy to work with despite the potentially complex mechanical work that has to be done to get it working. If you look around, you'll see that most beginner robot designs do not use DC motors; instead they use modified *radio-control* (R/C) servos modified to turn continuously to provide a drivetrain and motor driver electronics in a simple package. I will present how R/C servos are used in mobile robots as well as a utility that you will need to be able to efficiently use them in your robot.

For this experiment, as well as the others that use servos, I am assuming that you are going to use standard and not Futaba servos. Standard servos are controlled by a 1 ms to 2 ms pulse (with 1.5 being the center), whereas Futaba servos are controlled by a 1 ms to 1.5 ms pulse. If you are using Futaba servos with the application, make sure that you change the data values accordingly.

Earlier in the book, I provided a block diagram for a R/C servo. The basic servo will only move 90 degrees (45 degrees in each direction off center) and must be modified to allow the output shaft to turn continuously. To make this modification, three changes to the servo must be made:

1. Disconnect the position feedback potentiometer and replace it with either two resistors of the same value as a voltage divider or a "trimmer" potentiometer, which is wired identically to the potentiometer in the servo. By disconnecting the potentiometer and passing the signal to a voltage divider set to half the input potentiometer voltage, you have a servo that will not move when a 1.5 ms pulse is passed to it. This servo will turn in one direction when a pulse less than 1.5 ms is passed to it and turn in the other when a greater than 1.5 ms pulse is passed to it. This is ideal for commanding the servo to move forward or backward or stop.

Figure 16-14 shows the top removed from a Hitec Model 322 servo. To disconnect the potentiometer, clip the leads running to it and wire in two 2.2k resistors wired as a voltage divider. The 322's potentiometer's total resistance is 4.7k, so the two 2.2k resistors are a reasonable approximation.

Instead of the two resistors, you can glue a "trimmer" potentiometer to the outside of servo. The advantage of having an external potentiometer is that the servo can be matched to the software instead of the other way around. The disadvantage of doing this is that the external potentiometer does take some space and it must be a trimmer potentiometer (which is more expensive than a

Figure 16-14   *Servo top removed, showing the gearing inside the servo. The motor drive is on the right side and the potentiometer is on the left.*

standard panel or PCB mount potentiometer). A standard potentiometer only has one turn for its full motion, whereas a trimmer potentiometer can have 10 or 20 turns. The increased number of turns will allow you to more precisely center the potentiometer to the software.

2. Remove any mechanical stops inside the servo that prevent it from turning a full 360 degrees. Often the position feedback potentiometer will only allow 90 degrees of movement, which will require that it is either removed or reworked to turn continuously. Along with this, you will probably find tabs on the inside of the servo that will restrict the output shaft's motion that will have to be removed with a knife. Figure 16-15 shows a "tab" built into a Hitec Model 322 servo that will have to be cut away with a sharp knife or a rotary (Dremel) tool.

    On some servos, you will have to take apart the feedback position potentiometer and remove the pot wiper, which restricts the range the pot can turn.

3. Replace the standard output arms, wheels, horns, and other actuator parts with a wheel. Some wheels available on the market are designed for use in robots, but I prefer using a model aircraft 2.5-inch diameter wheel that has been 5-minute epoxied onto a cut-down control arm. The wheel's hub may have to be drilled out to access the servo arm holding the screw.

The only other thing that you will need is the breadboard-to-servo connector adapter that you made for the earlier servo experiment in the book. This adapter isn't absolutely necessary (you could use 22-gauge wires between the breadboard and the servo connector), but it is very reliable and will not damage the servo's connector or the breadboard. When you make up this connector adapter, I suggest that you make a half dozen or so; they are easy parts to lose (see Figure 16-16).

I have not included instructions for modifying any specific servos (even the photographs and instructions for the Hitec Model 322 are pretty sketchy). After doing a quick look on the Internet, I found instructions for modifying almost 50 different servos; when you select a servo to use, do a quick Google™ search to make sure instructions are available for modifying the servo. Although most servos can be modified for continuous rotation, for a number of them the effort required to do so is prohibitive. I should caution you that most instructions for modifying servos warn you about making sure that you

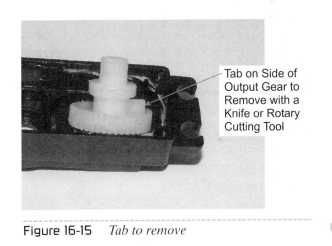

Tab on Side of Output Gear to Remove with a Knife or Rotary Cutting Tool

Figure 16-15   *Tab to remove*

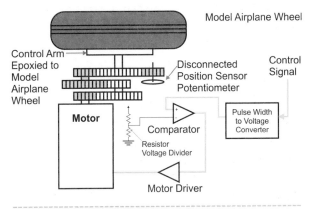

Model Airplane Wheel

Control Arm Epoxied to Model Airplane Wheel

Disconnected Position Sensor Potentiometer

Control Signal

Motor

Comparator

Resistor Voltage Divider

Pulse Width to Voltage Converter

Motor Driver

Figure 16-16   *Modified servo innards*

know what you are doing and make the operation sound quite ominous, but you will find that it is actually quite straightforward and will take you less than 15 minutes for each servo. When modifying a servo, it is important to keep track of the parts (taking pictures of them before they are removed using a digital camera is not a bad idea).

For this experiment, I am going to assume that you are going to replace the position feedback potentiometer with two 2.2k resistors. Although the two resistors are fairly close to an exact value of 2.2k, they are actually off by a few percent. Unless you are very, very lucky, these two resistors will be of different values, and the voltage divider voltage will not be exactly half the total voltage. This difference means that you cannot pass a 1.5 m signal (using a "pulsout 750" statement) and guarantee that the servo will be stopped. You will need an application to find the center or calibrate the value that will hold the servo still. To find this calibration value once you have modified a servo to turn continuously, set up the circuit shown in Figure 16-17 and then enter in the following BS2 program.

If the servo does not move when you put in a value such as 500 or 1000, then check your power supply and the servo's wiring. Although the wiring shown in Figure 16-17 should be correct for the

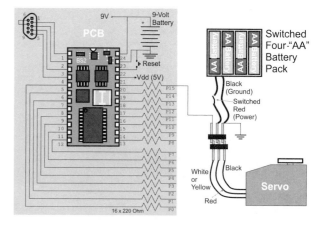

Figure 16-17   *Servo calibration circuit*

servo you're using, you might find that some servos are wired differently. Miswiring the servo should not damage it unless you are passing more than 6 volts to it.

Do not power a servo from the BS2 power supply! The BS2 power supply cannot supply enough current for the servo to operate, and if you drive the servo from the BS2, you *will* burn out the BS2's built-in power supply. When you are running this experiment, or any other application using servos, it is a good idea to provide a separate 4.8-volt to 6-volt battery power supply capable of 300 mA of current output for the servos.

```
'  Calibrate - Find the Center/Not Moving Point for Servo
'{$STAMP BS2}
'{$PBASIC 2.50}

'  Variables
Servo            pin 15
CurrentDelay     var word         '  Servo Center Point
i                var byte

'  Initialization/Mainline
  low Servo                       '  Set Servo Pin Low
  CurrentDelay = 750              '  Start at 1.5 ms

  do                              '  Repeat forever
    debug "Current Servo Delay Value = ", dec CurrentDelay, cr
    for i = 0 to 50               '  Output servo value for 1 second
      pulsout Servo, CurrentDelay
      pause 18                    '  20 msec Cycle Time
    next
    debug "Enter in New Delay Value "
    debugin dec CurrentDelay
    do while (CurrentDelay < 500) or (CurrentDelay > 1000)
      debug "Invalid Value, Must be between 500 and 1,000", cr
      debug "Enter in New Delay Value "
      debugin dec CurrentDelay
    loop
  loop
```

This application will wait for you to specify a center value and then pass it to the servo for a second so you can see whether or not it causes the servo to turn. In the application, I use the "debugin" statement to allow you to enter numeric values from your PC's keyboard to find the "pulsout" value that holds the servo motionless. Once you have found the center pulsout statement value, I suggest that you write it on the servo.

Along with being used for centering the servo, this application code can be used to observe the operation of the modified servo when different values are passed to it. Remember that the values passed to the servo via the "pulsout" statement *must* be in the range of 500 (1 ms) to 1000 (2 ms).

Experiment 114
Controlling Multiple Servos

Setting up and calibrating robot servos with the BS2 (or any other microcontroller) is not very difficult. What can be confusing is how to set up a useful application for a robot that has more than one servo built into it. The BS2 is more than fast enough to effectively control two (or more) servos at the same time, but you will have to sit down for a few minutes to figure out how you can control two servos simultaneously while computing how the robot is to move. Fortunately for you, I have done this work and in this experiment, I will present it to you.

The obvious solution to the problem of controlling the servos is to send a pulse to each of them once every 20 msecs while in a loop, as I show in Figure 16-18. This probably seems quite easy to do because the BS2 is rated as executing one statement every 250 μs. The problem comes in when you look at the *actual* execution speed of the various statements, and of the "pulsout" statement and compound statements in particular. The code read and initiation portion of the pulsout statement takes about 250 μs, but you have to add on the time that it is active to get the actual time it is executing. If you have a compound mathematical statement, such as

$$A = (B + C) \times D$$

you have to remember that each operator requires about 250 μs, so this statement would take about 500 μs to execute.

To avoid the variable timing issue of the pulsout statement, the simple solution is to execute two pulsout statements with a set maximum time. When I work with servos, I assume that the maximum time

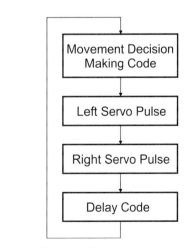

Figure 16-18   *Multiservo program*

for the pulse is 2.25 ms, which means that the first pulseout statement is sent to the servo while the second picks up the difference. The general case form for the servo pulse output that I use is

```
pulseout  SelectedServo, ServoValue
pulseout DummyServo, 1125 - ServoValue
```

The "DummyServo" pin is used to provide a place to put the leftover delay so that the pulseout executes a full 2.25 ms pulse and keeps the statement timing equal. The DummyServo pin should be an I/O pin on the BS2 that is not used for any other purpose.

The granularity provided by the pulseout statement in the BS2 is 2 μs; so to get a total delay of 2250 μs using the pulsout statement, a total delay of 1125 is used. In these two statements, notice that there will actually be three instruction delays (the "1125—ServoValue" code requires an extra 250 μs). When it is added to the 2.25 μs delay of the pulse output, the total delay for the servo pulse output is 3 ms. This makes it quite easy to time out a loop controlling two or more servos to execute within 20 msecs.

To demonstrate the operation of these instructions in the control of a robot with more than one servo, I came up with the following application that moves the robot randomly using servos (which can operate at the same time).

```
'   Servo Random Movement - Move Randomly
About the Room
'{$STAMP BS2}
'{$PBASIC 2.50}

'   Variables
LeftServo       pin 15
RightServo      pin 0
DummyServo      pin 1
LeftServoVal    var word
RightServoVal   var word
LeftStop        con 750
RightStop       con 750
LeftForward     con 600
LeftBackward    con 900
RightForward    con 900
RightBackward   con 600
CurrentStep     var word
RandomValue     var word
i               var byte

'   Initialization/Mainline
  low LeftServo: low RightServo: low
DummyServo
  RandomValue = 1000
  i = 1
  do
    i = i - 1
    if (i = 0) then
      random RandomValue
      select ((RandomValue / 4) & 3)
```

```
      case 0:
        LeftServoVal = LeftForward
      case 1:
        LeftServoVal = LeftForward
      case 2:
        LeftServoVal = LeftStop
      case 3:
        LeftServoVal = LeftBackward
    endselect
    select ((RandomValue / 16) & 3)
      case 0:
        RightServoVal = RightForward
      case 1:
        RightServoVal = RightForward
      case 2:
        RightServoVal = RightStop
      case 3:
        RightServoVal = RightBackward
    endselect
    i = ((RandomActive & 3) + 1) * 120
    else
      pause 4
    end if
    pulsout LeftServo, LeftServoVal
    pulsout DummyServo, 1125 -
LeftServoVal
    pulsout RightServo, RightServoVal
    pulsout DummyServo, 1125 -
LeftServoVal
'   Statements Above Take 11 msecs in total
    pause 9
  loop
```

The application should be quite easy to follow; the same pulse will be sent to the servos unless the counter ("i") is decremented (one taken away from it) to zero, at which point random movement values will be saved in the variable assigned with the servo's pulse width. In this application, I have assumed that full-speed rotation will take place with a 1,200 μs or 1,800 μs pulse. The value that stops the pulse will have to be found using the calibrate application.

To test out this application, I just wired the two servos according to Figure 16-19 and then loaded in the application. I didn't feel that a schematic was necessary for this application because of its simplicity and reliance on the book PCB. When downloading the application, remember to keep the servo power switch off as the robot may start moving immediately following download of the application code.

To test how well I understood the timing requirements for this application, I looked at the servo outputs supplied to the two servos, as well as the "DummyServo" using an oscilloscope (Figure 16-20). The total loop time is almost exactly 20 ms, which is exactly what is required by the servos. In case you are suspicious, I can honestly say that I did not "cook" the results shown in Figure 16-19. Although I did spend a bit of time figuring how long each section of

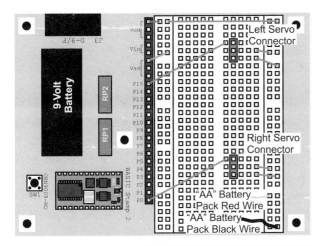

Figure 16-19   *Multiservo wiring*

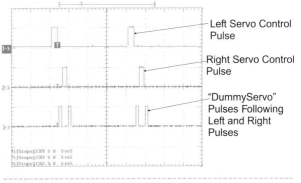

Figure 16-20   *Multiservo scope*

the loop would take to execute, I really lucked out when I combined them and ended up with the ideal 20 ms loop. Even if you are within 10 percent of the

specified loop time, your application should still run fine.

When you have finished with this experiment, don't disassemble it. The servo platform will be required for the next two experiments.

Experiment 115
Robot Artist

Parts Bin

Assembled PCB with
    breadboard and BS2

Assembled plywood base
    with servos and four-
    AA-battery pack and
    power switch

Elastic band (see text)

Magic Marker

Tool Box

Wiring kit

Carpenter's protractor
    (see text)

Machines for making art or patterns have been manufactured over the past two or three hundred years. This is probably surprising to you because you can't think of any valuable works of art that have been made by machine. Actually, to find an example of machine-drawn art that has been in use for centuries, you don't have to look any further than your wallet. The swirls, loops, and patterns were made by a machine built from a collection of gears and cogs that scratched the pattern on the printing plate used for the bill. As a young child, you may have played with a toy called a "Spirograph," which was a collection of gears that you could pin to a piece of paper with one

free gear that you would put a pen tip through. The spirograph worked on exactly the same principle as the etching of the patterns on a bill; the gears of the spirograph would put the pattern on a sheet of paper in an almost identical (but slightly different) pattern.

You can replicate this pattern with a piece of graph paper, a pencil, and a ruler, as I have shown in Figure 16-21. To make the curve that you see in this picture, I started with two lines 90 degrees apart and drew a line from one extreme to the origin and moved along by a few millimeters. The resulting series of intersecting lines appears to describe a curve. This figure could be repeated with different

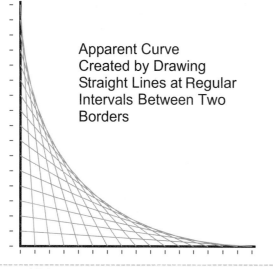

Apparent Curve
Created by Drawing
Straight Lines at Regular
Intervals Between Two
Borders

Figure 16-21   *Line curve*

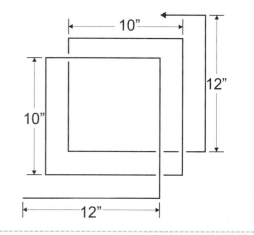

Figure 16-22   *Proposed robot art*

length lines with different angles. When I was a kid, a popular hobby was to place a series of nails into a piece of wood and then run string between the nails, trying to see what kind of patterns we could come up with.

If you get a chance to go to one of the large robot competitions, you will have the opportunity to watch (and maybe compete in) the robot art competitions. In these competitions, robots are programmed to create their own patterns either as a preestablished pattern or as a response to the environment they are operating in. If the robot is programmed to respond to the environment, it uses either light sensors (usually in the form of CDS cells) or a microphone and amplifier.

For this experiment, I wanted to see how close I could get a robot to draw a repeating pattern like the one shown in Figure 16-22. In this pattern, the robot moves in repeating squares with two sides longer than the other two. For the robot to draw the line, I simply wound an elastic band around the rear supports on the servo robot base and used it to hold a Magic Marker.

In terms of the robot itself, I used the same circuit for controlling the two servo motors as was presented in the previous experiment. Before attempting to create the program, I had to come up with some way of determining how long the robot would have to run in order to move 12 inches, then move 10 inches, and turn 90 degrees. The program I came up with for doing this is listed here and has the stop value that I

found I required for the servos that I was using with the robot base. This program delays 5 seconds before moving for 200 20-millisecond loops (which take 4 seconds). The program can by restarted simply by pressing the reset button on the PCB.

```
'   Servo Distance Calibrate - Figure Out
Robot Speed/unit time
'{$STAMP BS2}
'{$PBASIC 2.50}

'   Variables
LeftServo        pin 15
RightServo       pin 0
DummyServo       pin 1
LeftServoVal     var word
RightServoVal    var word
LeftStop         con 777
RightStop        con 770
LeftForward      con 600
LeftBackward     con 900
RightForward     con 900
RightBackward    con 600
i                var word

'   Initialization/Mainline
  low LeftServo: low RightServo: low
DummyServo
  LeftServoVal = LeftForward
  RightServoVal = RightForward
  i = 200
  pause 5000
  do while (i <> 0)
    i = i - 1
    pulsout LeftServo, LeftServoVal
    pulsout DummyServo, 1125 -
LeftServoVal
    pulsout RightServo, RightServoVal
    pulsout DummyServo, 1125 -
LeftServoVal
'   Statements Above Take 7 msecs in total
    pause 13
  loop
  end
```

After the robot stopped, the distance traveled was recorded, and the operation was repeated five times to get the average and used a set number of 20 ms loops before stopping. I repeated the operation five times and averaged the resulting value. Table 16-2 lists the five values as well as the average.

This average distance was then divided into 200 to get the number of cycles per inch (7.1 cycles per inch). To go 12 inches, I would require 85 cycles, and to go 10 inches, 71 cycles are required.

Measuring the angle required the same program (but with "LeftServoVal" loaded with "LeftReverse" rather than "LeftForward," and the number of cycles it turned was reduced to 25 so that the turn would be less than 90 degrees). Before starting the program, I aligned the robot to a sheet of paper, ran it, and measured the difference in direction between the starting angle and the finishing angle. Amazingly enough, the robot turned 90 degrees consistently with 25 servo cycles.

So, with the number of cycles for moving 12 inches and 10 inches, as well as turning 90 degrees, I was ready to test out the program to see how well it could draw what I was calling a "wandering square."

The program I came up with to draw the wandering squares follows. Note that after moving, I stopped the robot for a quarter-second. This was because the turning angle began and ended with a stopped robot—the acceleration of the robot in the term was part of the value. Rather than using a table to determine what the robot was to do next, I calculated the responses algorithmically. It actually has 16 positions; each even one is a stop, and each one that is equal to one after the step number modulo 4 is found is a straight line. Finally, if it is a straight line movement and the step number is greater than 8, then it is a

Table 16-2 Robot distance travelled in five seconds

Trial	Distance
1	28.00"
2	28.13" (28 1/8")
3	28.38" (28 3/8")
4	28.13" (28 1/8")
5	28.25" (28 1/4")
Average	28.178"

short line. Looking through the program, you can see that I went to some lengths to make sure that the loop cycle time was 20 msecs.

```
'   Robot Artist - Try to draw the
"Wandering Square"
'{$STAMP BS2}
'{$PBASIC 2.50}

'   Variables
LeftServo        pin 15
RightServo       pin 0
DummyServo       pin 1
LeftServoVal     var word
RightServoVal    var word
LeftStop         con 777
RightStop        con 770
LeftForward      con 600
LeftBackward     con 900
RightForward     con 900
RightBackward    con 600
i                var word
j                var word
StepNum          var byte

'   Initialization/Mainline
  low LeftServo: low RightServo: low
DummyServo
  LeftServoVal = LeftStop
  RightServoVal = RightStop
  i = 250
  StepNum = 0
  do
    i = i - 1
    if (i = 0) then
      StepNum = (StepNum + 1) // 16
      j = StepNum // 2
      if (j = 0) then
        LeftServoVal = LeftStop
        RightServoVal = RightStop
        i = 25
        pulsout DummyServo, 250
      else
        RightServoVal = RightForward
        j = StepNum // 4
        if (j = 1) then
          LeftServoVal = LeftForward
          if (StepNum > 8) then
            i = 71
          else
            i = 85
          endif
        else
          LeftServoVal = LeftBackward
          i = 25
          j = 16
        endif
      endif
    else
      pulsout DummyServo, 875
    endif
    pulsout LeftServo, LeftServoVal
    pulsout DummyServo, 1125 -
LeftServoVal
    pulsout RightServo, RightServoVal
    pulsout DummyServo, 1125 -
LeftServoVal
'   Statements Above Take 8.5 msecs in
total
    pause 11
    pulsout DummyServo, 125
  loop
```

The result of this program is shown in Figure 16-23, and you should notice two things in the photograph. The first is the funny arc that is drawn at each curve. The pen is not located at the center of the turn, so it follows the path of the part of the robot that it is attached to. Based on my previous experience with robots, this was not unexpected.

The second problem is that although I measured 90 degrees using the "Servo Distance Calibrate" program for 25 cycles, it obviously wasn't correct for the actual application. The actual turn should have been something less than 25 cycles.

Despite these two issues, I believe that the resulting figure is actually quite attractive, and if it were allowed to continue for more than four circuits, it could have produced quite an interesting figure. If you are interested in true precision, then you would want to add a compass to the robot (to make sure the turns are absolutely sharp) along with some way of measuring distance, and put the pen at the center of the robot's turn. Even without these extras, the picture the robot produced after just a few hours of

Figure 16-23 *The robot did not produce the expected "wandering square" but came up with its own unique pattern that could definitely be called art.*

work is quite interesting, and I'm sure that when you repeat this experiment, your robot will come up with something just as unique and interesting.

Experiment 116
Parallax's "GUI-Bot" Programming Interface

Parts Bin

Assembled PCB with breadboard and BS2

Dual servo robot base with four-AA-battery pack with power switch connected

Two radio-control (R/C) servos

Three-pin R/C servo connectors (see text)

Two 10k resistors

Two microswitch whiskers with long actuating arms

Solid core 24-gauge wire

Tool Box

Wiring kit

Five-minute epoxy

Clippers

Wire stripper/knife

Soldering tools

If you have been looking at different hobbyist robots, then I'm sure that you will be familiar with the Parallax BOE-Bot, which is a dual servo robot controlled by a BS2, and it has some built-in sensors and a vari-

ety of different methods for output. The robot is built around Parallax's Board of Education PCB, which includes a BS2 socket, power input, a breadboard, and interfaces to servos and Parallax's line of App-

Mod BS2 adapters. The PCB included with this book wasn't based in a large measure on the Board of Education PCB to allow you to take control of the BOE-Bot software and development tools like the GUI-Bot.

The servo differentially driven BOE-Bot is an excellent way to get into robotics for people who don't want to do the cutting, drilling, gluing, painting, and parts finding that are required for the robots presented in this book. Along with the parts necessary to build the robot, the kit comes with an excellent manual (written by BASIC Stamp expert Jon Williams), along with a wide variety of parts that will allow you to learn more about electronics and the BASIC Stamp 2.

One feature of the BOE-Bot that makes it very attractive for beginners is the GUI-Bot software that is designed for use with the robot. This tool will allow you to create applications for the BOE-Bot that will have it running around and performing basic operations in no time. In this experiment, I would like to show you how easy it is to work with the GUI-Bot software and give you some idea of how graphical programming can be implemented for robots.

Before you can work with GUI-Bot, you must have a BOE-Bot equivalent robot. For this case, I will show you how to make a robot that is functionally equivalent to the BOE-Bot using the R/C servo base that you built earlier in this section. This base can be used almost exactly as built; you will just have to add two microswitch whiskers, similar to the ones you added to the DC motor base. The microswitches should have wires soldered to them so that when the switch actuators are pressed, the connection is closed. Once the wires are soldered to them, the microswitches can be soldered to the servo base as shown in Figure 16-24.

The wiring necessary to convert the PCB and robot into a BOE is quite simple. Figure 16-25 shows the circuit elements that are necessary and how they are wired together. The wires connecting P0 and P8 to Vcc are needed to simulate an infrared remote control receiver, which is used on the BOE-Bot as an object detector. For this robot, just the microswitch whiskers are used for object detection.

Once you have the robot completed, you can download the "GUI-Bot" from the Parallax web site

Figure 16-24 *Servo base with microswitch whiskers added as object sensors to make the servo base compatible with the BOE-Bot*

(www.parallax.com) and install it the same way as you did with the BASIC Stamp Windows Editor Software (left-click on the link on the web page and open the application to install it). I suggest that you open and print out the readme file. The first time you open the application, click on "Beginner Mode" of operation, at which point you will be greeted with a dialog box similar to Figure 16-26.

The big difference between Figure 16-26 and what you will see when you first run the application is the program that I have entered under "Actions to Be Performed." The seven actions that I have written out cause the robot to describe a right-angle, triangle as it moves. The robot goes forward for three seconds, turns, goes forward, and so on until it finishes, at which point it jumps back to the start to try again.

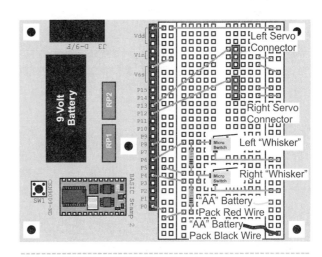

Figure 16-25 *BOE-Bot circuit*

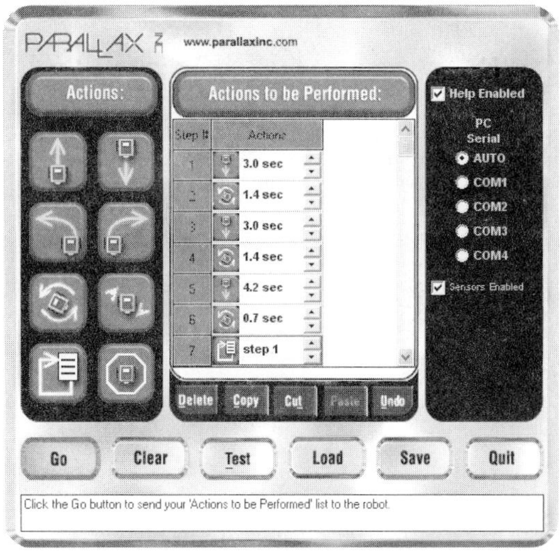

Figure 16-26   *Simple GUI-Bot*

You will have three issues with the GUI-Bot software when you first use it. The first is learning how long actions should be performed for. This is found empirically (also known as trial and error). To get the correct timings for the triangle, I spent about a half-hour getting the values so that the robot would move about reasonably correctly. This isn't a difficult process, but it means you will run back and forth to your computer as you work on your program.

The second issue is that when you stop the robot for any length of time, you will find that the servos will be turning because the "Stop" value isn't correct for the actual servo hardware. To eliminate this, click the Test button on the GUI-Bot Interface. This brings up the Test Mode dialog box, which can be used to test your whiskers as well as calibrate the stop position of the servos that you are using. This is the same function as the PBASIC program I presented earlier for finding the stop position of a servo. When you have adjusted the application so that the servos are stopped, click "Calibrate Servos" to save this information in the application.

The last issue is that once you program the BS2 with your program (by clicking Go), you will discover that the robot will start moving. To prevent this from happening, I strongly recommend that you turn off the power from the four AA batteries to the servos (this is why I keep the BS2 and motor power separate) and suspend the robot with the wheels not touching anything. Doing both is probably overkill, but hopefully you'll remember to carry out at least

one of these operations and not watch your robot run off a table and smash on the floor while under program control.

If you would like to test out the robot on the floor, press the reset button on the PCB and then turn on the servo power. After the BS2 is reset, you have a couple of seconds before the program starts executing.

Once you have run a few basic programs commanding your robot to move about the floor, you are ready for the Advanced Mode, which incorporates the whiskers and other sensors as part of the program. Quit the GUI-Bot application, restart it, and select, "Advanced Mode" and you will see something like Figure 16-27, which is a simple wall follower and to implement it, I added some wire to the right whisker so that the microswitch would be closed when it was touching a wall.

In this program, the robot normally turns right (into the wall), but if the right sensor detects an object, it turns away. This is a very simple application and not one that is very useful because it can only circle an object in a clockwise direction. To develop an application, you first set down a nominal program and then create smaller programs (called sets) that are responses to different sensor actions. In this example, Set1 turns the robot in the opposite direction if the sensor detects the wall. You will find creating Advanced Mode applications to be fairly easy, but you will find it quite limited when you want to monitor the operations of your robot (by LEDs, LCDs, or audio), and it will not allow you to implement applications that require variables or advanced decision making.

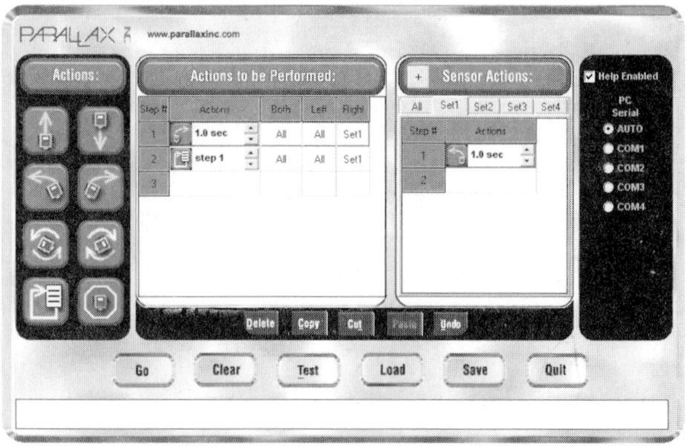

Figure 16-27   *Advanced GUI-Bot*

Experiment 117
Stepper Motor Control

Parts Bin

Assembled PCB with breadboard and BS2

Four-AA-battery pack

Five-volt bipolar stepper motor

Four-pin stepper motor/breadboard connector (see text)

Eight 1N4148 (1N914) silicon diodes

Paper

Tool Box

Wiring kit

Scissors

Krazy Glue

A number of different chips will interface to a stepper motor and update its position when a clock is passed to it. These chips usually have three inputs, an output enable, direction, and clock (which initiates a change in the position of the stepper motor). These chips normally do not connect directly to the stepper motor itself because of the differing voltage and current ratings of different motors; instead their output is passed to a driver circuit that has been designed for the motor that is being controlled. These chips are quite easy to work with, but they are really not necessary, as I will show in this experiment.

When I introduced stepper motors in the book, I neglected to indicate that two different types are commonly used. The unipolar stepper motor consists of a motor with four coils, wired as pairs with a common center connection. The unipolar stepper motor's drive electronics are very simple, as I show in Figure 16-28. To energize a coil of the unipolar's stepper motor, its connection is simply pulled to ground using a transistor.

Although the control wiring of the unipolar's stepper motor is quite simple, it does not provide as much torque as the bipolar stepper motor (which was demonstrated earlier in the book). This motor has four coils, just like the unipolar stepper motor, but it does not have the center connection and has to be driven using half-H-Bridges as shown in Figure 16-29.

I prefer working with the bipolar stepper motor because of the extra torque it provides for small robots (two coils are always active in the X and Y direction instead of the single coil of the unipolar stepper motor). To drive a stepper motor, I use the 754410 motor driver, and to demonstrate its operation, I would wire it to a BS2 as shown in Figure 16-30. The control code for turning the bipolar stepper motor is surprisingly simple and is listed here. After building the circuit, glue a paper arrow onto the stepper motor (just like in the earlier experiment) so that you can see the motor turning easily.

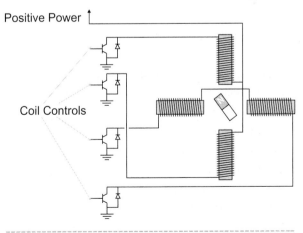

Figure 16-28   *Unipolar stepper motor control*

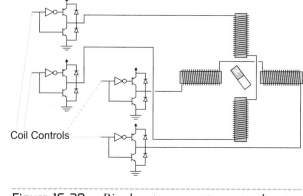

Figure 16-29   *Bipolar stepper motor control*

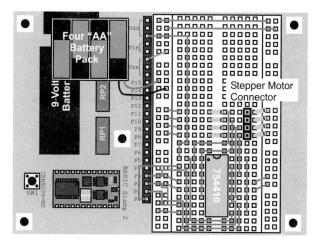

Figure 16-30   *Stepper control circuit*

When you have the circuit built, enter the application code listed and save it as "Stepper Motor Control" in a Stepper Motor folder located in the Evil Genius folder.

```
'   Stepper Motor Control - Turn the
Bipolar Stepper Motor
'{$STAMP BS2}
'{$PBASIC 2.50}

' Variables
MotorDIRS            var DIRS.NIB0
MotorCtrl            var OUTS.NIB0
i                    var byte

'  Initialization/Mainline
  MotorCtrl = %0000
  MotorDIRS = %1111
  i = 0
  do
    pause 100
    lookup i, [%0111, %0101, %0001, %1001,
%1000, %1010, %1110, %0110], MotorCtrl
    i = (i + 1) // 8
  loop
```

These lines of code are deceptively simple, and I should point out a few features to you. The first feature to note is that I take advantage of the lookup PBASIC statement to convert a countervalue to commands for the motor coils' half-H-Bridge drivers and save the value directly in the BS2's I/O pins. PBASIC's capability to set labels as different I/O bits can simplify and speed up your application when used in situations like this. Second, after incrementing the position counter ("i"), I find its modulo with 8. This keeps "i" in the range of 0 to 7, and although I normally eschew placing multiple operations in a single PBASIC statement, this is a case where the left-

to-right order of operations results in a statement that is easily readable. For controlling the motor speed, I have entered the "pause 10" statement, and this statement's value can be changed to alter the speed of the motor. I started with a delay of 100 msecs (using the "pause 100" statement), and the motor I used made a full revolution in two or three seconds. With a 10 ms pause, the pointer glued to the motor's output shaft becomes a blur.

I should make a few points about stepper motor movement. The first is that every change in the coils moves the motor just a few degrees (the full eight-coil values will not turn the motor a full revolution), so the actual turning speed of the motor is significantly less than the update speed. If a motor moves 1.5 degrees per step and you are changing the steps 1,000 times per second, this works out to about 4 turns per second. Also note that the stepper motor armature can only turn so fast; the motor will not be able to respond to a certain rate of coil change, and it will result in the motor stopping (although the shaft will probably vibrate back and forth very rapidly). The maximum speed of the stepper motor is usually specified by the manufacturer and should never be exceeded.

In Table 16-3, I have reviewed three different types of motors for driving mobile robots. It covers their different features so that you can best decide which one to use for your robot.

Table 16-3 Characteristics of different motors used in robots

	DC Motor	Stepper Motor	R/C Servo
Size	All ranges.	All ranges.	Generally small robots.
Torque	Fair to good depending on gearing.	Very good. Also will hold position well.	Good. High torque servos available.
Battery consumption	Good. PWM can be used to lessen current consumption.	Poor. At least one coil energized at any time.	Fair. Servo electronics can require relatively high current while motors idle.
Speed	Fair to very good depending on gearing.	Fair but usually good enough for robot applications.	Poor to fair but usually good enough for robot applications.
Ease of physical installation in robot	Can be difficult. There are some kits (notably by Tamiya) that can make installation of DC motor and drivetrain easier.	Fairly easy. Stepper motors often have mounting flanges to facilitate bolting the motor into the robot.	Very easy. Servos can be mounted using either two-sided tape or lugs built into servo case.
Ease of controller programming	Easy to difficult depending on requirements and controller. Can be very difficult to implement PWM in some controllers.	Quite easy.	Fair. Implementation of timed pulses for the servos can be difficult, especially if servos do not maintain current position if no pulse train sent.
Ease of position measurement	Odometry sensors must be added to robot.	Each change in coil values results in an easily determined robot movement.	Odometry sensors required.
Scalability	Good, larger motors (with different drivers) can be used for "growing" applications.	Very good. Motors and drivers can be changed easily with little impact to software.	Very difficult once robot is larger than what the standard servos can drive easily.
Hazardous environment usage	Poor. Chance of arcing/sparks within motor make DC motor not recommended for this type of application.	Very good. No changing connections in motors.	Good if cases have gas-tight seals.
Cost	Low although drivetrain and gearing may be costly.	Moderate. Note that stepper motors are normally geared so drivetrain gearing normally not required.	Low to high. Low-cost servos are competitive with other motor solutions, but do not have ball-bearing outputs, metal gears, and other features desired for robot applications.

Experiment 118
Infrared Two-Way Communications

Parts Bin

Assembled PCB with breadboard and BS2

Separate BS2

Separate breadboard

Nine-volt battery and power clip

Two 555 8-pin timer chips

Two 74LS74 dual D flip flops

38 kHz IR TV remote control receivers

Two IR LEDs

Two visible light LEDs, any color

Two 670 Ω resistors

Four 470 Ω resistors

220 Ω resistor

Two 100 Ω resistors

Two 10 µF electrolytic capacitors

Six 0.01 µF capacitors, any type

Tool Box

Wiring kit

A good portion of this book has been devoted to showing you how to interface output devices (such as the LEDs, speakers, and LCDs) to the BS2 so that you can get some kind of feedback as to what the robot is sensing and what decisions are being made. As I have been told often by people that work on robots professionally, it isn't unusual to see an engineer following behind the robot reading the LCD on the top of the robot. I don't like monitoring the performance of a robot for the simple reason that there is the question of whether or not the robot's developer is part of the environment the robot is working in. The question has to be asked, will the robot operate and respond the same way in the same environment if the developer is or isn't present?

Unfortunately, for many robots, the answer to this question is no. I have been to more than one show where a robot engineer explains that his creation works perfectly, only to discover that when they put it on the floor it either does nothing or something that is completely unexpected and undesirable. The developer usually wanders away cursing or tries the operation again, this time hovering over it like an expectant father, only for the robot to behave correctly. It's disappointing, but very few people realize what is happening. I liken this situation to Heisenberg's Uncertainty Principle working at the macro (rather than micro) level; the developer observing the robot becomes part of the robot's environment and affects its operation.

This is why I like to design my robots with bright LEDs and audio output so that I can observe the behavior of the robot as well as monitor its inputs and decision making remotely. To facilitate this, I will put LEDs on the robot's motor controllers, any object sensors, and line sensors, so that I can observe from a safe distance what is happening. Of course, situations occur where more is happening than can be comfortably observed, so I need some method of sending and receiving data to the robot.

The most obvious way of doing this is to use a radio receiver and transmitter. The problem with this method is that it is quite costly, and it can be difficult to get two-way communications working. Instead of RF, I went back to infrared and came up with the circuit shown in Figure 16-31 for providing two-way communications between robots. As I indicated previously, R/C infrared communications are accomplished using Manchester encoding, which cannot be easily implemented into the BS2, so I decided to see how well standard *Nonreturn to Zero* (NRZ) communications would work in this environment.

NRZ communications are known colloquially as asynchronous serial communications. The data format is shown in Figure 16-32 and consists of an initial low start bit, followed by the data, an optional parity error detection bit, and finally the high stop bit. The high stop bit is where asynchronous communications gets its name—at the end of each packet of data (in which there is one byte) the line is returned to a high voltage level. The most popular data format is 8-N-1, which indicates that the packet contains eight data bits, one stop bit, and no parity. Data speeds range from 110 *bits per second* (bps) to 115,200 bps in modern PC systems. The BS2 works best at data speeds ranging from 300 bps to 4800 bps (other models of BS2 can comfortably handle higher speeds).

You may be familiar with the term RS-232 as the electrical standard for sending asynchronous serial data. RS-232 is something of a strange electrical standard that goes back to the earliest days of telegraph communications. I will discuss RS-232 and show how you can interface the BS2 to a true RS-232 device, but for this experiment, I will just use standard CMOS/TTL levels to and from the IR LED and TV R/C receiver.

This experiment will consist of a BS2 that interfaces to your PC via the debug dialog box. It will send the key character that you press to a remote BS2 via modulated infrared light as shown in Figure 16-31. To modulate the infrared light, I will use our old friend the 555 timer chip to produce a stable timebase for the circuit. When you are working with the IR remote controls, you will discover that it is very sensitive to the duty cycle of the received IR light. To make sure that I get as close to a 50 percent duty cycle as possible, I have set up the 555 timer to produce a frequency twice of what I need. Then I have divided it by two using a D flip flop wired as a T flip flop. Both the master and slave devices are wired according to Figure 16-33.

I take advantage of the IR LED's operation and use the BS2 to connect it to ground while the 555 and 74LS74 provide the positive voltage for its operation.

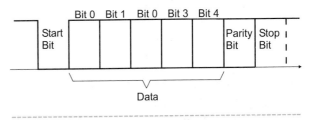

Figure 16-32   *Asynchronous data*

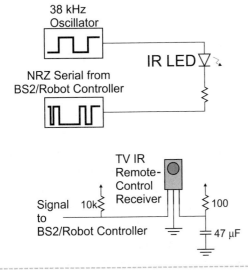

Figure 16-31   *IR serial communications*

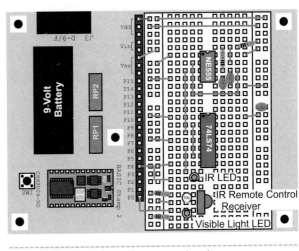

Figure 16-33   *Asynchronous circuit*

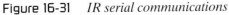

This positive voltage is modulated, so when the asynchronous data is sent from the BS2, the LED's output is also modulated. You should be aware of two other features in this circuit. The first is that the IR R/C receiver produces a low voltage on its open collector line when data comes in that is identical to the signal sent (and expected) by the BS2. Secondly, that I have put a 470 Ω resistor and visible LED on the IR R/C's output. This was done to give a visual indication when data is being received by the IR R/C receiver.

When I first built the circuit, I used two prototype book PCBs that I had on hand (the circuit is identical for both the master and the slave). Although it's somewhat unreasonable to expect you to get two copies of this book to get two PCBs, by getting a simple, long PCB, you can repeat the wiring with a BS2 at one end with a 9-volt battery and clip. The 555 and 74LS74 are powered by the 5-volt regulator on the BS2.

With the hardware designed and in place, I came up with the following program to send out an infrared ping packet with a character that you enter from the debugin statement. Save this program as "Master Comms:"

```
'    Master Comms - Send "Ping" to Slave
BS2
'{$STAMP BS2}
'{$PBASIC 2.50}

'  Variables
SerialOut        Pin 4
SerialIn         Pin 0
i                var byte
j                var byte
Retn             var byte(5)
Flag             var bit

'  Initialization/Mainline
  high SerialOut
  do
    debugin str i\1
    j = i ^ $ff
    debug "Sending Character '", str i\1,
"', Hex $", hex i, cr
    serout SerialOut, 3313, ["Ping ", str
i\1, str j\1, cr]
'     serout SerialOut, 3313, [str i\1, str
j\1]
    pause 50
    Flag = 0
'  Use Flag to Indicate Response
    serin SerialIn, 3313, 1000,NoResponse,
[WAIT("ACK"),str Retn\5]
    Flag = 1
'  Indicate Data Found
NoResponse:
'  Timeout - No Response
```

```
    if (Flag = 0) then
      debug "No Response from Remote", cr
    else
      j = "N" ^ $FF
      if (Retn(0) = "N") and (j = "N")
then
        debug "Message not properly
received", cr
      else
        j = Retn(2)
        debug "Response to Message was '",
str j\1, "'", cr
      endif
    endif
  loop
```

This application waits for an "ACK" message from the following "Slave Comms" program and responds accordingly.

```
'  Slave Comms - Responde to "Ping"
'{$STAMP BS2}
'{$PBASIC 2.50}

'  Variables
SerialOut        Pin 4
SerialIn         Pin 0
i                var byte
j                var byte
k                var byte
Message          var byte(3)

'  Initialization/Mainline
  high SerialOut
  do
    serin SerialIn, 3313, [WAIT("Ping "),
str Message\3]
    pause 75
    i = Message(0): j = Message(1) ^ $FF
    if (i = j) then
      i = i + 1
      j = i ^ $FF
      k ="Y" ^ $FF
      serout SerialOut, 3313, ["ACKY", str
k\1, str i\1, str j\1, cr]
    else
'  Bad Message Received
      k = "N" ^ $FF
      serout SerialOut, 3313, ["ACKN", str
k\1, str i\1, str j\1, cr]
    endif
  loop
```

Slave Comms, if the ping is received properly, responds with an "ACKY" message and returns the byte passed to it after incrementing it. In both programs, note that when I send a data byte, I follow it by its "1's complement" (each bit inverted using the XOR operator) as a checksum. In Master Comms, if no response is received within a second or data is corrupted, it prints out an error message. To make these operations simpler, note that I have taken advantage of the PBASIC "serin" statement's "WAIT" and "TIMEOUT" parameters.

I found that with this simple setup, the two BS2's could communicate successfully across the length of my basement. I suspect that in the real world, you may want to have multiple IR LEDs and IR R/C receivers pointing in different directions in an effort to have a direct line between the transmitter and operating receiver. This shouldn't be a problem by putting LEDs and receivers in parallel (the open collector output of the receivers makes doing this quite simple).

Section Seventeen

Navigation

One of the most difficult things to implement with a mobile robot is giving it the ability of knowing where it is. Robots do not naturally have our capability of being able to look around and find out where they are. Humans naturally develop the ability to sense where they are relative to other objects using sight and sound. When you first look at adding this capability to a robot, you will probably be stumped with no idea from what direction to attack the problem. Anytime you are faced with a problem that you don't know how to solve, I recommend that you look at situations where people have solved similar problems historically.

I am saying "historically" because with the advent of technology, we rely on very technologically complex solutions that can be difficult to implement in a small robot. An example would be using the *Global Position System* (GPS) satellites in Earth's orbit; they can be used for navigating a robot to within just a few feet (less than a meter), but they require an unobstructed view of the sky and can be costly. GPS is an incredibly useful tool as it will return the current position (in terms of latitude and longitude as well as altitude) and velocity (with direction). Unfortunately, GPS is somewhat impractical for the robots that you will start out working with. Because GPS is a relatively new invention and people have been traveling all over the world for centuries, the question that you should be asking is, how did people do it without becoming lost?

A situation where people, like robots, did not have any reference points, yet had to navigate is on the oceans. As you are no doubt aware, locations on Earth are specified by the use of latitudes and longitudes as shown in Figure 17-1. *Longitudes* are lines that run from the North Pole to the South Pole and are numbered from -180 to 180. Zero latitude is

Greenwich in Great Britain, and latitudes increase in number as you travel to the east, each unit of latitude being one degree. *Latitudes* are concentric circles starting at the equator (zero longitude) and extending to the poles as either north longitudes or south longitudes to a maximum value of 90, and they are given the units of degrees like latitude.

Early navigators found their position on the Earth using three tools. The first was to use a very accurate clock (called a *chronometer*) kept at Greenwich Mean Time or Zulu Time ("Zulu" being the phonetic name for the letter "Z," the first letter of *zero*). To find the current latitude, when the sun reached its highest point in the sky (noon), the current time was recorded and the latitude was calculated as the difference between when noon took place at Greenwich and when it took place for the navigator. Because there are 360 degrees of latitude and 24 hours in a day, then there are 15 degrees for every hour of difference between Greenwich noon and local noon.

The second tool was the compass. Aside from being important for keeping track of the direction of travel, it was needed to determine when the sun was at its highest point in the sky. Noon can be defined as the time when the sun is in a direct line with the navigator and North.

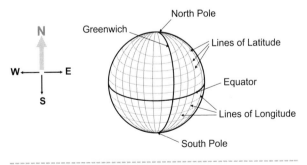

Figure 17-1   *Earth mapping*

The last tool was the *sextant*. This device was built from a collection of protractors and mirrors that were used to measure the position of the sun (or stars) over the horizon. Longitude could be calculated as 90 degrees with the angle of the sun when it is at its highest point in the sky subtracted from it.

The use of three tools can be seen in Figure 17-2, in which the angle of the sun at its highest point in the sky is used with knowledge of the time at Greenwich and the direction of Earth's North Pole. This method is a simplified model for celestial navigation and is *much* simpler than what is required for actual navigation; I have ignored the fact that the Earth's axis is tilted with respect to its orbital plane around the sun. Along with this, I have also ignored the fact that the magnetic North Pole is a surprisingly large distance away from actual point the Earth turns about.

From this example, you can take the idea that three angles are required to find your location in space. This can be used by robots in the situation like the one I have shown in Figure 17-3. Three lights can be placed outside the robot's expected movement area or envelope. By continually sighting the three points, measuring the angles between them, and knowing their actual positions, you can calculate your actual position with a great deal of precision quite easily. As I have laid out the movement envelope in Figure 17-3, the robot can only navigate within a certain area in which the three lights can be observed easily.

In World War II, navigating bombers to their targets in Europe was a major problem. This was due to the often inclement weather the bombers would have to fly in, the need to fly at night (because they were easy targets during the day), and the expense and time required training the thousands of navigators that were needed. Even if navigators were available, the time required to calculate the aircraft's position from reading the positions of the stars was a problem because by the time the calculations were finished,

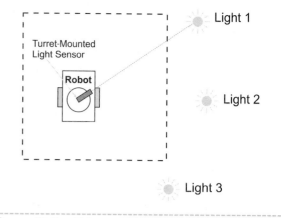

Figure 17-3   *Trig navigation*

the bomber could conceivably be a good distance away from the point where the measurements were made. The solution to this problem was quite inspired. Using radio-direction-finding equipment already aboard the bombers, powerful transmitters were set up, and the bombers were instructed to keep the transmitter behind them while flying at a continuous heading as shown in Figure 17-4.

Robots can use a similar system by marking a black line on the surface they run on that is sensed by an infrared LED and phototransistor combination as I showed earlier in the book.

Another method of navigation is taken from how a blind person walks through a room. In this case, rather than trying to avoid objects, the blind person tries to find them to help understand his or her location within the room. A cane is used as a touch sensor to identify objects in front of the person to help him or her navigate. This method can be implemented similarly in a robot by moving it from object to object

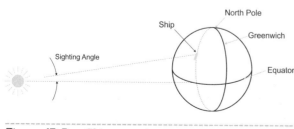

Figure 17-2   *Ship mapping*

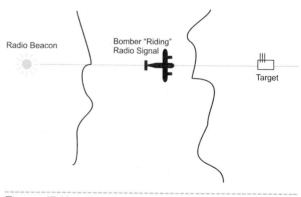

Figure 17-4   *Bomber navigation*

as shown in Figure 17-5. Normally, the robot's object sensors are used to help it avoid objects, but in this case objects are actively searched out, a complete reversal of the purpose you would expect for the object sensors.

Many other historical methods of navigation can be used in a robot. In the previous examples, I did not mention dead reckoning, in which the direction of movement is known, as well as the speed of movement and the length of time since leaving a known point. To get the current position, the time is multiplied by the speed with the product being the distance traveled.

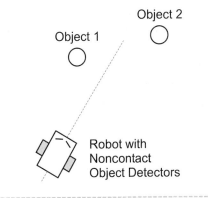

Figure 17-5   *Blind object detection*

Experiment 119
Line-Following Robot

Parts Bin

Assembled PCB with breadboard and BS2

DC robot motor base with four-AA-battery clip and switch

Two Opto-Interupters mounted on sheet metal from experiment 48 with infrared optointerrupter halves glued to it

LM339 quad comparator

Two ZTX649 NPN bipolar transistors

Two XTX749 PNP bipolar transistors

Two LEDs, any color

Two 100k resistors

Six 10k resistors

Two 470 Ω resistors

Two 100 Ω resistors

Two 10k breadboard-mountable potentiometers

Twenty-two-inch by 28-inch white Bristol board with path marked on it (see text)

Tool Box

Wiring kit

I've tried not to repeat experiments in this book, but I would like to revisit one case, and that is the line-following robot that I first presented in the "Optoelectronics" section of the book. In this experiment, I created a simple methodology for following a line; each wheel would be controlled by a sensor on its side. When the sensor detected white underneath it, the motor would be turned on, and when the sensor detected black, the motor would be turned off. By doing this, if a sensor on one side detected black beneath it, the robot would turn away from the black section by shutting off that motor until it was back in the white. With a bit of work, you probably came up with a set of values that worked for your robot, and it ran over the path quite well except in one case: when both sensors are over a black section of the path. In this case, the robot would just stop there. The solution to this problem is actually quite simple: If both sensors are over black, then the robot should move forward until one of them is over white, at which point the robot could turn to the sensor that was still over black, hopefully center itself, and continue going over the path. The problem with implementing this strategy back in the "Optoelectronics" section was that I had not yet discussed the concept of digital logic and decision making, so I had to go with this simple strategy.

With digital logic, I could have created a truth table in Table 17-1 to plot out how the robot should move.

In the table, I have assumed that when a "1" is returned, the sensor is over a white part of the paper, and a "0" indicates a black. Using this table data, I could have determined that the left motor would be controlled by the formula:

$$\texttt{Left Motor = !(!Left \cdot Right)}$$

and the right motor by the formula:

$$\texttt{Right Motor = !(Left \cdot !Right)}$$

Table 17-1 Wall-following robot logic truth table

Left Sensor	Right Sensor	Left Motor	Right Motor
White (1)	White (1)	On (1)	On (1)
White (1)	Black (0)	On (1)	Off (0)
Black (0)	White (1)	Off (0)	On (1)
Black (0)	Black (0)	On (1)	On (1)

This could have been implemented using a single 74C00 (Quad dual-input NAND gate), but I decided to use the BS2 to implement this function instead of using digital logic. The reasons for going with the BS2 were to take advantage of the delay ("pause") and *pulse width modulation* (PWM) functions built into the controller, as well as its easy programmability to allow for smoother and more accurate operation.

The BS2-based line-following robot takes advantage of the two optointerrupters mounted on a piece of sheet metal that you made back in Experiment #48, along with the sheet of Bristol board that you marked up with a simple path for the robot to follow. The circuit that I used for this experiment uses the same parts as Experiment #48 (with the inclusion of the BS2) and looks like Figure 17-6.

Once you have wired the robot and calibrated and tested the function of the optointerrupters (by plac-

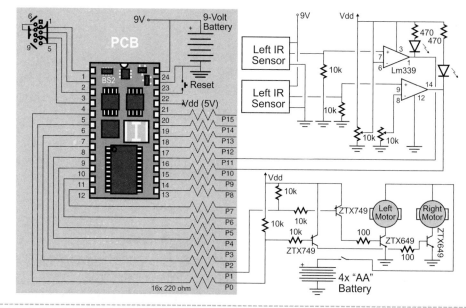

Figure 17-6   *Line-following circuit*

ing them over a white and black sheet of paper and seeing the LEDs light when they are over the black), you are ready to test it out with the following program:

```
'   Line Follower - Follow the Line
'{$STAMP BS2}
'{$PBASIC 2.50}

'   Variables
LeftSensor      Pin 11
RightSensor     Pin 10
LeftMotor       Pin 1
RightMotor      Pin 0

'  Initialization/Mainline
  high LeftMotor: high RightMotor
  input LeftSensor: input RightSensor
  do
    if (LeftSensor = 0) and (RightSensor =
1) then
        high LeftMotor
    else
      low LeftMotor
    endif
    if (RightSensor = 0) and (LeftSensor =
1) then
        high RightMotor
'  Stop the Right Motor
    else
        low RightMotor
'  Else, Right Motor can Run
    endif
    pause 50
```

```
  high LeftMotor: high RightMotor
  pause 100
loop
```

Like the original line-following robot, you are going to have to adjust the "Run" and "Stop" times for best performance. One nice feature of using the BS2 is that you can take advantage of the PWM command built into the PBASIC language. After spending a bit of time experimenting with the different on and off values, I found that I could run my robot around the ring at about twice the speed as the original, LM339-based robot with much better accuracy.

When you run the robot, you might notice that the robot bounces back and forth depending on the amount of torque available when the motors stop and start. In my own case, I found that I had to add another castor to the front of my robot to prevent the front from "diving" down when the robot paused and incorrectly sensed the lines in front of it. If you don't want to do this, you can add to the pause time (decreasing the overall speed of your robot) to allow the robot to stop bouncing before the next time the infrared sensors are polled.

Experiment 120
Wall-Following Robot

Parts Bin

Assembled PCB with breadboard and BS2

Servo robot motor base with four-AA-battery clip and switch

Two Sharp GP2D120 IR object detectors

LM339 quad comparator chip

Two LEDs, any color

Two 470-ohm resistors

Two 10k breadboard-mountable potentiometers

Two servo-connector-to-breadboard adapters

Sheet metal mounting plate

Twenty-four-gauge red solid core wire

Twenty-four-gauge green solid core wire

Twenty-four-gauge black solid core wire

Tool Box

Wiring kit
Five-minute epoxy
Small clamps
Soldering iron
Solder

Although many people consider the "classic" robot application to be the light seeker, I always look toward the wall-following application as being my favorite. The reason for my fondness for this application goes back to when I first started working on the "TAB Electronics Build Your Own Robot Kit," and I had three applications (random movement, light following, and light avoiding) built in and I was looking for a fourth. I had the good fortune to ask my daughter what she would like to see a robot do and she said that the robot should be able to solve a maze. Coincidentally, I had just read a *Flash* comic book in which an evil villain had captured the Flash's wife and was holding her hostage. To keep the Flash at bay, the villain placed a giant maze between them with the thought that, even though the Flash can move very quickly within the maze, it would still take him enough time that the villain could get away with his hostage. What the villain did not count on was Flash's knowing that you can solve a maze by simply keeping one hand on a wall as you go through it—eventually you will travel from the entrance to the exit. In the end, mere seconds after the villain explained to the Flash the purpose of the maze, the Flash apprehended the villain and got a big hug from his wife.

I realized that by coming up with a wall-following robot, I also had, in effect, a maze-solving robot. So my daughter, who normally expects her ideas for robot projects (such as make one of her dolls walk around and stomp, Godzilla-style, on smaller dolls) to be rejected, was surprised to see how enthusiastically I embraced the idea.

By placing two IR object detectors at an angle to the front of the robot, it is actually very easy to have it follow a wall and work through the perimeter of a room as I show in Figure 17-7. In this example, the robot will have its right wheel turning forward until the left sensor detects an object, at which point the

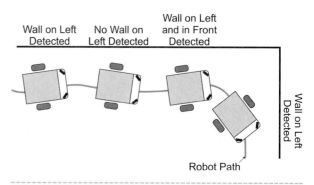

Figure 17-7   *Wall-follow waddle*

Experiment 120 — Wall-Following Robot (vertical side text)

right wheel stops and the left wheel turns. When both sensors detect an object in front of it, the right wheel reverses and the left wheel turns forward to tightly turn the robot away from the obstruction. I call this motion a "waddle;" the reason why will become obvious when you see the robot in operation.

Earlier in the book, I introduced you to the Sharp GP2D120 IR object detectors and showed how they could be used with a comparator and a voltage-dividing potentiometer to indicate when there was an object at a specific distance from it. By gluing on two of these sensors to the leftover sheet metal mounting plate that you made for the line-following robot (Figure 17-8), you will have noncontact object sensors for your robot that can be used for normal operations, such as solving a maze or following a wall as I show in this application. Because of the weight and awkwardness of the wires, you will have to clamp the GP2D120s to the sheet metal while the glue is curing.

The actual circuit for implementing the line-following robot is quite simple (Figure 17-9). I chose to use the servo base for this experiment, but the DC motor base can be used just as easily. The only surprise that you should be aware of is the connection of the GP2D120's power to the four-AA-battery clip (which also powers the servos) because I found their current draw was more than the 5-volt regulator on the BS2 could handle.

Figure 17-8 *Front of wall-following robot with Sharp GP2D120s glued to a sheet metal mounting plate attached to the front of the robot*

When writing application code for the servo base, it is important to remember that the left wheels turn in the opposite direction of the right wheels when the same pulse train is passed to it. For this reason, in the PBASIC "wall follow" application that follows, I have defined pulse values for moving forward and backward for each servo.

```
'  Wall Follower - Follow the Perimeter
of a Wall
'{$STAMP BS2}
'{$PBASIC 2.50}

'  Variables
```

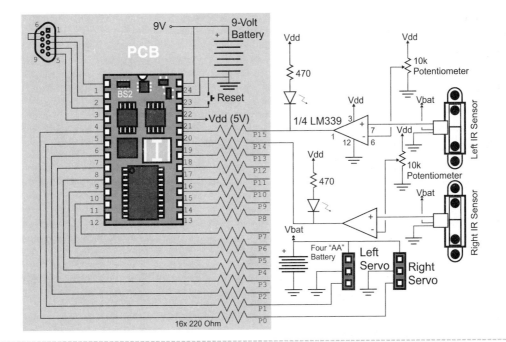

Figure 17-9 *Wall-following circuit*

```
LeftSensor        Pin 15
RightSensor       Pin 14
LeftServo         Pin 1
RightServo        Pin 0
LeftForwards      var word
LeftBackwards     var word
RightForwards     var word
RightBackwards    var word
i                 var byte

'  Initialization/Mainline
   low LeftServo: low RightServo
   LeftForwards = 500: LeftBackwards = 1000
   RightForwards = 1000: RightBackwards =
500
   input LeftSensor: input RightSensor
   do
     if (LeftSensor = 0) then
```

```
      for i = 0 to 5
        pulsout LeftServo, LeftBackwards
        pause 18
      next
    endif
    if (RightSensor = 0) then
      for i = 0 to 5
        pulsout RightServo, RightForwards
        pause 18
      next
    else
      for i = 0 to 5
        pulsout LeftServo, LeftForwards
        pause 18
      next
    endif
    pause 100
loop
```

Experiment 121
Ultrasonic Distance Measurement

Parts Bin

Assembled PCB with
   breadboard and BS2

Polaroid 6500 Ultrasonic
   Distance-Measuring
   Unit

74LS 123 Dual One Shot

74LS74 Dual D flip flop

10k resistor

2.2k resistor

1,000 µF electrolytic capacitor

1 µF capacitor, any type

Two 0.01 µF capacitor, any type

Tool Box

Wiring kit

Six-volt lantern battery
   (see text)

If you are a fan of old submarine movies, you might want to watch for a mistake that is very commonly made in them. While the submarine is "rigged for silent running" and the captain (Clark Gable, Cary Grant, or someone else of their stature) is hovering over the sonar operator asking if the destroyer that is attacking them is leaving the area, you will hear the distinctive "pinging" sound that is used in movies to indicate the sub's sonar is working. This is actually embarrassing because the submarine would never send these signals—instead they would be produced by the destroyer as it scanned the ocean for the sub. Once the reflected sound waves were received by the destroyer, it would then use the time it took for the signals to return and the direction they were coming from to identify where the submarine was and what depth it was running at. Once the destroyer identified where the sub was, it was then a race to move over the submarine to drop depth charges in an effort to destroy it. As any Tom Clancy reader would know, the game of searching for a submarine is not this simple; sound waves in water, only travel so far before they are attenuated below the microphone's ability to pick them up, and by monitoring the water temperature and salinity (salt content), the sound waves can be reflected away from the submarine.

Earlier in the book, I introduced you to IR LEDs and phototransistors and how they can be used to detect objects when modulated light is reflected off them. Ultrasonic "sonar" works in the same way and can be easily added to your robot. The most popular device used for this purpose is the Polaroid 6500 Sonar Ranging Module (Figure 17-10). This module has two primary issues that you should be aware of if you would like to use it in your robot; it takes a bit of work to interface to a BS2, has a very narrow field of view, and requires a *lot* of power to operate (1 amp while the ultrasonic pulses are being transmitted).

The first concern is somewhat alleviated by desoldering the connector that comes with the 6500 and adding individual wires as shown in Figure 17-10. This nine-pin connector has six signals that you will have to be familiar with if you are going to interface to the module (see Table 17-2).

Normally, the 6500 will have power connected to it with "BLNK" and "BINH" tied to ground with "INIT" tied to a output driver and "ECHO" pulled up and connected to a receiver. Figure 17-11 shows normal operation of the 6500 with INIT driven high and ECHO going high upon the reflected pulse.

To find out more information about the Polaroid 6500 Sonar Ranging Module, you can download its datasheet from the Internet by first doing a Google™ search on it. The 6500 was originally designed for use as a camera's range finder (point the black and gold transducer at the subject to find its range). This was an excellent application for it as ultrasonic sonar modules have very narrow fields of view (the 6500 is most sensitive for the four degrees within perpendicular), but one that can make them somewhat diffi-

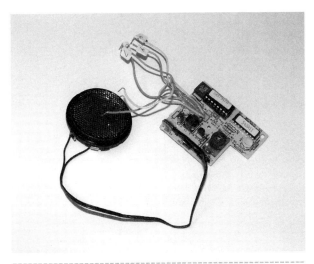

Figure 17-10 *Polaroid 6500 Sonar Ranging Module with standard connector removed and breadboard wiring added*

Table 17-2 Polaroid 6500 power and control wires

Pin	Label	Function/Comments
1	Ground (Gnd)	
2	BLNK	When driven high, any reflected signal is blanked out.
4	INIT	Driven high to initiate sonar ranging. When INIT is driven low, the operation of the 6500 stops, even if an echo has not yet been received.
7	ECHO	Open collector signal held low until reflected signal received.
8	BINH	Driven high to disable 2.38 ms internal masking.
9	Vcc	4.5 to 6.4 volts, with up to 1 amp drawn. Should have 1,000 µF capacitor across it and Gnd.

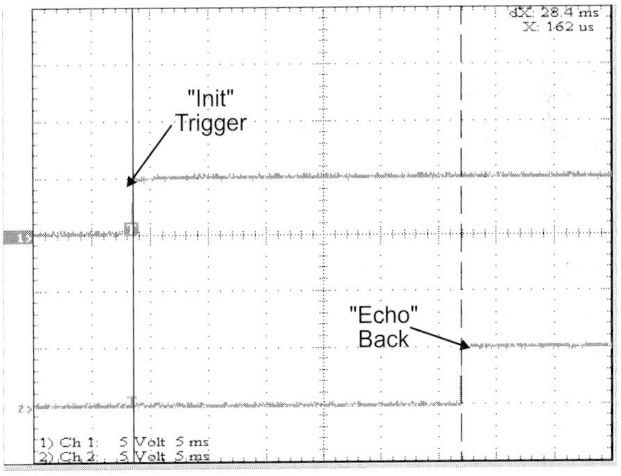

"Init" Trigger

"Echo" Back

Figure 17-11 *Ultra echo*

cult to use in robots. A possible solution is to mount the sensor on a servo driven "turret." By swivelling the turret, the sonar distance sensor will give you bearing and distance to the objects around the robot.

Using the distance and angle information, the position of the robot can be fairly easily determined; the complexity of the mechanical installation of the sonar range finder and the data that is returned is why I consider it to be a navigation tool and not a sensor. For the range finder to be used as a sensor, it would have to scan the area around it continuously, and to do it effectively, this would significantly slow down the movement of the robot.

Having to continuously scan further exacerbates the final issue of sonar range finders. When they are active, they use a lot of power. The 6500 requires a full amp of current when it is operational, much more than the PCB's 9-volt battery and 5-volt BS2 supply can source. For this experiment, I bought a large 6-volt lantern battery. The 6500 can be driven from the battery directly, but if a higher voltage battery is used then its output would have to be regulated down to 4.5 to 6.4 volts. For many large robots, the requirement of an amp of current at 6 volts is not a significant issue, but for small robots, supplying an amp of current is a significant issue (when the cameras that first used the sonar ranger were designed, they provided special batteries that were discarded with the empty film carriers).

To interface the BS2 to the 6500, I came up with the circuit shown in Figure 17-12. This circuit was developed to allow the BS2 to use the "pulsin" statement by changing the time between the INIT line going high and the ECHO line going high into a single, negative pulse. This was done by using a 74LS123 to "Pulse" when the these two events occurred, changing the state of a D flip flop (one-half of a 74LS74) and producing the pulse on the bottom line of the oscilloscope picture shown in Figure 17-13. To make sure that the D flip flops of the circuit are in the correct initial state, I added an extra line (which I call "InitSetupPin") that is pulsed low to set the two flip flops. This is one of the advantages of having a

microcontroller; it can be used to initialize external circuit hardware more efficiently than coming up with a reset circuit on your own.

Coming up with this circuit is not a trivial exercise. The BS2 is not well suited for this type of operation for two reasons. The first is that it does not have a method of understanding how long a statement takes to execute. The time from when the INIT signal goes high (the "pulsout" statement) to when the BS2 starts polling the ECHO signal (pulsing) is not known; if it were, this time could be added to a statement polling the ECHO signal to see when it comes high. This ability is normally present in a traditional microcontroller and the extra circuitry needed for this experiment would not be required. The second problem is the ability to carry out more than one task at a time, or the ability to "interrupt" the current task. Again, this capability is available in more traditional microcontrollers. After wiring the circuit on the breadboard, I entered in the following program and saved it as "Ultra Sonic Ranging" in the Sonar folder, which is located in the Evil Genius folder.

```
'   Ultra Sonic Ranging - Use Polaroid
6500 Sonar Ranging Module
'{$STAMP BS2}
'{$PBASIC 2.50}

'   Pin/Variable Declarations
InitPin         pin 15
InitSetupPin    pin 13
FlightPin       pin 0
SoundFlight     var word
```

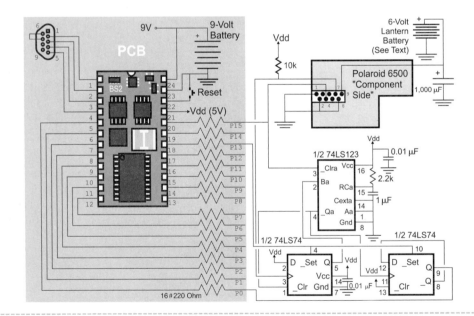

Figure 17-12   *Sonar circuit*

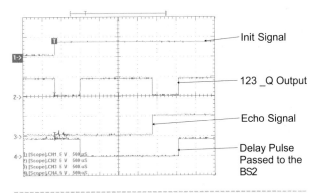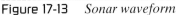

Init Signal

123 _Q Output

Echo Signal

Delay Pulse
Passed to the
BS2

Figure 17-13   *Sonar waveform*

```
SoundIn        var word
SoundFt        var word

' Initialization/Mainline
  low InitPin
  high InitSetupPin
  input FlightPin

  do
    pulsout InitSetupPin, 10
```

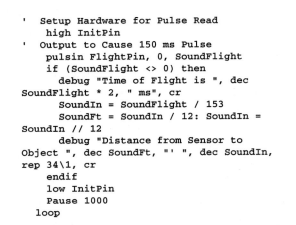

```
' Setup Hardware for Pulse Read
  high InitPin
' Output to Cause 150 ms Pulse
  pulsin FlightPin, 0, SoundFlight
  if (SoundFlight <> 0) then
    debug "Time of Flight is ", dec
SoundFlight * 2, " ms", cr
    SoundIn = SoundFlight / 153
    SoundFt = SoundIn / 12: SoundIn =
SoundIn // 12
    debug "Distance from Sensor to
Object ", dec SoundFt, "' ", dec SoundIn,
rep 34\1, cr
  endif
  low InitPin
  Pause 1000
loop
```

To finish off the discussion on different types of object detectors, I have summarized the three primary methods (whiskers, IR proximity, and ultrasonic ranging) used by robots in Table 17-3. It is important to note that there is not one method that will be optimal for all applications (unless you tightly control the environment the robot is operating in).

Table 17-3   Different object detection methods used by robots

Object Detection Method	Advantages	Disadvantages	Comments
Physical whiskers	Detects all objects Inexpensive Variety of different ways to manufacture	Easily damaged Require constant attention Require software debounce Limited range may result in damage to robot Potential for static electricity charge build-up Generally small field of view	Best for "worst case" collision detection Best designed with robot's mechanical design
Infrared proximity	Reliable Relatively inexpensive No debouncing required Prepackaged modules available Wide field of view	Difficult to get range of object May be difficult to set up hardware/software for detector Wide field of view may pick up objects that are not dangerous or interesting to robot	Good general purpose object detection mechanism Can be built on electronics PCB or put remotely on robot
Ultrasonic ranging	Reliable Narrow field of view allows "plotting" of objects around robot	Requires large amounts of power Limited field of view requires some method of scanning to find all nearby objects Most expensive option	Often most difficult method to get working Should only be considered for advanced robot applications

Experiment 122
Hall Effect Compass

In choosing the experiments for this book, I have used two primary criteria. The first is to look at the theory and practice of very basic robot components. To a large extent, I believe I have succeeded. You should be able to come up with a unique robot design on your own and choose the major subsystems required and specify the parts. Secondly, when presenting the different parts, I have tried to present them in a nontraditional way, looking for experiments, circuits, and structures that aren't generally thought of when the topic of robots is discussed. I am bringing this up because I have not yet discussed a number of useful robot components.

One of the more useful devices that you should be aware of is the Hall effect switch (Figure 17-14). In operation, this device passes a current through a piece of silicon, and if no external magnetic field is operating on it, the current passes straight through it. If there is a magnetic field, then the current is defected and passed to another sensor.

This magnetic deflection of current through a semiconductor is known, not surprisingly, as the Hall effect and is often used in robots to implement

odometry. Rather than setting up an optointerrupter, as I did earlier in the book, you can simply glue a magnet or two to a wheel, gear, or axle and count its rotations using a Hall effect switch. Hall effect switches are very often used in automotive antilock braking circuits as wheel turn sensors—being magnetically actuated, they are very robust and do not require any cleaning, as an opto-interrupter would.

Most Hall effect switches come in a three-pin package, with a positive voltage (power in) pin, a negative voltage pin (ground), and a signal pin. The signal pin provides output in a variety of different formats (such as open collector/drain, totem pole, or analog output). Generally, Hall effect switches will indicate the presence of a "South Pole" of a magnetic field.

An interesting application for the Hall effect switch is the one that I am going to show you in this experiment using two Hall effect switches to indicate which direction is south. I originally discovered this application in a book of Texas Instruments data sheets. I photocopied the datasheet and tucked it away in my files, looking for an opportunity to use it. As I discussed at the start of this section, for early navigation, knowing which direction was north and south was of great importance. I have always wanted to try out this circuit to see how well it works in relation to robots. The circuit itself is very simple (Figure 17-15) and will just take you a few moments to wire.

Using the specified resistor values gives me an amplification of the differential signal of about 700 times. This op-amp application is excellent for situa-

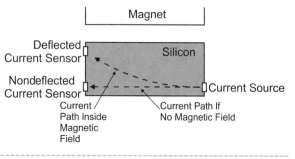

Figure 17-14   *Hall effect switch*

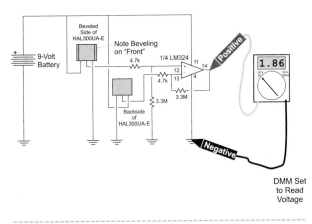

Figure 17-15   *Compass circuit*

tions like this one where the difference between two signals is to be observed.

Before coming up with an interface to the BS2, I wanted to test the operation of this circuit. It consists of two Hall effect sensors turned in opposite directions and their (differential) output greatly amplified using an op-amp. When the Hall effect switch connected to the positive input of the op-amp was pointing directly south, its output was at a maximum, whereas the op-amp pointing north was at its minimum output. The differences in the output would be very small because of the relatively small value of the Earth's magnetic field that this circuit is detecting. By passing these voltages to the inputs of an op-amp and amplifying the difference between them, the circuit should be at its highest voltage when the positive input Hall effect switch is pointing south and the negative input Hall effect switch is pointing north.

After wiring the circuit, I tested it out to see how effective it was in determining north and south. In my house, the circuit did not produce any kind of noticeable change when I turned around. Going outside, I found that the circuit would repeatedly show an output voltage of 1.86 volts when the end with the battery was pointing toward north, and this would drop to 1.83 volts when the circuit was turned in any other direction. This is a difference of 1.6 percent, which would be very difficult to observe within a simple computerized circuit.

I would have pursued the compass circuit (and looked for different or more efficient Hall effect sensors) except that the compass output, once the rotation of the circuit stopped, took 15 seconds or more to settle down into a constant value. Along with this long time, I found that the range of north was about 15 degrees, much too wide to be accurate for any kind of practical robot.

Along with these issues, I also wondered how effective the compass would be on a robot chassis with magnetic devices, many of which produce fields that are larger than the Earth's magnetic field. You can test this last supposition by moving a permanent magnet close to the two Hall effect switches.

So in conclusion, the electronic compass circuit that I have come up with here does not provide a significant enough signal to be easily measured, the range in which it would indicate north is larger than would allow effective navigation, and the output settling time is very long.

Experiment 123
NMEA GPS Interface

Parts Bin

Assembled PCB with breadboard and BS2

GPS unit with RS-232 interface

Maxim MAX232 RS-232 interface chip

Five 1.0 μF electrolytic capacitors

Solderable 9-pin "Male" D-Shell connector

Twenty-four-gauge solid core black wire

Twenty-four-gauge solid core red wire

Twenty-four-gauge solid core green wire

Tool Box

Wiring kit

Soldering ron

Solder

Before ending this book, I realized that I missed presenting to you what I consider to be the most important computer-to-computer interface that you will have an opportunity to work with. In the previous section, I introduced you to the concept of *Non-return to Zero* (NRZ) serial communications, and you have been using RS-232 to program the BS2, but I did not show the RS-232 electrical interface that is, by far, the most popular way of connecting two computer systems. I'm always discouraged by the number of graduate engineers that I get to work with who have never successfully implemented their own RS-232 connection between two computers. It's not hard; you just have to go through it at least once to understand how it is done.

RS-232 goes back to the very early days of electronic computers when there was more than one computer and there was a need for them to communicate with one another. These early systems were not built from transistors (let alone integrated circuits) so their operating parameters were somewhat unusual compared to today's computers. Today, we would probably want to use 5 volts for communications and let two computer systems "talk" to each other with just some logic gates to buffer (protect and repower) the signals between them. This is actually the path that was taken with the first computers,

but they did not run with simple +5 and ground logic levels. They used vacuum tubes for logic, and their operating voltages were considerably higher and different. To facilitate simple communications between the two systems, the voltage level of −3 to −15 volts was used to indicate a "1" (or a mark) and +3 to +15 volts indicates a "0" (or a space). The 6-volt region in between the two voltage ranges (from –3 volts to +3 volts) was called the *switching region*, and the voltages of the communications passing between the two systems should never be in this range. The standard specified a maximum of 20 mA of current to be passed between the two systems—this value goes back to the first teletypes.

When the first two systems were connected by RS-232 they were not in the same room; you must remember that these systems generally took up as much space as a high school gymnasium and required as much electrical power as a small subdivision. This is *not* an exaggeration. Because of the space and power requirements, it was not practical for the two systems to be in close proximity, so they had to communicate over the phone lines using a device called a *modem*. I'm pointing this out because there had to be some way of connecting the computer systems to the modem; this was done by specifying standard connectors with different pin functions. These connectors are known as D-Shell connectors and were given a standard wiring pattern, which exists to this day. Figure 17-16 shows the original 25-pin standard and the newer 9-pin standard, which was created for personal computers.

The connectors that are built into computers are male (which means they have pins, as in the photograph of the connector in Figure 17-16), and are known as *data terminal equipment* (DTE) connectors. Modems have the mating (female) connectors, which are known as *data communications equipment* (DCE). DTE (namely personal computers) generally uses straight-through connectors to connect them to modems and other peripherals (like the PCB included in this book).

If you look around on the Internet or read some introductory texts on RS-232, you will see some pretty strange circuits used to allow a standard 5-volt (or 3.3-volt) circuit to interface with an RS-232 line. These circuits will work anywhere from most of the time to some of the time, with your ability to figure out communications problems being very limited. To avoid the problems of trying to debug RS-232 failures in circuits that have taken shortcuts with voltage levels, I am going to recommend that you avoid these circuits and use something like the Maxim MAX232 to provide you with an RS-232 interface that operates at the correct voltage levels. The MAX232 is a very common and popular chip and is actually quite easy to wire into a circuit. When you are buying the chip, you must decide between the MAX232 (which uses 1.0 µF capacitors) and MAX232A (which uses 0.1 µF capacitors). Most people go with the original MAX232 because 1.0 µF electrolytic capacitors can be purchased quite inexpensively.

For this experiment I would like you to create an RS-232 interface between the book PCB and a peripheral device. The peripheral device that I have chosen is a *global positioning system* (GPS) receiver that can interface to other devices using the NMEA 0183 communications standard. NMEA comes from the National Marine Electronics Association and is an RS-232 connection running at 4800 bps and transferring navigation data as a series of "sentences." You are probably familiar with GPS; it is a number of satellites in orbit around the Earth that provide signals to help aircraft, boats, and motor vehicles to navigate.

You can buy surprisingly sophisticated GPS receivers such as my Garmin eTrex (Figure 17-17) for just a few hundred dollars. When choosing a GPS receiver, you should be looking for its ability to communicate with other systems via NMEA 0183 (RS-232) and transmit both position (GPS) as well as heading (compass) data. Depending on the amount of money you have to spend, along with features like position (returned as latitude and longitude) and compass heading, a GPS unit can provide you with a moving map display, showing you exactly where you are and where you are headed.

The NMEA "sentences" consist of data in the following format in Table 17-4.

The NMEA data may look like garbage and is always coming in at a rate (each sentence is separated by a 0.8- to 5-second pause) that you can't handle, but it is quite easy for a controller such as the

DB-25 (Male)   D-9 (Male)

Pin Name	25 Pin	9 Pin	I/O Direction
TXD	2	3	Output (O)
RXD	3	2	Input (I)
Gnd	7	5	
RTS	4	7	O
CTS	5	8	I
DTR	20	4	O
DSR	6	6	I
RI	22	9	I
DCD	8	1	I

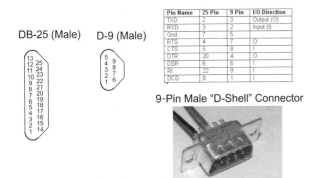

9-Pin Male "D-Shell" Connector

Figure 17-16   *RS-232 connectors*

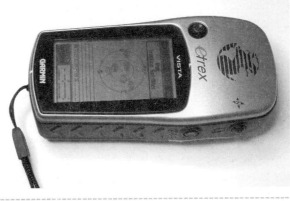

Figure 17-17   *The Garmin eTrex provides position and heading information via an NMEA interface as well as a moving map display.*

Table 17-4 NMEA 0183 communications "sentence" features

Character	Position	Example—Comments
Header	1	Always "$."
Identifier	2-3	"GP"—GPS/"HC"—Compass heading.
Formatter	4-6	"GSV"—GPS Satellites in view/"HDG"—heading with deviation.
Data	7 . . .	ASCII data separated by commas. Different for each Formatter.
Sentence ending	Last two Characters	Carriage return and line feed ASCII characters.

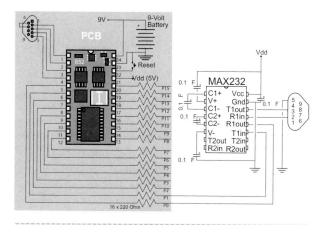

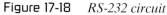

Figure 17-18 *RS-232 circuit*

BS2 to interpret the data and find some useful information from it.

The final experiment of this book is to create a simple BS2 application that monitors the incoming NMEA data stream and extracts the compass heading information from it. The sentence format for the compass heading is the following:

```
$HCHDG,Heading,,,Deviation,W*0A,cr-1f
```

It can be received using the hardware shown in Figure 17-18 and extracted from the stream of data (which can be seen in Figure 17-17) using the following application, which should be stored in the RS-232 folder in the Evil Genius folder:

```
'  GPS Receiver - Receive Data from eTrex
'{$STAMP BS2}
'{$PBASIC 2.50}

'  Variables
SerialInput     Pin 0
SerialOutput    Pin 1
```

```
CompassHead     var byte(6)
i               var byte

'  Initialization/Mainline
  high SerialOutput
  do
    serin SerialInput, 188,
[WAIT("HCHDG,"), STR CompassHead\6]
    i = 5
    do while (CompassHead(i) <> ".")
      i = i - 1
'  Display to Decimal Point
    loop
    debug "Heading ", str CompassHead\i, "
degrees.", cr
  loop
```

In this application, I take advantage of PBASIC's "serin" (serial input statement) and wait for the string "HCHDG" (it is limited to six characters long in PBASIC) before saving the next six characters. This capability of PBASIC to wait or search data makes applications like this very easy to write. Once I have the six characters that follow, I read back through them, looking for the decimal point in the angle (a tenth of a degree is better accuracy than I need for most applications), and then pass it back to the programming PC using the "debug" statement.

PBASIC Reference

BS2 Specifications

Memory Size: 2K *electrically erasable programmable read-only memory* (EEPROM), 26 bytes of variable RAM

Programming Language: PBASIC

Number of I/O Pins: 16 + 2 dedicated RS-232 lines

Speed: Approximately 4,000 statements per second

Current Draw: 8 mA running, 100 uA sleep

Current Supply:

Pin Current Parameters: Source 20 mA, sink 25 mA (typically)

Onboard Regulator Output: 5 V +/-5 percent, 50 mA from a 12-volt source, and 150 mA from a 7.5-volt source

Pin	Name	Function
1	SOUT	RS-232 serial output from BS2. Serial pin 16.
2	SIN	RS-232 serial input to BS2. Serial pin 16.
3	ATN	RS-232 attention pin. Connected to host PC DTR line (DB9 pin 4).
4	VSS	BS2 ground. Same as pin 23.
5-20	P0-P15	General-purpose I/O pins.
21	Vdd	Regulated 5 volts. If external power source on VIN, then +5 volts out (up to 90 mA). Can have +5 volts applied to drive BS2.
22	RES	Pulled-up, negative active reset pin.
23	VSS	BS2 ground. Same as pin 4.
24	VIN	Unregulated power in the range of 5.5 to 15 volts.

PBASIC Header Statements

The following statements must be in place before any application code statements. Both statements must be used to take advantage of all the built-in PBASIC functions:

```
' {$STAMP BS2}
' {$PBASIC 2.5}
```

PBASIC Built-in I/O Port Labels

Name	Function
INS	Sixteen-bit word that returns the value of P0-P15.
INL	Eight-bit value returned from P0-P7.
INH	Eight-bit value returned from P8-P15.
IN#	Single-bit value returned for P#.
OUTS	Sixteen-bit word used to set the output values of P-P15.
OUTL	Eight-bit value passed to P0-P7.
OUTH	Eight-bit value passed to P8-P15.
OUT#	Single bit to set the output value of P#.
DIRS	Sixteen-bit word used to set the I/O state of P0-P15. "1" written to an I/O pin will put it into "output mode." "0" written to an I/O pin will put it in "input mode."
DIRL	Eight-bit value setting I/O state of P0-P7.
DIRH	Eight-bit value setting I/O state of P8-P15.
DIR#	Single-bit value setting I/O state of P#.
W#	Sixteen-bit variable (W0 to W12 available).
B#	Eight-bit variable (B0 to B25 available). Note that the space used for 8-bit variables is shared with space used for 16-bit variables. Divide the # in B# by 2 and round down to determine which 16-bit variable the 8-bit variable is located in.

Courtesy of Parallax, Inc. of Rocklin, CA (http://www.parallax.com).

PBASIC Constant, Variable, and Pin Declarations

Constant values are defined using the statement format:

```
ConstantName con value
```

Constant values can consist of other values manipulated with mathematical operators or constants with their type modifiers consisting of the following:

Modifier	Data Type
Nothing	Decimal value
$##	Hex value
%## . . .	Binary value
#	ASCII value returned

Variables can be given names and space dynamically within the application using the declare statement:

```
VariableName var Type
```

where Type can be bit, nib (four bits), byte, and word (16 bits).

Type can also be another variable to share the space or can be subdivided into highbyte (or byte1), lowbyte (or byte0), highnib (or nib3 or nib1 depending on the variable type), lownib (or nib0), and bit# (as well as highbit and lowbit). For example, an LED on Output pinp can be defined as

```
LED var OUTS.bit0
```

Single-dimension arrays (numbered from 0 to size 1) are defined as

```
VariableName var Type(size)
```

To define pins, the declaration statement

```
PinName Pin #
```

is used with # being the number of the pin being defined (from 0 to 15).

Arithmetic Operators

The following table lists the operators available in PBASIC. The operators only available for use in constant declarations are indicated in the Constant column.

Symbol/Format	Constant	Function
SQR A	n	Returns square root of A.
ABS A	n	Returns absolute value of A.
COS A	n	Returns 16-bit cosine value for 8-bit (0 to 255) angle A.
SIN A	n	Returns 16-bit sine value for 8-bit angle A.
DCD A	n	2**n power decoder (DCD 4 = %0000000000010000).
NCD A	n	Priority encode of 16-bit value (NCD 12 = 8).
A MIN B	n	Returns smallest value.
A MAX B	n	Returns largest value.
A DIG B	n	Returns digit B from decimal number A.
A REV B	n	Reverses B bits (starting at LSB) of A.
A ^ B	Y	Bitwise XOR A and B.
A	B	Y	Bitwise OR A and B.
A & B	Y	Bitwise AND A and B.
~A	Y	Returns the bitwise complement of A.
A >> B	Y	Shifts A right B bits.
A << B	Y	Shifts A left B bits.
A / B	Y	Divides A by B.
A // B	n	Returns the remainder of A divided by B.
A * B	Y	Multiplies A by B.
A ** B	n	Multiplies A by B and return high 16 bits of 32-bit result.
A */ B	n	Multiplies A by B and return the middle 16 bits of the 32-bit result.
A - B	Y	Subtracts B from A.
-A	Y	Returns negative A.
A + B	Y	Adds A to B.

Debug and Serout Data Formatter Commands

Output strings are specified within double quotes, and numeric and ASCII data can be displayed by using the formatters listed in the following table. Multiple data (with optional formatters) can be output from one statement by separating them with commas.

Formatter	Function
?	Displays "Symbol = #" and carriage return. Can be combined with other formatters to display data in different formats (such as hex ? hexVariable).
ASC ?	Displays "Symbol = #" and carriage return where # is the ASCII representation for the symbol.
DEC{1…5}	Decimal display. The number of digits is displayed optionally specified.
SDEC{1…5}	Signed decimal display. The number of digits displayed optionally is specified.
HEX{1…4}	Hexadecimal display. The number of digits displayed is optionally specified.
SHEX{1…4}	Signed hexadecimal display. The number of digits displayed optionally specified.
IHEX{1…4}	Indicates (with proceeding $ character) hexadecimal display. Number of digits displayed is optionally specified.
ISHEX{1…4}	Indicates (with proceeding $ character) signed hexadecimal display. Number of digits displayed is optionally specified.
BIN{1…16}	Binary display. The number of digits displayed is optionally specified.
SBIN{1…16}	Signed binary display. The number of digits displayed is specified.
IBIN{1…16}	Indicates (with proceeding % character) signed binary display. The number of digits displayed is optionally specified.
ISBIN{1…16}	Indicates (with proceeding % character) signed binary display. The number of digits displayed is optionally specified.
STR bytearray{\#}	Displays ASCII string in array until NUL ($00) character is encountered or an optionally specified number of characters is displayed.
REP byte\#	Displays the specified ASCII character # times.

For display control with the debug statement, the special control characters listed in the following table are used. For special control characters, use "rep #\1" where # is the ASCII value listed in the table.

When using the special control characters with "serout," just bell, bksp, tab, Line Feed, and cr are recommended as these will be supported by different ASCII terminal devices.

Special Control Characters	Symbol	ASCII	Function
Clear Screen	cls	0	Clears screen and places cursor at home position of Debug Terminal.
Home	home	1	Places cursor in home position (top left of Debug Terminal).
Move to (x,y)		2	Moves cursor to specified location on Debug Terminal. Must be followed by two values (x and then y).
Cursor Left		3	Moves cursor one column to the left in Debug Terminal.
Cursor Right		4	Moves cursor one column to the right in Debug Terminal.
Cursor Up		5	Moves cursor one row up.
Cursor Down		6	Moves cursor one row down.
Bell	bell	7	Beeps the PC's speaker.
Backspace	bksp	8	Deletes character to the left of cursor and moves the cursor to its position.
Tab	tab	9	Moves cursor to the next tab column.
Line Feed		10	Moves the cursor down one line.
Clear Right		11	Clears the line to the right of the cursor.
Clear Down		12	Clears the Debug Terminal below the cursor.
Carriage Return	cr	13	Moves cursor to the start of the next line in Debug Terminal. In other devices, the cursor may be returned to the start of the current line and a line feed character will be needed to move to the next line.

Serin and Serout Baudmode Bit Value Definitions

The Baudmode value is defined using the following bits:

Bits	Function
15	Serout *only*. Reset with data driven out. Set (add 32,768) with open drain output.
14	Reset with data *not* inverted. Set (add 16,384) with data inverted.
13	Reset with data in the 8-N-1 format (eight data bits, no parity, one stop bit). Set (add 8,192) with data in the seven-bit format with even parity.
11–0	Data rate defined using the formula Bits 11–0 = INT (1,000,000/desired baud rate) − 20.

PBASIC Functions and Statements

In the following table, braces ({ and }) indicate optional parameters and "..." indicates that parameters can be repeated.

Statement	Example	Comments/Parameters
Comment	'	Everything to the right of a single quote (') is ignored.
Assignment	A = B + C	Assigns a variable the result of a single arithmetic operation. If multiple operations are specified in the statement, they are executed from left to right. The order of operations in a multiple-operator assignment statement can be "forced" by enclosing higher-priority operations within parentheses.
Stop BS2	end	Stops the BS2 from executing any more code. BS2 goes into a low-power mode.
Stop BS2	stop	Stops the BS2 without placing the BS2 in a low-power mode.
Loop code	do ... loop	Executes the code between do and loop repeatedly. Can exit using a goto or exit statement.
Loop while	do while condition ... loop	Repeats the code between the do and loop statements while condition is true.

Statement	Example	Comments/Parameters
Loop until	do ... loop until condition	Repeats the code between the do condition and loop statements until the condition is true.
Address Label	Label:	Specific address within application.
Jump to Label	goto Label	Changes execution to continue after Label: statement.
Jump to subroutine	gosub Label	Executes instructions after Label: statement until a return statement is encountered.
Return from subroutine	return	Returns to statement after gosub Label statement.
Conditionally execute statement	if condition then else statement	If condition is true, it executes the first statement. If condition is false, it executes the statement after "else." "Else" and the statement that follows it are optional. Multiple statements can be executed if the condition is true or false if they are separated by a colon (:).
Conditionally execute multiple statements	if condition then ... else ... endif	If the condition is true, then execute the statements immediately following the if statement. If the condition is false, execute the statements following the else statement. Else and the statements following it are optional.
Conditionally jump	if condition then Label	The test condition is in the format "A cond B" where cond is = (equals), <> (not equals), < (less than), <= (less than or equal to), > (greater than), or >= (greater than or equal to). A and B can be variables, constants, or expressions. Parentheses are allowed in this statement. Logical operators (AND, OR, and NOT) can be used to combine multiple condition expressions.
Repeat code	for Variable = InitialValue to StopValue {step StepValue} ... next	Loads Variable with the InitialValue and executes code to the next statement. At the next statement, increment Variable either by 1 or Step Value if it is defined. When Variable is greater than StopValue, execute the first statement after next.
Execution branch	branch Offset, [Address0, Address1, ...]	Goes to Address# for a branch value of #.
Execute code according to data value:	case condition ... case condition ... case else ...	select expression

Statement	Example	Comments/Parameters
Low on Pin	low Pin	Pin will be changed to output and a low voltage will be output from it.
Pin "input"	input Pin	Pin mode is changed to an input.
Pin "output"	output Pin	Pin mode is changed to an output.
Change Pin mode	reverse Pin	Pin changes between input and output modes.
Change Pin value	toggle Pin	The output value for Pin is complemented.
Find list value	lookdown Target, [ComparisonOp], [value0, value1[,...]], Variable	Returns the offset of Target in the list of values in Variable. Note that values can be a string enclosed in double quotes. ComparisonOp is an optional conditional operator for specifying how the search is to be done (equals [=] is the default).
Find list offset	lookup Index, [value0, value1[,...]], Variable	Returns the Index value from the list in Variable. Note that the values can be a string enclosed in double quotes.
Power down	nap Value	Power down for Value length of time where, if Value is 0, the nap time is 18 msecs. It is 36 msecs for a Value of 1.72 msecs for 2,144 msecs for 3,288 msecs for 4,576 for 5, 1.15 secs for 6, and 2.30 secs for 7. Note that actual delays can be different by as much as 50 percent.
Power down	pause Value	Stops for the number of msecs specified by Value. This mode does not save as much power as nap or sleep.
Power down	sleep Value	Power down for Value times 2.3 seconds.
Measure pulse	pulsin Pin, State, Variable	Waits for Pin to be at the specified State and times how long it is at this state. It then returns the time (in 2-usec increments) in Variable.
Drive out pulse	pulsout Pin, Period	Pulses Pin for the specified Period times 2 usecs.
PWM output	pwm Pin, Duty, Cycles	Drives out a *pulse width modulation* (PWM) cycle on the specified Pin for the Cycles' number of times. The PWM period is 255 msecs, with the Duty being the number of msecs, and the PWM output is active high.

(continued)

Statement	Example	Comments/Parameters
	endselect	Executes according to the value of the expression. Condition consists of a constant value with an optional comparison operator (=, <>, >, >=, < or <=) and executes the code statement if the condition is true. Case else is for expression values that do not match any of the set conditions.
Button debounce	button Pin, Downstate, Delay, Rate, Workspace, TargetState, Address	Polls Pin and increments the byte variable Workspace if Pin is in a downstate each time the Button statement is encountered. It jumps to Address if the number of times specified by Delay is equal to the current value of Workspace and the TargetState is met. If the button is released, Workspace is reset. If Delay is equal to 0, no debounce or autorepeat takes place. If Delay is equal to 255, debounce is performed, but with no autorepeat. Autorepeat is executed if the Button statement is executed with Delay plus Rate * n times (and Delay does not equal 0 or 255).
Count pulses	count Pin, Interval, Counter	Delays 1 msec times the number of Interval and counts the number of times Pin is pulsed. The result is saved in Counter.
Return status	debug {formatter} Data/String/ Constant, ...	Returns a string of data to the host PC in the Debug Terminal after BS2 programming to indicate what is happening in the application. Note that if no formatter exists, variable values are printed in decimals. The different formatters available are listed and explained in the previous table along with the special control characters that can be used to control the operation of the Debug Terminal.
Touch-tone output	dtmf Pin, {onTime, offTime,} [tone[,...]]	Generates a touch-tone phone. onTime and offTime are optional values, setting the length for the tone in msecs. Tone is the touch-tone button to send data. If onTime and offTime are not specified, the BS2 defaults to 200 msecs and 50 msecs, respectively.
Output frequency	freqout Pin, Period, Freq1 {,Freq2}	Outputs Freq1 on Pin for the specified Period (in msecs). If Freq2 is specified, it will be mixed with Freq1. Freq1 and Freq2 are specified in Hz and the maximum value is 32,767 Hz.
High on Pin	high Pin	Pin will be changed to output and a high voltage will be output from it.

Statement	Example	Comments/Parameters
Get random value	random Variable	Returns a 16-bit pseudo-random number in Variable.
Measure RC	rctime, Pin, State, Variable	Measures the charge/discharge time (specified by State) and saves the result in Variable, which is required for an RC network on a Pin. The recommended method of using this circuit is to place a capacitor at Vdd with the resistor grounded at the other end and use a State of 1. The discharge time returned is in 2 usec units.
Preload EEPROM	data [@Starting Address,] Value[, "string"]	Saves the specified value or string in EEPROM at the first available address. The starting address can be specified by placing an integer value proceeded by a @ character before the data that is to be saved.
Write EEPROM	write Location, [word] Value	Writes a byte value at the specified location in EEPROM memory. Location can be any value from 0 to 2,047. To ensure application code is not overwritten, use the Ctrl-M command from BASIC Stamp Editor to check application locations. The optional word parameter indicates that Value is 16 bits instead of the default 8 bits.
Read from EEPROM	read Location, [word] Variable	Read the 8 bits in EEPROM at Location into Variable. The optional word parameter indicates that Value is 16 bits instead of the default 8 bits.
Serial input	serin Pin [\Ffpin], Baudmode, =[Plabel,] [Timeout, Tlabel,] [InputData]	Gets serial input data from I/O pin. Note that Pin 16 is the programming serial port. fpin is the flow control output pin (indicating if data can be received). Baudmode has been defined previously. The Plabel and Tlabel jumps to addresses on error (parity and timeout, respectively) are optional. If the Tlabel option is used, then the Timeout interval (units are msecs) and the flow control pin (fpin) must be specified. InputData consists of a destination variable along with formatting variable. If waiting for a numeric string of some kind, the DEC, HEX, or BIN filters are used along with the S or I prefixes, indicating that a signed value or integer, respectively, is required. Using the STR filter stores a string into an array until the number of specified characters has been received (using the \# parameter). A second \# parameter can be used to indicate a string-ending character. The WAIT(#,#...) filter will not return any values until the specified values have been

Statement	Example	Comments/Parameters
		received first. WAITSTR will wait for a series of bytes to match the contents of a specified array (which can be limited to # characters). Finally, using SKIP #, a set number of characters can be ignored before reading the input data.
Serial output	serout Pin [\Ffpin], Baudmode, [Pace,] [Timeout, Tlabel,] [Data]	Outputs serial data onto the specified pin. fpin is an optional pin used for flow control. Baudmode has been defined earlier. Pace is the time between byte outputs and is in the units of msecs. Timeout and Tlabel operate exactly the same way as serin. Data and its optional formatting options are identical to the ones used with debug and are listed in the previous table.
X-10 output	XOUT Mpin, Zpin, [House/Command [\Cycles] {,...}]	Output X-10 outputs to other devices. Mpin is the modulation source pin to the power-line interface device. Zpin is the zero-crossing signal from the power-line interface device. House is the X-10 "house code" and Command is the command to send. Cycles is the number of times the House and Command values are to be sent.
Shift data in	shiftin Dpin, Cpin, Mode, [Variable {\Bits} {,...}]	Synchronous shift of data on Data and Clock pins. The Mode values are MSBPRE (value 0 is the most significant bit first and sample before clock), LSBPRE (value 1 is the least significant bit first and sample before clock), MSBPOST (value 2 is the most significant bit first and sample after clock), and LSBPOST (value 3 is the least significant bit first and sample after clock). Data is shifted into the specified variable with an optional number of bits (the default is eight). The clock is high for 14 usecs and low for 46 usecs.
Shift data out	shiftout Dpin, Cpin, Mode, [OutputData {/Bits} {,...}]	Synchronous shift of data out on Data and Clock pins. The Mode values are LSBFIRST (value 0 is the least significant bit first) and MSBFIRST (value 1 is the most significant bit first). Data is taken from OutputData and can have a set number of bits shifted out (eight is the default). The clock is high for 14 usecs and low for 46 usecs with data available 14 usecs before the clock goes high.

Acknowledgments

If anybody suggests that you write a book consisting of a number of small projects or experiments, run away. This book, although very satisfying and fun, was an *awful* lot of work. There is no way I could have completed it without the help and support of the following individuals and groups (and some others I've forgotten):

- The Tabrobotkit, Basicstamp, and PICList list servers. Normally, I am very active on these lists, but as I got closer to (and past) my deadline, the time spent on these great resources decreased, but I still spent a lot of time lurking and learning. These three groups are probably the best resources experts and novices alike can use to learn about robots, programming, and electronics.

- Ken Gracey and the people at Parallax have been a great help and inspiration for this book. They have a great product line that they are constantly adding to and they are committed to customer support. I consider myself lucky to have received as much of Ken's time and support that I did. Their product line is so good that if you watch closely you will find a BASIC Stamp 1 in *The X-Files: Fight the Future* movie.

- My regular employer, Celestica. Although not being a robotics company per se, it is an outstanding technology company with the best people in the industry. I am continually humbled by how much I don't know and what kind of resources the company has.

- Blair Clarkson and the people at the Ontario Science Centre here in Toronto. The Centre started off as a robot workshop for Celestica, and it has been a lot of fun talking about and planning robots for people to learn with. I hope our relationship continues for a long period of time.

- Ben Wirz, who is a great resource for bouncing ideas off of and for discovering that my great ideas are not unique (or often workable) in any way. Over the years, Ben and I have worked together on three hobbyist robots and I look forward to doing many more with him.

- Judy Bass is my long-suffering editor and I appreciate her work with me on the unique format for the book as well as her patience despite how far past the deadline the manuscript ended up being. I owe you a big lobster dinner for putting up with me.

- My wife Patience, who always lives up to her name, and daughter Marya, who does her namesake proud. Nobody could ask for any more support than I get from the two of you. Even when smoke and curses are rising from the basement, I know that I can get a smile and hug from you.

- Lastly, to everyone that has been involved in the production of the many, many TV shows and movies I have watched for entertainment, inspiration, and relaxation for a good fraction of my life. You're rarely right on how you have portrayed technology, but I feel like I am a richer person for questioning, sneering at, and trying to reproduce what you've created.

Index

Note: Boldface numbers indicate illustrations.

A

B

Index

D

Index

N

Index

Index

About the Author

Myke Predko is a new technologies test engineer at Celestica in Toronto, Canada. He is the author of McGraw-Hill's *Programming and Customizing PICMicro Microcontrollers*, Second Edition, and is the principal designer of the *TAB Electronics Sumo-Bot*.